BIOCHEMISTRY RESEARCH TRENDS

ALKALOIDS

BIOSYNTHESIS, BIOLOGICAL ROLES AND HEALTH BENEFITS

BIOCHEMISTRY RESEARCH TRENDS

Additional books in this series can be found on Nova's website
under the Series tab.

Additional e-books in this series can be found on Nova's website
under the e-book tab.

BIOCHEMISTRY RESEARCH TRENDS

ALKALOIDS

BIOSYNTHESIS, BIOLOGICAL ROLES AND HEALTH BENEFITS

EDUARDO SOBARZO-SÁNCHEZ
EDITOR

nova publishers
New York

Library of Congress Cataloging-in-Publication Data

ISBN: 978-1-63482-074-5
Library of Congress Control Number: 2015930589

Published by Nova Science Publishers, Inc. † New York

CONTENTS

PREFACE

This book discusses a group of natural compound s that is referred to in many bibliographic references for its multiple medical and therapeutic applications, which have been carried out by civilizations in the past and continue to be used in the present. Thus, the alkaloids have been isolated from marine and terrestrial sources and human beings have had the aptitude to determine the chemical structure of many derivatives of simple and big complexity as well as observing the biological effects of every compound in the living organism. Different natural sources as well as the synthesis of many alkaloids of big therapeutic activity have been the basis for the hundreds of drugs that are applied successfully in the scope of the health and combating diverse diseases. Alkaloids ' low cytotoxicity in many cases and versatility in transforming into stable salt have generated diverse drugs of easy administration in the organism without the side effects associated with the ingestion of organic and inorganic salt of difficult tolerance.

In this sense, this contribution covers several chapters which include: mechanisms and strategies against cancer, wherein certain types of alkaloid take control of important and selective form; the use of boldine as the alkaloid of current reference in the traditional medicine and used actively as natural antioxidant; alkaloids from vegetable origin as coming from the Amaryllidaceae; curious brominated alkaloids from marine sources between several outstanding examples; alkaloids derived from the Erythrina including the synthesis and pharmacological applications; the technological approaches of some derivatives originated from Tropane; an interesting contribution of the application of Trabectedin as alkaloid of clinical use in the treatment of ovarian cancer; the mention of a small group of alkaloids called oxoisoaporphines as the big medical tool in the treatment of mental disorder s such as depression; and finally a complete review on the Daphniphyllum alkaloids.

This complete review of chapters that includes diverse nitrogenated compounds such as marine and terrestrial alkaloids that cover the future development of new vaccines as well as the clinical treatment of diseases of diverse nature, will help us to know and understand the importance and amplitude of the alkaloids and their role in our lives. The challenges that they generate towards the future, principally in the feeding and in the scope of the health will be mentioned and discussed in each chapter.

In: Alkaloids
Editor: Eduardo Sobarzo-Sánchez

ISBN: 978-1-63482-074-5

Chapter 1

TARGETING MITOCHONDRIAL BIOENERGETICS OF CANCER CELLS WITH ALKALOIDS: MECHANISMS AND ANTI-CANCER STRATEGIES

Félix A. Urra[1,2,*] and Ramiro Araya-Maturana[1,†]
[1]Department of Organic and Physical Chemistry,
Faculty of Chemical and Pharmaceutical Sciences,
University of Chile, Santiago de Chile, Chile
[2]Laboratory of Bioenergetics and Cancer,
Molecular and Clinical Pharmacology Program,
ICBM, Faculty of Medicine, University of Chile, Chile

ABSTRACT

Mitochondria play a central role in cancer cells, participating in high proliferative capacity, resistance to cytotoxic drugs and metastasis. Several mutations allow for cancer mitochondrial bioenergetics to be different counterpart from non-malignant cells, offering the possibility to explore new targets to obtain anti-cancer effects. Currently, some few alkaloids with well-defined clinical uses act as anti-cancer drugs, exerting their actions on microtubule dynamics or DNA replication and topology. However, several alkaloids isolated from different natural sources and their derivatives have been identified as anti-cancer agents, affecting selectively mitochondrial bioenergetics by inhibition of oxidative phosphorylation through interaction with electron transport chain or inhibition of Adenine Nucleotide Translocator (ANT), inhibition of Tricarboxylic acid (TCA) cycle and transport of metabolites. These events can trigger activation of energetic sensor AMPK favoring anti-migratory effects or mitochondria -mediated cell death by intrinsic pathway apoptosis, autophagy or programmed necrosis. Strategies such as metabolic reprogramming, enhancing the mitochondrial metabolism of the more glycolytic and aggressive cancer cells, through pharmacological inhibition of Pyruvate Dehidrogenase Kinase (PDK) and the use of lipophilic cations to obtain mitochondrial potential-dependent accumulation could enhance anti-cancer actions of alkaloids. This chapter

* E-mail: felix.urra@qf.uchile.cl.
† E-mail: raraya@ciq.uchile.cl.

shows the role of mitochondria in the action of anti-cancer alkaloids, which offer an attractive target for the design of new alkaloid derivatives, considering organelle -specific delivery strategies.

Keywords: alkaloids, mitochondria, cancer cells, cell death, drug delivery

I. INTRODUCTION: ALKALOIDS ACTING ON MITOCHONDRIA OF CANCER CELLS

Alkaloids are a vast group of natural N-containing organic compounds produced by plants, bacteria, fungi, marine sponges and slugs, arthropods, amphibians, birds and mammals. In contrast with other classes of natural compound s, alkaloids have practically unlimited frameworks, many of which exhibit relevant bioactivities [1, 2]. A great variety of structures of alkaloids exhibits stereogenic centers, heteroatoms, several H-bond donors and acceptors and a low ratio between the aromatic rings and total heavy atoms generate. As a consequence of this, alkaloids have the chance to interact with a wide spectrum of molecular targets [3]. Regarding this subject, two pharmacological approaches coexist in medicinal chemistry: the goal of developing highly selective molecules acting on single targets [4] is facing to the objective to obtain molecules that interfere simultaneously with multiple targets, which might be more effective than single target agents [5].

Alkaloids are an important component of the current cancer chemotherapy and are widely used as lead compounds by medicinal chemistry researchers [6]. Alkaloids with clinical uses affect the progression of cell cycle by alteration of microtubule polymerization or inhibition of topoisomerases [3]. However, it has been showed that an increasing number of alkaloids, including the classical anticancer alkaloids, can present several cellular targets, including mitochondria, affecting selectively to cancer cells [7, 8]. On the other hand, the severe toxicity and development of chemoresistance have driven to search strategies or new targets to enhance tumor selectivity [6, 9]. In this sense, selective targeting is currently an active research field, whose findings allow fully exploiting the pharmacological potential of this type of molecules [10, 11]. Furthermore, it offers the possibility to recover compounds discarded in previous pharmacologic studies.

In recent years, an increased interest to find new pharmacological targets allowed to identify several mitochondrial functions that offer a possibility to develop the mitochondrial pharmacology [12]. It is now clear that mitochondria are central to cell death, cell differentiation, hypoxia sensing, innate immune system, metabolism of amino acid s and calcium, iron sulfur center and heme biosynthesis [13]. Disruption to these processes contributes to several pathologies, making mitochondria an important therapeutic target [6, 7, 9, 14]. In cancer cells, mitochondria have a main role in the neoplastic phenotype, resistance to apoptosis, metabolic reprogramming and limitless proliferative potential [15, 16]. It has been suggested that efficient mitochondrially-targeted therapies would affect mitochondrial processes that participate in several of these features [15]. Besides, mitochondria control the programmed cell death by several mechanisms that can be activated through alteration of cellular redox balance, disruption of mitochondrial bioenergetics, and release or activation of pro-apoptotic factors [8, 9, 17, 18].

Although several compounds are known to act on mitochondria, it is necessary to develop particular strategies to target them selectively [19]. These strategies must take into account that mitochondria are located inside the cell and, in turn, the targets of drugs are located inside mitochondria. This imposes many physicochemical requirements for crossing several membranes and achieves selective accumulation of the molecules in these organelles [20, 21, 22]. It is very likely that many potential drugs do not exhibit adequate tumor and mitochondria-specific accumulation. Therefore an effort to develop tumor-targeted mitochondria-specific approaches might be an important aid towards exploiting mitochondrial targets for cancer therapy, enhancing mitochondria-specific accumulation of the anticancer molecules. Some specific approaches have been described. One of the most used strategy is to link a triphenylphosphonium cation to structurally different compounds, allowing a mitochondrial potential ($\Delta\psi m$)-dependent accumulation and delivery them into the mitochondria [23]. Also, the leader sequences of some matrix proteins have been used to mediate the mitochondria-specific accumulation of molecules [17]. Besides, pharmaceutical nanocarriers like liposomes, micelles and solid nanoparticles have also been used for this purpose [17, 22, 24]. With this in mind, we proposed that the anticancer activity of many alkaloids could be increased using this strategy [25].

A new class of anti-cancer drugs that affect the mitochondrial function has been defined as "mitocans" (an acronym for mitochondria and cancer), existing at least eight classes of these agents [26]. As far as we know, none anti-cancer alkaloid have been considered in this classification. In this chapter, we present alkaloids and their derivatives that have the mitochondrial bioenergetics as a central component for their anti-cancer actions (Figure 1).

II. Mitochondrial Bioenergetics as Target for Anti-Cancer Alkaloids

1. Inhibition of OXPHOS by Interaction with Electron Transport Chain and Transport of ADP /ATP

Oxidative phosphorylation (OXPHOS) involves electron transport chain (ETC) and ATP synthesis. ETC is composed by four respiratory complexes: NADH -ubiquinone oxidoreductase (complex I), succinate-ubiquinone oxidoreductase (complex II), ubiquinol-cytochrome -c oxidoreductase (complex III) and cytochrome-c oxidase (complex IV). Complex I and complex II oxidize NADH and $FADH_2$ to reduce molecular oxygen to water by complex IV (process known as mitochondrial respiration). Moreover, this process involves electron transfer, which is coupled to the pumping of protons to the intermembrane space by complexes I, III and IV forming the mitochondrial potential membrane ($\Delta\psi m$). ATP synthase utilizes this proton-motive force to generate ATP at the mitochondrial matrix (Figure 3)[27]. On the other hand, the maintaining of $\Delta\psi m$ allows ATP synthesis for cellular energetic requirements and control of signaling of cell death (e.g., certain types of regulated necrosis and intrinsic pathway of apoptosis). In cancer cells, mutations in several genes such as p53, RAS and c-Myc affect metabolic pathways [28, 29] producing mitochondria -related metabolic changes that participate in cancer progression. It exhibits low activity of some respiratory complexes and high $\Delta\psi m$ [30]. Moreover, alterations in ETC promote migration and invasion, [31] and elevated mitochondrial ROS production at complex III level promotes anchorage -independent growth [32].

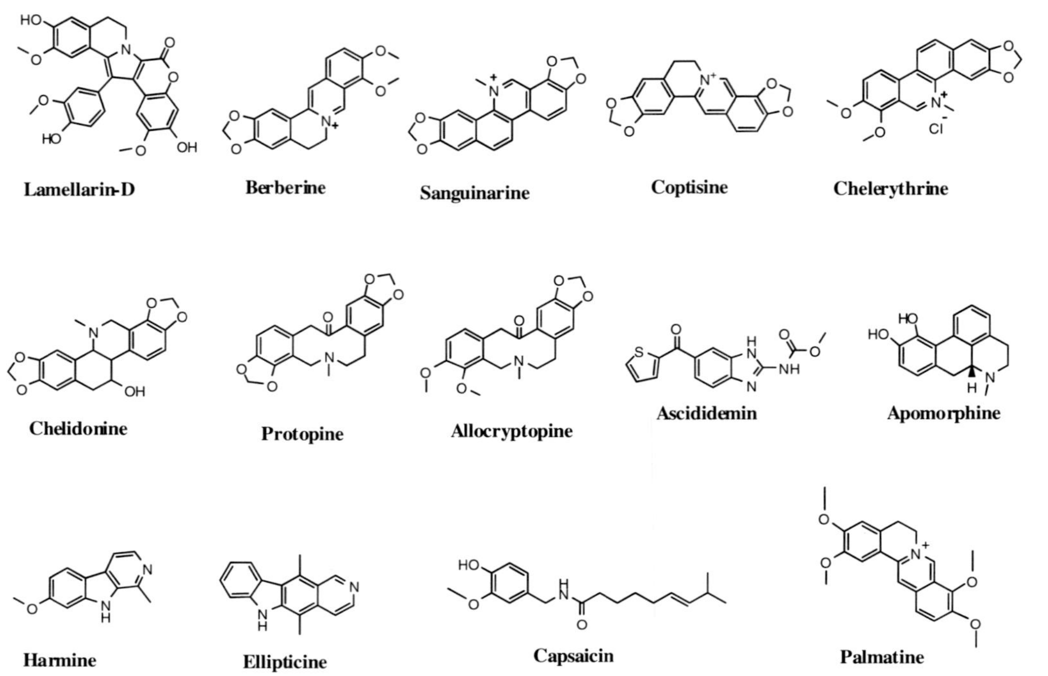

Figure 1. Chemical structures of alkaloids that affect mitochondrial bioenergetics in cancer cells.

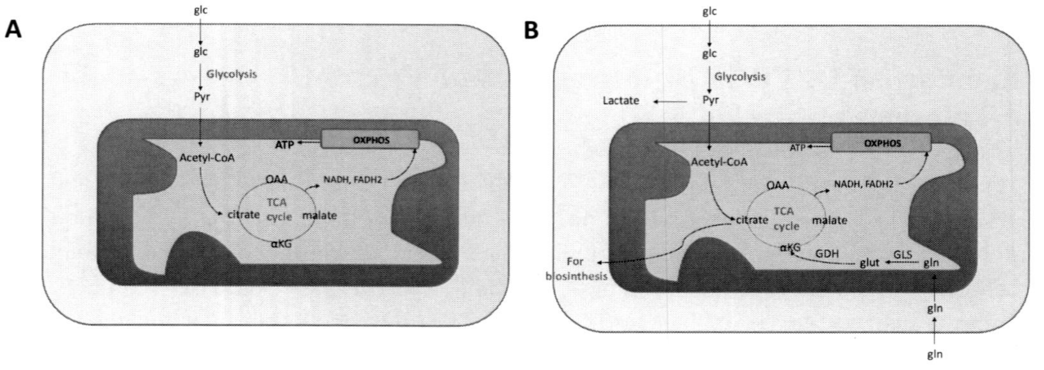

Figure 2. Participation of metabolic pathways in normal cells (A) and cancer cells (B). In normal cells, oxidative phosphorylation (OXPHOS) maintains the ATP production required by cellular energetic demand. In contrast, cancer cells (**B**) exhibit an elevated glucose consumption and consequently high glycolytic rate. Moreover, intermediates of TCA cycle are used for *de novo* synthesis of proteins, lipids and nucleic acid s, which are required for high proliferation rate. The replenishment of intermediates of TCA cycle is obtained through anaplerotic reactions such as glutaminolysis. Abbreviations: glc: glucose, pyr: pyruvate; OAA: oxaloacetate, αKG: α-ketoglutarate, OXPHOS: oxidative phosphorylation, glut: glutamate, gln: glutamine, GLS: glutaminase, GDH: glutamate dehydrogenase.

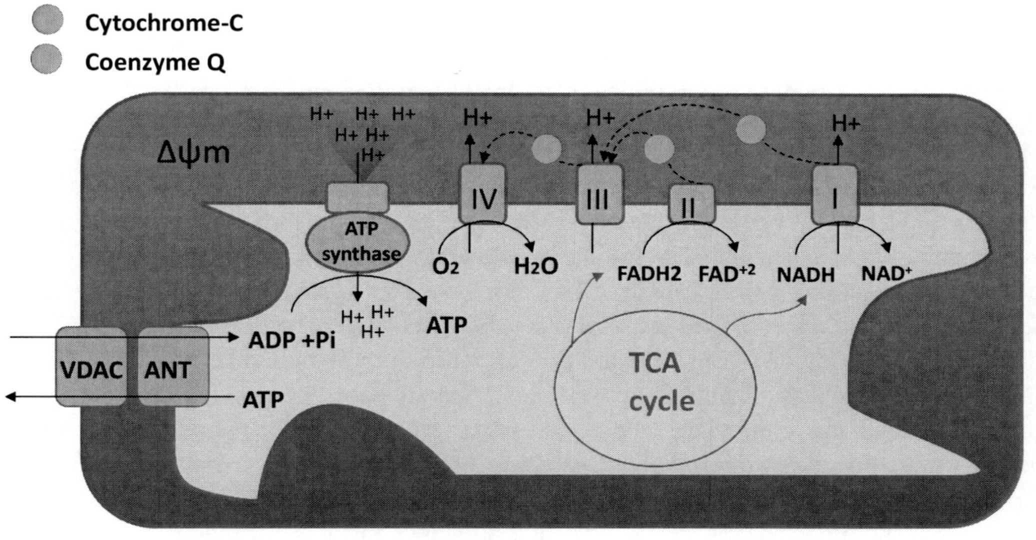

Figure 3. Oxidative phosphorylation (OXPHOS). It is composed by electron transport chain (ETC) and ATP synthase. In ETC, oxidation of reducing equivalents (NADH and $FADH_2$) allow the electron transport from complex I and II to complex III by ubiquinone (violet circle) and from complex III to complex IV by cytochrome -c (pink circle). This process finishes in reduction of molecular oxygen, occurring mitochondrial respiration. **Mitochondrial potential membrane (Δψm) generated by** ETC is used for ATP synthesis by ATP synthase. Adenine nucleotides are transported by voltage-dependent anion channel (VDAC) and adenine nucleotide translocator (ANT).

These characteristics have allowed the development of a broad group of mitochondriotoxic agents known as mitocans [33, 34]. Interestingly, alterations in different levels of OXPHOS, especially inhibition or uncoupling of respiration or inhibition in the transport of ADP /ATP, are events that produce ATP depletion with induction of selective death of cancer cells. Different mechanisms of inhibition of OXPHOS have been described for anti-cancer alkaloids. Two types of interactions can be identified: 1) direct interaction with ETC, specially with complex I and III, affecting the mitochondrial respiration and Δψm and 2) inhibiting OXPHOS (and ATP synthesis) by inhibition of Adenine Nucleotide Translocator (ANT), producing diminished adenine nucleotide transport and alteration in mitochondrial ADP/ATP levels.

Interaction of many alkaloids with mitochondrial bioenergetics in cancer cells have been described: Lamellarin-D, a hexacyclic pyrrole alkaloid, is a pro-apoptotic agent insensitive to P-glycoprotein-mediated drug efflux [35]. It induces apoptosis by two mechanisms of action: topoisomerase I inhibition and induction of apoptosis intrinsic pathway of a p53 nuclear signaling-independent manner [36, 37]. However, some reports indicate that lamellarin-D primarily produces changes in mitochondrial bioenergetics in several cancer cells [38, 39], reducing complex III activity with decrease of mitochondrial respiration, ATP synthesis, and Δψm dissipation. These alterations result in MPTP opening and apoptosis in human melanoma cells [40]. Recent evidence indicates that topoisomerase I inhibition by Lamellarin-D produces senescence and apoptosis which can be due to inhibition of mitochondrial bioenergetics [41].

Berberine is an isoquinoline alkaloid with several pharmacological properties [42, 43] that affect different levels of the mitochondrial functions [44, 45]. In mouse melanoma cells, it selectively accumulates in mitochondria, depending of mitochondrial potential. Moreover, it induces oxidative stress and inhibition of complex I that result in ATP depletion, alterations in mitochondrial structure and reduction in mtDNA copy number [44]. It has also been suggested that the induction of MPTP opening and respiration inhibition are events due to interaction with ANT [46]. This protein facilitates the exchange of ADP /ATP across the mitochondrial inner membrane and several isoforms can be involved in the progression of cancer, being a regulator of acquisition of resistant to apoptosis [47, 48].

Sanguinarine is a benzophenanthridine alkaloid that exhibits anti-proliferative activity on several human tumor cells without affect normal cells [49, 50]. In mouse metastatic melanoma cell line, this alkaloid presents a concentration-dependent dual action: it primarily accumulates in cell nuclei, causing inhibition of cell proliferation and cell death, but at low concentrations, induces alteration in the mitochondrial bioenergetics [51]. In contrast to berberine, sanguinarine produces $\Delta\psi m$ dissipation without to induce oxidative stress. In isolated mitochondria, this alkaloid inhibits preferentially the complex I-dependent respiration, increasing the state 4 and state 4+oligomycin and decreasing state 3 and uncoupled respiration [51]. Moreover, it induces cyclosporin A-sensitive opening of MPTP and presents a mixed effect on calcium loading capacity. At low concentrations, sanguinarine decreases calcium loading capacity however, at high concentrations increases this capacity [51]. Moreover, other mechanisms of apoptosis have been reported in cancer cells, including mitochondrial alterations such as a ROS-mediated mitochondrial pathway [52, 53].

Interestingly, isoquinoline alkaloids isolated from *Chelidonium majus* can inhibit complex I and complex II-dependent respiration [54]. In isolated mitochondria, alkaloids with a positive charge (due to a quaternary nitrogen atom) inhibit complex I-dependent respiration (berberine and coptisine) or complex I- and complex II-dependent respiration (sanguinarine and chelerythrine). No-cationic alkaloids chelidonine, protopine and allocryptopine do not exhibit effect on mitochondrial respiration; however, in submitochondrial particles produce inhibition of NADH dehydrogenase activity. Taken together this evidence, it has been suggested that presence of cationic charge confer to some alkaloids permeability across mitochondrial membranes to produce inhibition of respiration and minimal structural requirements for the inhibition of complex I can be the presence of at least two adjacent aromatic rings. In no-cationic alkaloids (protopine and allocryptopine) can be the presence of carbonyl group, which can possibly react with sulfhydryl groups in complex I [54].

2. Inhibition of TCA Cycle and Transport of Metabolites

Tricarboxylic acid (TCA) cycle, also known as Kreb´s cycle, has a central role in mitochondrial oxidative metabolism. TCA cycle produces the reducing equivalents (NADH and $FADH_2$), which are oxidized by ETC to generate mitochondrial ATP (Figure 2A) [17]. In cancer cells, several mutations in key enzymes such as succinate dehydrogenase, fumarate hydratase and isocitrate dehydrogenase promote the tumorigenesis [55, 56]. In highly proliferative cells such as cancer cells, TCA cycle provides metabolic intermediates to *de novo* synthesis of proteins, lipids and nucleic acid s. This requires the replenishment of intermediates of TCA cycle through anaplerotic reactions such as glutaminolysis (Figure 2B) [57]. Moreover, dependence on glutamine in several cancers suggests that this anaplerotic pathway can promote tumor growth and invasion [58, 59, 60].

In this view, changes in intermediates of TCA cycle or inhibition of enzymes involved in anapletoric pathways can be an unexplored target to obtain anti-cancer effects by alkaloids. A recent research, using an approach by proton NMR spectroscopy -based metabolomics, identified new mitochondrial metabolic targets for two marine alkaloids with known anti-cancer activities [61] Ascididemin, a pyridoacridine alkaloid, inhibits topoisomerase II and induces apoptosis, involving mitochondrial ROS production and JNK-dependent activation of caspases [62]. Recent evidence indicates that this alkaloid produces accumulation of citrate, an initial product of TCA cycle and decrease of metabolites related with bioenergetic pathways as adenosine phosphate s, acetate, lactate and *N*-acetylaspartate in MCF7 breast cancer cells [61]. Alterations in the citrate metabolism induced by ascididemin are due to inhibition of ATP -citrate lyase, a common enzyme for cholesterol and *de novo* fatty acid synthesis and inhibition of cytosolic and mitochondrial aconitases (Aco1 and Aco2, respectively), which catalyze dehydrogenation of citrate into isocitrate [61]. Aconitases inhibition could be by an oxidative mechanism. On the other hand, lamellarin-D, which exhibits anticancer action by direct action on ETC, can also inhibit the cycle malate-aspartate shuttle (MAS) in MCF7 breast cancer cells, producing glutamate and aspartate accumulation [61]. In addition, lamellarin-D does not affect the mitochondrial ROS levels, indicating that has no effect on ROS-dependent aconitase inhibition [61].

On the other hand, it has suggested that chelerythrine and apomorphine are inhibitors of glutaminase [63], which has a central role in the proliferation of cancer cells via glutaminolysis, catalyzing the hydrolysis of glutamine to glutamate (Figure 2B) [64]. Chelerythrine and apomorphine could be competitive inhibitors, being more efficient that the well-known glutaminase inhibitors DON (6-diazo-5-oxo-L-norleucine) and BPTES (bis-2-(5-phenylacetimido-1,2,4-thia-diazol-2-yl) ethyl sulfide). Identification of glutaminase as a new target for anticancer alkaloids increases the question if anti-cancer effects could be secondary events to inhibition of this enzyme [63]. The effect of these alkaloids on cancer cells with glutamine-dependent growth requires to be investigated.

3. Activation of Energetic Sensor AMPK

AMP-activated protein kinase (AMPK) is a sensor of cellular energetic status, regulating the metabolism of proteins and lipids in response to changes in substrate availability [65]. AMPK is a heterotrimer with a catalytic α-subunit and regulatory β and γ subunits [65]. It is activated by several physiological and pathological stresses that increase the intracellular AMP/ATP ratio by increase of ATP consumption or decrease of ATP production [66]. Complete activation of AMPK requires phosphorylation in Threonine 172 of catalityc α-subunit and exchange of ATP by AMP in the γ subunit. Phosphorylation in α-subunit is mediated by LKB1, which is constitutively active, by Ca^{2+} calmodulin-dependent kinase kinase β (CAMKKβ) which actives to AMPK in presence of high Ca^{2+} levels, independently of changes in AMP/ATP ratio. In response to energetic stress, AMPK up-regulates catabolic pathways to produce ATP and down-regulates anabolic pathways, inhibiting ATP consumption [65]. It induces translocation of glucose transporters to plasma membrane, increasing glucose uptake and glycolysis [65]. Moreover AMPK increases fatty acid oxidation by phosphorylation of Acetyl-CoA carboxylase-2 (ACC2), increases mitochondrial biogenesis and decreases the sterol and glycogen synthesis.

In cancer, AMPK signaling is down-regulated [67, 68]. LKB1, considered a tumor suppressor, is muted in patients with autosomal dominant inherited Peutz-Jegher syndrome, who develop hyperpigmented macules of the oral mucosa and intestinal hamartomatous polyps and are predispose to develop several types of cancers [69, 70]. Consequently, AMPK activation can produce anti-cancer effects such as cell cycle arrest [71] and cell death in lung, prostate, liver and breast cancer cells [72]. Therefore, it has been suggested that AMPK activation through small molecules by indirect or direct mechanisms is a promising strategy to prevent and inhibit tumor cell growth [66, 72].

Anti-diabetic drug metformin is a complex I inhibitor that induces AMPK activation [73] and evidence from epidemiological studies with diabetic patients suggests that activation of AMPK can suppress tumorigenesis. In similar manner to metformin, berberine exhibits AMPK-mediated hypoglycemic effect with benefits in diabetic and insulin -resistant states [74], activating the glucose transport activity of glucose transporter (GLUT1) [75, 76]. Berberine appears to activate to AMPK by an indirect mechanism similar to metformin (Figure 4), involving inhibition of complex I in ETC and increase of the AMP/ATP ratio [77, 78]. This activation has been reported independent of LKB1 or CAMKKβ [79]. In cancer cells, berberine inhibits the metastatic potential of colon cancer and melanoma cells by AMPK activation [80, 81]. In B16F10 melanoma cells, berberine inhibits the migration, adhesion and invasion through activation of AMPK by increase in ROS production, decreasing ERK activity and protein levels of COX-2 [81]. In cancer progression, malignant cells acquire invasive properties favoring metastasis and involve the expression of mesenchymal markers and loss of epithelial markers. This phenomenon known as epithelial to mesenchymal transition (EMT) implies expression of several transcription factors such as Twist, Snail, Slug and Zeb, which participate in the establishment of metastatic phenotype in cancer cells [82]. Interestingly, AMPK activation induced by berberine decreases expression of EMT-related genes: Snail, Slug, Zeb and N-cadherin [81].

On the other hand, berberine exhibits anti-migratory effect on SW480 and HCT116 human colon cancer cells, producing down-regulation in the levels of integrin β1 [80]. This protein is one subunit of α/β heterodimeric cell surface receptors, which mediate cell-cell and cell-matrix adhesion [83]. Binding of components of extracellular matrix (e.g., fibronectin, proteoglycans, collagen) to integrins produces the recruitment of several proteins involved in proliferation, motility and survival and in cancer cells favor cancer progression and metastasis [83, 84, 85].

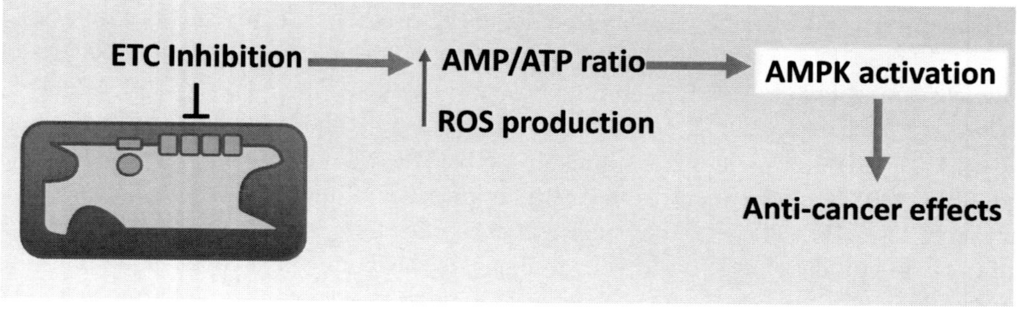

Figure 4. Indirect mechanism of AMPK activation by ETC inhibitors. Metformin and berberine are inhibitors of ETC, producing decrease of mitochondrial ATP synthesis. Berberine increases AMP/ATP ratio and ROS production, activating AMPK in cancer cells. This event produces anti-cancer effects.

Berberine induces activation of AMPK by increase of ROS production, affecting post-translationally the levels of integrin β1 and its downstream targets Src, FAK and p130Cas with consequent decrease of migration in human colon cancer cells [80].

III. Induction of Mitochondria -Mediated Cell Death by Anti-Cancer Alkaloids

Extensive literature has been dedicated to define the types of cell death, considering morphological and biochemical aspects [86, 87, 88]. Induction of death of cancer cells by alterations of mitochondrial bioenergetics have been proposed as promising approach for synthetic and natural compound s [23]. Mitochondria participate in different types of cellular death, specially apoptosis (intrinsic pathway), programmed necrosis and autophagy [15, 16, 89], which can be triggered by anti-cancer alkaloids.

1. Mitochondria -Mediated Cell Death: Intrinsic Pathway Apoptosis

Apoptosis is composed by the extrinsic and intrinsic pathways, which act autonomously at the beginning and involve caspase activation. Extrinsic pathway is triggered by stimulation of cell surface death receptors, such as TNFα (tumor necrosis factor -α), Fas (CD95 /APO1) and TRAIL (TNF related apoptosis inducing ligand) receptors. The intrinsic pathway in turn, is triggered by $\Delta\psi m$ dissipation, accompanied with increase of ROS levels and mitochondrial swelling. Moreover, induction of MPTP opening or oligomerization of pro-apoptotic proteins (e.g., Bax, Bak) and inhibition of anti-apoptotic proteins (e.g., Bcl-2, Bcl-xl) produce mitochondrial membrane permeabilization, favoring the release of cytochrome -c and pro-apoptotic proteins (e.g., AIF and Smac/DIABLO) from mitochondrial intermembrane space with consequent caspases activation, chromatin condensation and DNA fragmentation.

Control of apoptosis by mitochondria is essential in cancer cells [90], allowing the development of resistance to cytotoxic agents [91, 92] and mechanism of control in cancer progression [93]. Alkaloids isolated from different natural sources have been described as inductors of mitochondria-dependent apoptosis (Figure 5) and their mechanisms of action involve $\Delta\psi m$ dissipation [8] in several cancer cell line s (Tables 1-3). This event produces release of cytochrome -c, activation of caspase-9 and caspase-3. In human endometrial cancer cells, ellipticine-induced apoptosis is characterized by increase of ROS production and activation of ERK and JNK signaling [94]. Similar effects have been reported for vinorelbine-induced apoptosis in human lung cancer cells [95]. This latter is a pyridocarbazole alkaloid with known anticancer activity mediated by induction of DNA damage, involving topoisomerase II inhibition and DNA adduct formation [96, 97, 98, 99, 100]. Several strategies of delivery based in self-assembling peptide have been designed for ellipticine. These peptides are organized in nanofibers and stabilize this hydrophobic anticancer agent [101, 102, 103]. This approach for a selective delivery can favor an increase in activity and reduce toxic events.

On the other hand, intrinsic pathway apoptosis accompanied of alteration on cell cycle progression for four alkaloids and derivatives have been reported. Aldisin-derivative B6 induces arrest in G1 phase and ellipticine, evodiamine and noscapine-derivative EM011 induce arrest in G2/M phase in cancer cells.

2. Mitochondria -Mediated Cell Death: Programmed Necrosis

Concept of necrosis has been considered such an accidental and uncontrolled process of cell death. Morphologically, necrotic cells exhibit swelling of organelles (such as endoplasmic reticulum and mitochondria), permeabilization of plasma membrane and consequent cellular lysis with a intact nucleus [86].

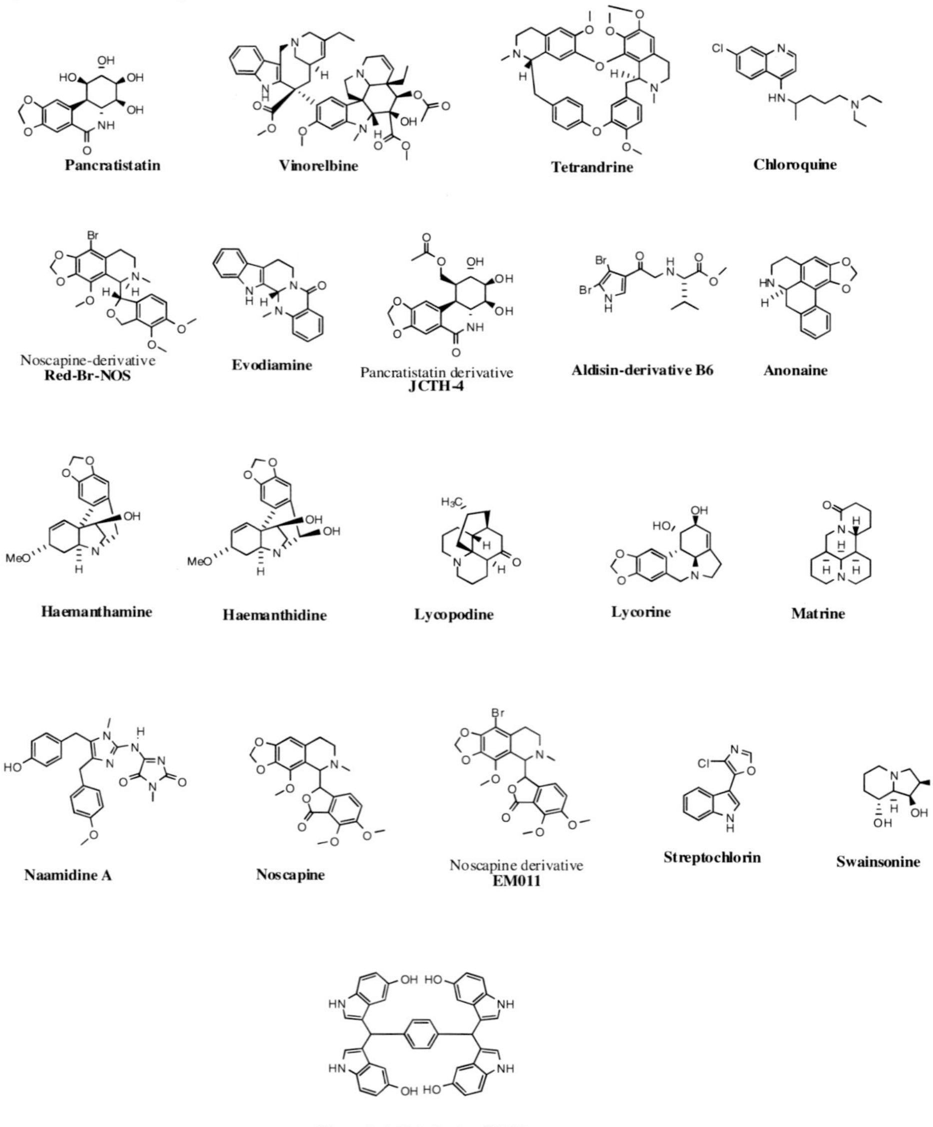

Figure 5. Alkaloids that induce mitochondria -mediated cell death in cancer cells.

Table 1. Anticancer alkaloids that induce mitochondria -mediated apoptosis (Intrinsic pathway) in cancer cells

Alkaloid	Mechanism of apoptosis	Cell type	Comments	Ref.
Aldisin-derivative B6	$\Delta\Psi_m$ dissipation, release of intracellular Ca^{2+}, activation of caspase-3 and -9, arrest in G1 phase of cell cycle.	Several human cancer cells	B6 is a synthetic molecule based on pyrrole alkaloid aldisin isolated from marine organism.	[128]
Anonaine	$\Delta\Psi_m$ dissipation, increase of intracellular nitric oxide, ROS, glutathione depletion, caspase activation and PARP cleavage	Human cervical cancer cell Hela	Alkaloid isolated from plant *Michelia alba* (Magnoliaceae)	[129]
Chelerythrine	Arrest in S phase, $\Delta\Psi_m$ dissipation, release of Cyt-*c*, activation of caspase-3, cleavage of PARP, up-regulation Bid and Bax, down-regulation Bcl-xl.	Human Hepatoma SMMC-7721 cells	Benzo[*c*] phenanthridine alkaloid present in several plant species (e.g., *Chelidonium majus*).	[130]
Ellipticine	Arrest G2/M phase of the cell cycle, release of cyt-*c* and AIF, activation of caspases, $\Delta\Psi_m$ dissipation, increase intracellular ROS levels, activation of ERK and JNK.	RL95-2 human endometrial cancer cells	Alkaloid isolated from Apocyanaceae plants with several anticancer mechanism reported: topoisomerase II inhibition and formation of DNA adducts [131].	[94]
Evodiamine	Up-regulation of Bax and p53, down-regulation of Bcl-2, $\Delta\Psi_m$ dissipation, activation of caspase-3, arrest in G2/M phase of cell cycle.	Human Colorectal Carcinoma cells COLO-205	Alkaloid isolated from Chinese herbal medicine *Evodia rutaecarpa* with antioxidant, anti-inflammatory activities reported.	[132]
Harmine	Nuclear fragmentation and chromosomal condensation, $\Delta\Psi_m$ dissipation, caspase-3 and caspase-9 activation, down-regulation of Bcl-2, Mcl-1 and Bcl-xl, without change in Bax.	Human hepatocellular carcinoma cell line HepG2	Harmine is a major component isolated from *Peganumharmala L.* (Zygophyllaceae) seed extract. Other effect reported: antinociceptive and DNA topoisomerase inhibitor.	[133]

Table 2. Anticancer alkaloids that induce mitochondria -mediated apoptosis (Intrinsic pathway) in cancer cells

Alkaloid	Mechanism of apoptosis	Cell type	Comments	Ref.
Haemanthamine and haemanthidine	$\Delta\Psi_m$ dissipation, cell cycle arrest with increased p16 expression and Chk1 Ser345 phosphorylation	Human leukemic Jurkat cells	Alkaloids isotated from plants of Amaryllidaceae family. Haemanthidine is more active that haemanthamine as apoptotic inductor.	[134]
Lycopodine	Increase of ROS levels, $\Delta\Psi_m$ dissipation, release of cyt-*c*, caspase-3 activation, chromatin condensation, inter-nucleosomal DNA fragmentation	HeLa cells	Alkaloid isolated from plant *Lycopodium clavatum* [135].	[136]
Lycorine	Release of cyt-*c*, activation of caspase-3 and -9, degradation of PARP, down-regulation of Mcl-1.	Human leukemia cells	Amaryllidaceae alkaloid [137]. Its anticancer derivatives also act inducing apoptosis [138, 139].	[140]
Matrine	$\Delta\Psi_m$ dissipation, release of cyt-*c*, ratio Bcl-2/Bax protein decreased caspase-3 activation.	Human multiple myeloma cells	Alkaloid isolated from the traditional Chinese herb *Sophora flavescens*.	[141]
Naamidine-A	$\Delta\psi_m$ dissipation, activation of caspase-3, -8 and -9.	A431 epidermoid carcinoma cells.	Imidazole alkaloid identified in marine sponge *Leucetta chagosensis* [142].	[143]
Noscapine	Up-regulation of Bax and release Cyt-*c*, down-regulation of Bcl-2 protein, activation of caspase-3 and -9.	Several human gastric cancer cell lines.	Alkaloid with antitussive activity known.	[144]

Table 3. Anticancer alkaloids that induce mitochondria -mediated apoptosis (Intrinsic pathway) in cancer cells

Alkaloid	Mechanism of apoptosis	Cell type	Comments	Ref.
Noscapine-derivative EM011	$\Delta\psi_m$ dissipation, increase of PUMA, BAX, caspase-3 activation, cleavage PARP, arrest G2/M phase, activation of the mitotic checkpoint.	Human prostate cancer cell lines PC-3, LNCaP, C4-2 and C4-2B.	Brominated analog of the plant-derived alkaloid noscapine.	[145]
Pancratistatin	Increase ROS production, ATP depletion, $\Delta\psi_m$ dissipation, caspase activation.	Human breast cancer c ell lines MCF-7 and Hs-578-T.	Alkaloid isolated from Hawaiian spider lily [146]	
Streptochlorin	Up-regulation of pro-apoptotic Bax and FasL, $\Delta\psi_m$ dissipation, degradation PARP, caspase-3 activation.	Human leukemic U937 cells.	Alkaloid isolated from marine *Streptomyces sp.* with antiangiogenic activity reported.	[147]
Swainsonine	Up-regulation of Bax, down-regulation of Bcl-2 expression, release of cyt-*c*, caspase -3 and -9 activation, cleavage of PARP.	Human Lung Cancer A549 cells.	Alkaloid isolated from plant as the Australian legume *Swainsona canescens* and in many *Astragalus* and *Oxytropis* species (Leguminosae).	[148]
Tetrandrine	Up-regulation of Bax, Bak, and Bad; and down-regulation of Bcl-2 and Bcl-xl, release of cyt-c and activation of caspase-3 and -9.	Human gastric cancer BGC-823 cells	Alkaloid isolated from the dried root of Hang-Fang-Chi (*Stephania tetrandra*) with apoptotic action in vitro a nd in vivo [149, 150]	[149]
Vinca alkaloid-derivative SK228	$\Delta\Psi_m$ dissipation, increase of mitochondrial ROS levels, DNA damage, release of cyt-*c*, activation of caspase-3 and -9, down-regulation of Bcl-xl, up-regulation of Bad.	Lung cancer cells, Esophageal carcinoma cells.	SK228 exhibits cancer cell-specific cytotoxicity and inhibits also cancer cell invasion, affecting FAK/Paxillin signaling.	[151]
Vinorelbine	$\Delta\Psi_m$ dissipation, down-regulation Mcl-1, increase of ROS levels, activation of JNK. JNK-regulated DNA damage i n prometaphase.	AS2 human lung cancer cells.	Vinca alkaloid	[95]

However, these characteristics can be inhibited by pharmacological and genetic interventions, revealing a cell death by regulated necrosis. Several types of regulated necrosis have been defined: parthanatos, mitochondrial permeability transition-dependent regulated necrosis and necroptosis [104]. Necroptosis is a programmed cell death that is caspase-independent with receptor interacting protein kinases (RIPK and RIPK3) participating, and it is inhibited by small molecule necrostatin-1. Its cellular signaling shares some features with apoptosis, however the molecular mechanisms of necroptosis require further understanding. Parthanatos is a non-apoptotic cell death that occurs with hyperactivation of poly(ADP - ribose) polymerase 1 (PARP1). In this cell death, PARP1 hyperactivation produces depletion in the ATP and reducing equivalents levels, causing necrosis. A modality of regulated necrosis is dependent of mitochondria, specifically of mitochondrial permeability transition (MPT) [104]. This event occurs when increases the permeability of the inner mitochondrial membrane to small solutes into to mitochondrial matrix, producing mitochondrial swelling, $\Delta\psi m$ dissipation and releases of pro-apoptotic factor that participate in intrinsic pathway apoptosis. However, MPT can trigger regulated necrosis, which is dependent of mitochondrial cyclophilin-D (CYPD-D). CYPD-D and other mitochondrial proteins constitute a supramolecular entity, known as mitochondrial permeability transition pore complex (MPTP), which is stimulated by increase of ROS production and increase of cytosolic calcium concentrations, producing MPTP opening and consequently, MPT. Recently, it has been described a type of regulated necrosis induced by berberine in prostate cancer cells [105]. Berberine increases ROS production, which promotes p53 translocation to mitochondria, physically interacting with CYPD-D to induce mitochondrial depolarization and MPTP opening. Interestingly, ROS production and p53 activation are also involved in apoptosis by berberine, a cyclophilin D-independent event. Both types of cell deaths are induced in prostate cancer cells.

3. Mitochondria -Mediated Cell Death: Autophagy

Autophagy is a multistep lysosomal degradation process responsible for misfolded proteins, damaged organelles and maintain energy by recycling of nutrients for cell survival. Therefore, autophagy primarily functions as a cell survival adaptive mechanism during induced cellular stress conditions. However when this condition is persistent, it can promote extensive autophagy, triggering autophagic cell death [106]. Mitochondria have been recognized to play an important role in the establishment of oxidative stress, favoring autophagic process [16]. Consistently, the induction of oxidative stress and autophagic cell death constitute a cancer therapy strategy [107]. This view can be useful if it is considered that frequently cancers cells are resistant to apoptosis or develop drug resistance, offering autophagic cell death an effective alternative therapy [106, 108]. However, chemotherapeutic drugs and radiation can promote autophagy, which has a cyto-protective function and its inhibition can favor apoptosis [109]. This dichotomous view of the autophagy in cancer suggests that the modulation of autophagy could be a potential anti-cancer strategy specific to each type of cancer cell. In this section are discussed the anti-cancer alkaloids that induce autophagic cell death, playing mitochondria a key role. In the next section (Cross-talk between different types of mitochondria-mediated cell death) are discussed the alkaloids that induce protective autophagy (anti-apoptotic and anti-necrotic effects) or autophagy that participates with other type of cell death.

Some alkaloids with anticancer activities have been reported as modulators of autophagy (Figure 5). Interestingly, they can be classified as autophagic inhibitors or inductors.

Tetrandrine is a bisbenzylisoquinoline alkaloid that at 5 μM concentration induces autophagy in human hepatocellular carcinoma cells [110]. The mechanism involves mitochondrial dysfunction. Initially, tetrandrine produces mitochondrial depolarization that precedes the increase of intracellular ROS levels, which is blocked by cyclosporine A, an inhibitor of the MPTP opening. Moreover, it induces the expression of Atg7, LC3-II, the formation of acidic autophagolysosome vacuoles, and ROS-induced activation of ERK MAP kinase signaling, which all are involved in tetrandrine-induced autophagy. Similar results are produced *in vivo* [110]. Interestingly, another report has shown that this alkaloid induces apoptosis in the same carcinoma cells at high concentrations (>20 μM) by increase of ROS levels [111]. In contrast, chloroquine is an agent that inhibits the autophagic process through prevention of lysosome - autophagosome fusion and the degradation of autophagosomal content [112]. In solid tumor, hypoxia conditions allow the development of resistance to treatment, especially acute/cycling hypoxia accompanied of ROS production. In this condition, cancer cells can acquire adaptations to minimize ROS production, involving reduction of mitochondrial content by specific autophagy (known as mitophagy) or removal of intracellular aggregates. Consequently, moderate hypoxia can trigger mitophagy with reduced respiration as an adaptative process [113, 114]. In human cancer cells MCF-7, HT29 and U373, inhibition of autophagy by chloroquine prevents the tolerance to acute hypoxia that allows specific degradation of mitochondria to decrease ROS levels in reoxygenation [115]. It has been suggested that this action of chloroquine decreases the hypoxic portion within tumors, being an opportunity to enhance anti-cancer therapy [115].

4. Cross-Talk between Autophagy and Different Types of Mitochondria - Mediated Cell Death

Apoptosis and necrosis are the two extensively studied modalities of cell death that often exhibit similar initiator/effector molecules, signaling pathway s and involving of same organelles [86]. Both types of cell death can co-operate in cellular destruction or can be prevented by autophagy. It has been reported some anti-cancer alkaloids that induce cell death by activation of apoptosis or necrosis, where autophagy participates as a survival and protective pathway (anti-apoptotic or anti-necrotic effect). In these cases, mitochondria participate at least as a modality of cell death.

Recently, the anticancer mechanism of noscapine derivative Red-Br-nos, (*R*)-9-bromo-5-((*S*)-4,5-dimethoxy-1,3-dihydroisobenzofuran-1-yl)-4-methoxy-6-methyl-5,6,7,8-tetrahydro-[1,3]-dioxolo[4,5-g]isoquinoline, has been described in human prostate cancer PC-3 cells [116]. Red-Br-nos triggers protective autophagy to the induction of apoptosis and it is an early detectable event that involves double membrane vacuoles, acidic vesicular organelles, increase of expression of LC3-II and beclin-1. This noscapine derivative produces an increase of ROS levels and induces alterations in mitochondrial architecture, particularly in the intracristal compartment, accompanied of $\Delta\psi m$ dissipation. ROS production is a key event for autophagy and caspase-independent, ROS-dependent apoptosis induced by Red-Br-nos [116].

Capsaicin induces different modalities of cell death in human osteosarcoma G292 cells [117]. It increases cytosolic calcium concentration possibly by vanilloid receptor type 1 in endoplasmic reticulum and activates a mitochondrial caspase-dependent death signaling. Moreover, capsaicin induces autophagy with anti-apoptotic effects. Interestingly, inhibition of capsaicin-induced protective autophagy produces a second cell death signaling independent of caspase activation [117].

For evodiamine, it has been reported the participating of protective autophagy and mitochondria -dependent apoptosis in the anti-cancer effects on human cervix carcinoma HeLa and glioblastoma cells [118, 119]. In HeLa cells, evodiamine induces mitochondrial membrane permeabilization through increase of Bax/Bcl-2 or Bcl-xl ratio and changes the intracelullar redox status, which modulates the autophagy activation [118]. In gliobastoma cells, evodiamine induces an increase of cytosolic calcium and $\Delta\psi_m$ dissipation, producing calcium-mediated intrinsic apoptosis. Protective autophagy reaches maximum levels at 24h and it is activated by a calcium- c-Jun N-terminal kinases (JNK) signaling pathway [119]. In contrast, a report shows that evodiamine-induced cytotoxic effect is dependent of simultaneous induction of apoptosis and autophagy in SGC-7901 human gastric adenocarcinoma cells [120]. Similarly other alkaloids can induce mitochondrial apoptosis and autophagic cell death in a same cancer cell. Berberine induces cell death by induction of both pathways in liver cancer cells [121]. A pancratistatin alkaloid-derivative compound JCTH-4 also can induce apoptosis and autophagy in cancer cells [122, 123]. Decrease of mitochondrial membrane potential, ROS production, and appearance of autophagic marker LC3-II have been described in the JCTH-4-induced autophagy in p53 positive and p53 negative human colorectal cancer cells [122]. In addition, the parent alkaloid pancratistatin produces $\Delta\psi_m$ dissipation, inducing cell death by intrinsic pathway apoptosis and autophagy in human colon and metastatic prostate cancer cells [124, 125].

On the other hand, sanguilutine, a plant quaternary benzo[c]phenanthridine alkaloid, induces necroptosis, a caspase-independent non-apoptotic death in human melanoma cells. This alkaloid activates a protective autophagy, delaying the induction of necroptosis in melanoma cells [126]. These effects could have a therapeutic potential in the treatment of apoptosis -resistant melanoma tumors [127].

IV. Alkaloids with Clinical Uses that Affect Mitochondria of Cancer Cells

Mitochondria participate in several cellular functions and fine interactions with other cellular component (other organelles or cellular structures) are required to response to energy demand. Events of mitochondrial fission /fusion and biogenesis require coordinated gene expression and signaling [152, 153, 154, 155]. In cancer cells have been reported several features that promote the cancer progression such as increased mitochondrial biogenesis [156, 157] and increased mitochondrial fission in metastatic cancer cells [158]. Mitochondrial morphology and dynamics (mitochondrial fission/fusion) require a functional microtubules network and evidences indicate that the modification in microtubule network can induce intrinsic pathway apoptosis and altered mitochondrial dynamics [159, 160, 161].

Microtubule-targeting agents are amply used in chemotherapy and recent evidence suggests that they require of mitochondria for their anti-cancer effectiveness (Figure 6). Microtubule-destabilizing drugs such as alkaloids colchicine (used for gout treatment and familial mediterranean fever [162]) and nocodazole produce increase of free tubulin and in parallel decrease of $\Delta\psi m$ in human hepatoma cells [163]. In contrast, decrease of free tubulin and increase of $\Delta\psi m$ are produced by microtubule-stabilizing drug paclitaxel [163]. It has been reported that free tubulin blocks mitochondrial ANT and regulates respiration [164], resulting in a mechanism of control of mitochondrial bioenergetics from mitochondria-cytosol interface. Although the targets of microtubule-targeting agents are well-known in cancer cells, recently the impact on mitochondria has been reported [165]. Colchicine prevents the closure of cyclosporin-sensitive MPTP in tumor mitochondria [166]. On the other hand, microtubule-destabilizing drugs inhibit mitochondrial biogenesis through inhibition of mitochondrial mass increase and mtDNA replication [167]. Vinca alkaloids induce also mitochondria-dependent apoptosis. In melanoma cells, vinblastine triggers $\Delta\psi m$ dissipation, release of cytochrome -c and activation of caspases -9 and 3 [168] and in lung cancer cells, vinorelbine induces mitochondrial depolarization, increase of ROS production and activation of apoptosis [169]. Recently, it has been shown that mitochondrial ROS production contributes to efficacy of microtubule-targeting agents. Mitochondrial ROS modulate the microtubule dynamics through ROS-dependent End-Binding protein 1 (EB1) association with microtubules, being essential for cancer cell migration and proliferation [170]. Signaling involved in phosphorylation of EB1 participates in cytotoxic and anti-migratory effects of microtubule-targeting agents [170].

On the other hand, trabectedin (yondelis, ecteinaiscidia 743 (ET-743)), a tetrahydroisoquinoline alkaloid isolated from the marine tunicate *Ecteinascidia turbinate,* is a recent anticancer drug clinically in use [171]. It has been approved for treatment of advanced soft tissue sarcoma and ovarian cancer [172, 173, 174].

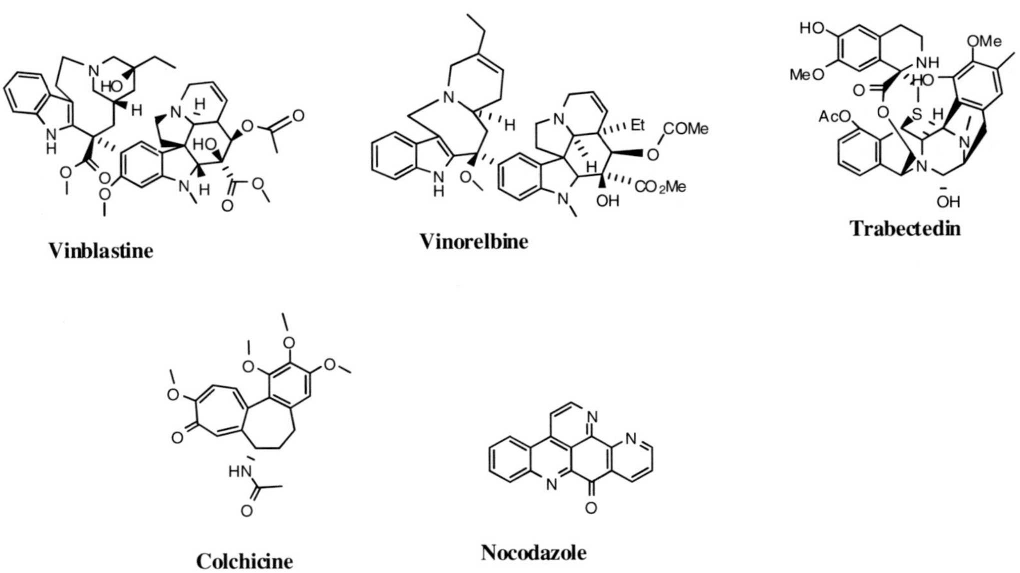

Figure 6. Alkaloids with clinical uses that affect mitochondria of cancer cells.

Interestingly, trabectedin appears to represent a new class of anti-neoplastic drugs acting on both cancer cells and the tumor microenvironment [175], directly affecting monocytes and tumor-associated macrophages or indirectly inhibiting the secretion of inflammatory signals involved in tumor angiogenesis [176]. Although its mechanism of action has been described as an alkylating agent [177], trabectedin can act by two different pathways: transcription -

dependent and transcription-independent [178]. In this latter pathway, continual JNK activation and mitochondria -dependent apoptosis, down-regulating the expression of Bcl-2, release of cytochrome -c and caspase-3 activation in MCF7 cancer cells has been described [179]. Similarly, it also has been reported intrinsic pathway apoptosis with increase of ROS production in MDA-MB -453 cancer cells. This evidence suggests that trabectedin causes selective activation of mitochondria-mediated apoptosis in two genotypically different models of breast cancer [180].

V. Strategies to Enhance Anti-Cancer Action of Alkaloids

1. Metabolic Reprogramming: Enhancing the Mitochondrial Metabolism in Cancer Cells

Cancer cells exhibit different metabolic organization during progression of cancer. Several cancer cells use mainly glycolysis as pathway for energy production and cell proliferation. In invasive and more aggressive cancer cells, this pathway can produce resistance to stimulus of cell death [181]. Moreover, metabolism of glucose and glutamine can compensate from one another to promote cancer cell survival [182, 183]. Metabolic organization can change to maintain the energetic requirements in cancer cells, up-regulating oxidative phosphorylation when glycolytic flux is inhibited [184], suggesting that in several glycolytic cancer cells the mitochondria remain functional [185].

In this scenario, PDK/PDH axis offers a promising option to enhance cytotoxic effect on cancer cells and overcome resistance to cell death. PDK/PDH axis is in a central position in the glucose oxidative metabolism. PDH complex catalyzes the rate-limiting oxidative decarboxylation of pyruvate into acetyl-CoA, being an important event to initiate the TCA cycle. PDH activity connects glycolysis with oxidative metabolism and is regulated by reversible phosphorylations. PDH kinase (PDK) phosphorylates and inactives PDH, in contrast, PDH phosphatase (PDP) activates PDH activity (Figure 7). Isoenzyme PDK3 is expressed in several type of cancers, including colon cancer [186], leukemia [187], melanoma s [185], conferring a glycolytic cellular phenotype. It has been suggested that PDK inhibition promotes oxidative metabolism, redirecting pyruvate into mitochondria and increase ROS production in cancer cells [188]. Moreover, it restores the sensitivity to anoikis, diminishing the metastatic potential [93], angiogenesis [188] and sensitizes to other cytotoxic agents [185, 189].

A pharmacologic strategy to enhance the oxidative metabolism in cancer cells is the use of dichloroacetate (DCA, Figure 7), a pyruvate mimetic that stimulates OXPHOS through inhibition of PDK. This compound is relatively inactive and induces apoptosis only at high concentrations, affecting selectively cells with defects in the mitochondrial ETC [190]. DCA synergizes the anti-cancer effects of existing chemotherapeutics that damage mtDNA (cisplatin and topotecan) [190], anti-diabetic complex I inhibitor [189] and several mitochondrial-targeted compounds and inductors of mitochondrial oxidative stress [191].

Metabolic reprogramming in cancer cells, specially drug-resistance cells, by combination of DCA and anti-cancer alkaloids is currently an unexplored area. Several anti-cancer alkaloids that affect mitochondrial respiration (e.g., berberine and lamellarin-D) or induce mitochondria -mediated apoptosis by increase of ROS levels (see Tables 1-3) could exhibit potent effects on several tumoral phenotypes with anti-proliferative and anti-metastatic actions.

2. Delivery System and Targeting to Cancer Cells

Selective delivery systems for cancer cells have been extensively explored in the recent years, offering selective pharmacologic actions, decrease of toxic side effect s and high potence of anti-tumoral agents [192].

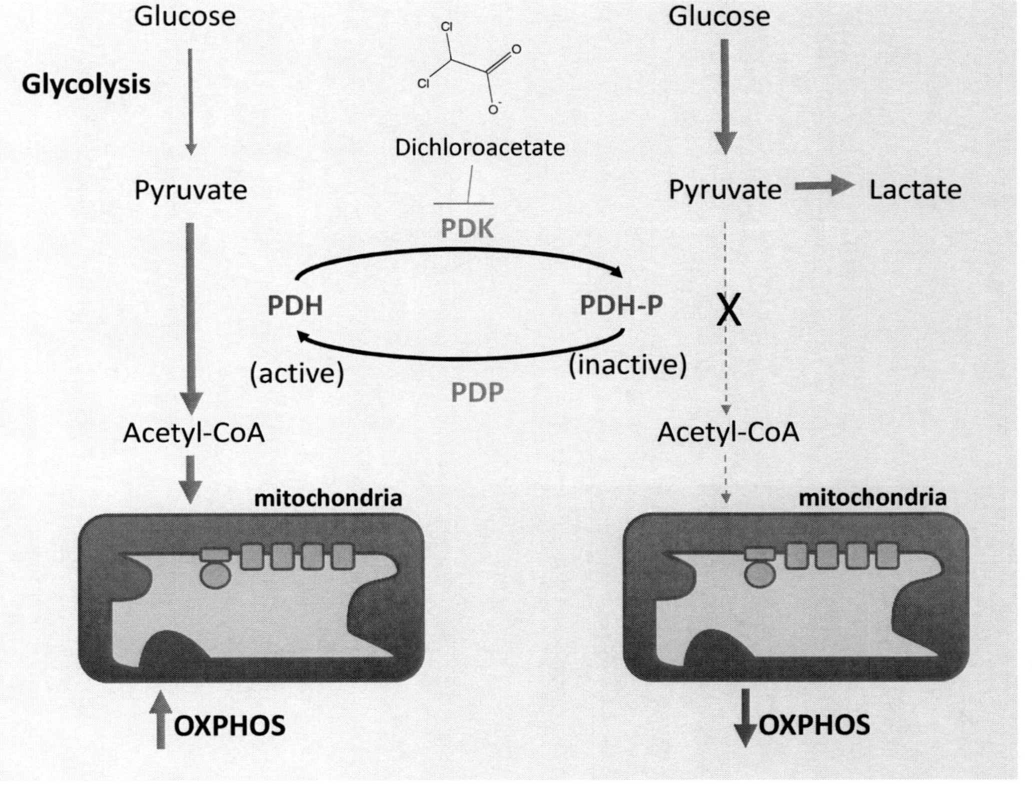

Figure 7. PDK/PDH axis and metabolic reprogramming in cancer cells. PDH activity is controled by PDK and PDP. When PDH is active, pyruvate (produced by glycolysis) is converted in Acetyl-CoA, which is used in TCA cycle and OXPHOS, increasing mitochondrial metabolism. In contrast when PDH is inactive, pyruvate is converted in lactate and mitochondrial metabolism, specially OXPHOS, is decreased. In this case, ATP demand is supplied by glycolysis in cancer cells. PDK inhibition by dichloroacetate (DCA) produces metabolic reprogramming toward mitochondrial metabolism, which increases OXPHOS and restores the sensitivity to anti-cancer compounds with mitochondrial targets.

Mitochondria of cancer cells exhibit differences compare to no-malignant cells such as mutations in mtDNA [193], different bioenergetic organization [37, 194], ultrastructural alterations, deficiency in β-F1 ATP synthase subunits, and have elevated mitochondrial membrane potential ($\Delta\psi m$) [195]. This latter characteristic has allowed the development of small molecules with selective anti-cancer action that incorporate a lipophilic cation (e.g., triarylphosphonium salts (TPP) and rhodamines) [196, 197]. Lipophilic cations (Figure 8A-B) are selectively accumulated within mitochondria by a $\Delta\psi m$ -dependent manner, leading to their several hundred-fold accumulation within mitochondria *in vivo* [198]. Interestingly, lipophilic cations (such as TPP) no produce specific delivery in mitochondrial compartments such as outer surface of the inner membrane, intermembrane space and outer membrane of mitochondria [12].

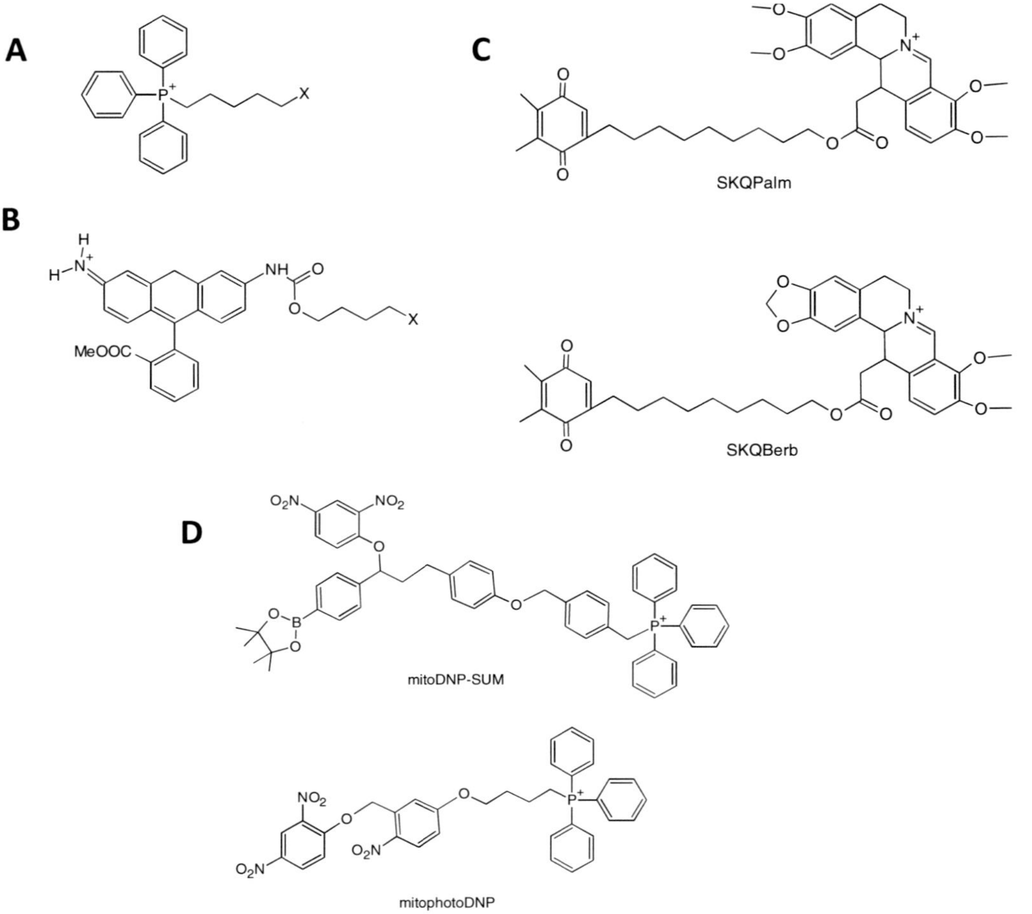

Figure 8. Mitochondrial delivery system based in lipophilic cations. (A) Triarylphosphonium salts (TPP) and (B) rhodamines are classic lipophilic cations used with chains of different lengths to bind a molecule of interest, targeting to mitochondria. (C) Berberine and palmatine, two cationic alkaloids that exhibit an accumulation in $\Delta\psi m$-dependent mitochondria, have been used to target antioxidant to mitochondria. (D) Two delivery systems based in TPP that allow to release an uncoupler (dinitrophenol, DNP) when mitochondrial ROS level is elevated (mitoDNP-SUM) or when a UV radiation favors the *in situ* cleavage of an uncoupler (mitophotoDNP).

Several natural compound s and derivatives have been conjugated with lipophilic cations, showing anti-cancer effects by a mitochondrial mechanism [199, 200, 201]; however, this strategy has not been used with anti-cancer alkaloids. Several strategies can be explored with anti-cancer alkaloids, using conjugated lipophilic cations: 1) covalent binding of alkaloid of interest to the lipophilic cation, a strategy well-studied to target small compounds, 2) photocleavable linker that binds to alkaloid of interest to the lipophilic cation. In this strategy, UV radiation produces photolysis *in situ* of the compound, releasing the active alkaloid within mitochondrial matrix. Recently, this strategy was described for a novel derivative of uncoupler DNP, mitophotoDNP (Figure 8D) [202], or 3) oxidative stress -sensitive linker that binds to alkaloid of interest to the lipophilic cation, a strategy that allows to release an active alkaloid when mitochondrial ROS level is elevated [203].

On the other hand, an useful extension of lipophilic cation such as delivery system to mitochondria with natural compound s has been investigated. Berberine and palmatine are cationic alkaloids that exhibit an accumulation in $\Delta\psi$m-dependent mitochondria [204]. This property has been used to design new mitochondria-targeted antioxidants, conjugating plastoquinone with berberine and palmatine [205]. These conjugates (SkQBerb and SkQPalm, Figure 8C) accumulate in mitochondria within 15-20 min and in absence of ADP, exhibit protonophoric effect, increasing the mitochondrial respiration. At higher concentrations, these compounds inhibit the rate of respiration, an event that is increased when the mitochondrial respiration is stimulated by NAD -dependent substrates [204]. This decrease of respiration is consistent with inhibition of complex I-dependent respiration by berberine, previously discussed. Moreover, protonophoric action of theses conjugates stimulate transport of fatty acid anions [206], which is mediated by ANT and is a event similar to the stimulation of ANT-dependent transport of fatty acids reported for tetraphenylphosphonium [207]. In intact cells, the accumulation of conjugates occurs within 1-1.5 h and left the cell without exhibit irreversible binding to cellular components [204]. In cervical cancer cells, SkQBerb and SkQPalm inhibit proliferation and partially restore normal phenotype of epithelium [204].

CONCLUSION

Mitochondria of cancer cells have a main role in the malignant phenotype, high proliferative capacity, resistance to cytotoxic drugs and metastasis. Because this, they are a valuable target for anti-cancer alkaloids. Some anti-cancer alkaloids selectively affect mitochondrial bioenergetics of cancer cells by inhibition of electron transport chain, inhibition of adenine nucleotide transport and TCA cycle. These effects can produce mitochondrial-mediated cell death. Therefore, these organelles offer many possibilities for the design of new alkaloid derivatives with selective actions. Additionally, several strategies to enhance anti-cancer action have been explored. Metabolic reprogramming, using a pharmacologic inhibition of PDK or organelle -specific delivery systems such as lipophilic cations are promising strategies to target subpopulations of cancer cells with resistance to apoptosis. Finally, because that between the anti-cancer compounds that act on mitochondrial functions, currently known mitocans (an acronym for mitochondria and cancer), anti-cancer alkaloids have not been included, this chapter contributes incorporating this type of compounds into this classification. In this chapter, we have summarized alkaloids and their derivatives that have the mitochondrial bioenergetics as a central component for their anti-cancer actions.

ACKNOWLEDGMENT

The authors gratefully acknowledge the financial support of CONICYT ACT 1107 grant, FAU PhD fellowship and FONDECYT grant 1140753.

REFERENCES

[1] S. Azimova, and M. Yunusov, Natural Compounds. Alkaloids. Plant Sources, Structure and Properties, 2013.

[2] C. Olano, C. Mendez, and J. Salas, Antitumor compounds from marine actinomycetes. *Mar. Drugs,* 7 (2009) 210-48.

[3] M. Wink, Molecular modes of action of cytotoxic alkaloids: From DNA intercalation, Spindle poisoning, Topoisomerase inhibition to Apoptosis and Multiple durg resistance. in: G. Cordell, (Ed.), The Alkaloids, Elsevier, 2007, pp. 1-47.

[4] C. Sawyers, Targeted cancer therapy. Nature, 432 (2004) 294-7.

[5] A. Petrelli, and S. Giordano, From single- to multi-target drugs in cancer therapy: when aspecificity becomes an advantage. *Curr. Med. Chem.*, 15 (2008) 422-32.

[6] C. Dumontet, and M. Jordan, Microtubule-binding agents: a dynamic field of cancer therapeutics. *Nat. Rev. Drug Discov.,* 9 (2010) 790-803.

[7] A. Rovini, A. Savry, D. Braguer, and M. Carre, Microtubule-targeted agents: when mitochondria become essential to chemotherapy. *Biochim. Biophys. Acta.*, 1807 (2011) 679-88.

[8] V. Kaminskyy, O. Kulachkovskyy, and R. Stoika, A decisive role of mitochondria in defining rate and intensity of apoptosis induction by different alkaloids. *Toxicol. Lett.,* 177 (2008) 168-81.

[9] M. Kavallaris, Microtubules and resistance to tubulin-binding agents. *Nat. Rev. Cancer,* 10 (2010) 194-204.

[10] N. Krall, J. Scheuermann, and D. Neri, Small Targeted Cytotoxics: Current State and Promises from DNA -Encoded Chemical Libraries. *Angew. Chem. Int. Ed.*, 52 (2013) 1384-1402.

[11] S. Rin Jean, D. Tulumello, S. Wisnovsky, E. Lei, M. Pereira, and S. Kelley, Molecular vehicles for mitochondrial chemical biology and drug delivery. *ACS Chem. Biol.,* 9 (2014) 323-33.

[12] R. Smith, R. Hartley, H. Cochemé, and M. Murphy, Mitochondrial pharmacology. *Trends Pharmacol. Sci.,* 33 (2012) 341-52.

[13] C. Desler, and L. Rasmussen, Mitochondria in biology and medicine -2012. *Mitochondrion,* 16 (2014) 2-6.

[14] A. S. Tan, J. Baty, and M. Berridge, The role of mitochondrial electron transport in tumorigenesis and metastasis. *Biochim. Biophys. Acta.,* 1840 (2014) 1454-63.

[15] P. Vandenabeele, L. Galluzzi, T. Vanden Berghe, and G. Kroemer, Molecular mechanisms of necroptosis: an ordered cellular explosion. *Nat. Rev. Mol. Cell Biol.*, 11 (2010) 700-14.

[16] J. Lee, S. Giordano, and J. Zhang, Autophagy, mitochondria and oxidative stress: cross-talk and redox signalling. *Biochem. J.,* 441 (2012) 523-40.

[17] A. Reichert, and W. Neupert, Mitochondriomics or what makes us breathe. *Trends Genet.*, 20 (2004) 555-62.

[18] Y. Chen, L. Fu, X. Wen, X. Wang, J. Liu, Y. Cheng, and J. Huang, Sirtuin-3 (SIRT3), a therapeutic target with oncogenic and tumor -suppressive function in cancer. *Cell Death Dis.*, 5 (2014) e1047.

[19] B. Chernyak, O. Pletjushkina, D. Izyumov, K. Lyamzaev, and A. Avetisyan, Bioenergetics and death. *Biochemistry (Mosc)*, 70 (2005) 240-5.

[20] R. Smith, R. Hartley, H. Cocheme, and M. Murphy, Mitochondrial pharmacology. *Trends Pharmacol. Sci.*, 33 (2012) 341-52.

[21] D. Wallace, W. Fan, and V. Procaccio, Mitochondrial energetics and therapeutics. *Annu. Rev. Pathol.*, 5 (2010) 297-348.

[22] L. Yousif, K. Stewart, K. Horton, and S. Kelley, Mitochondria -penetrating peptides: sequence effects and model cargo transport. *Chembiochem.*, 10 (2009) 2081-8.

[23] G. Chen, F. Wang, D. Trachootham, and P. Huang, Preferential killing of cancer cells with mitochondrial dysfunction by natural compound s. *Mitochondrion,* 10 (2010) 614-25.

[24] S. Ralph, S. Rodríguez-Enríquez, J. Neuzil, E. Saavedra, and R. Moreno-Sánchez, The causes of cancer revisited: "mitochondrial malignancy " and ROS-induced oncogenic transformation - why mitochondria are targets for cancer therapy. *Mol. Aspects Med.,* 31 (2010) 145-70.

[25] F. Urra, M. Cordova-Delgado, H. Pessoa-Mahana, O. Ramírez-Rodríguez, B. Weiss-Lopez, J. Ferreira, and R. Araya-Maturana, Mitochondria: a promising target for anticancer alkaloids. *Curr. Top. Med. Chem.*, 13 (2013) 2171-83.

[26] J. Neuzil, L. Dong, J. Rohlena, J. Truksa, and S. Ralph, Classification of mitocans, anti-cancer drugs acting on mitochondria. *Mitochondrion,* 13 (2013) 199-208.

[27] G. D'Souza, and V. Weissig, Subcellular targeting: a new frontier for drug-loaded pharmaceutical nanocarriers and the concept of the magic bullet. *Expert Opin. Drug Deliv.*, 6 (2009) 1135-48.

[28] C. V. Dang, A. Le, and P. Gao, MYC-Induced Cancer Cell Energy Metabolism and Therapeutic Opportunities. *Clin. Cancer Res.*, 15 (2009) 6479-6483.

[29] W. Ma, H. Sung, J. Park, S. Matoba, and P. Hwang, A pivotal role for p53: balancing aerobic respiration and glycolysis. *J. Bioenerg. Biomembr.*, 39 (2007) 243-6.

[30] A. Panov, and Z. Orynbayeva, Bioenergetic and antiapoptotic properties of mitochondria from cultured human prostate cancer cell line s PC-3, DU145 and LNCaP. *PLoS One,* 8 (2013) e72078.

[31] A. Santidrian, A. Matsuno-Yagi, M. Ritland, B. Seo, S. LeBoeuf, L. Gay, T. Yagi, and B. Felding-Habermann, Mitochondrial complex I activity and NAD +/NADH balance regulate breast cancer progression. *J. Clin. Invest.,* 123 (2013) 1068-81.

[32] F. Weinberg, R. Hamanaka, W. Wheaton, S. Weinberg, J. Joseph, M. Lopez, B. Kalyanaraman, G. Mutlu, G. Budinger, and N. Chandel, Mitochondrial metabolism and ROS generation are essential for Kras-mediated tumorigenicity. *Proc. Natl. Acad. Sci. U S A,* 107 (2010) 8788-93.

[33] S. J. Ralph, S. Rodríguez-Enríquez, J. Neuzil, E. Saavedra, and R. Moreno-Sánchez, The causes of cancer revisited: "Mitochondrial malignancy " and ROS-induced oncogenic transformation - Why mitochondria are targets for cancer therapy. *Mol. Aspects Med.,* 31 (2010) 145-170.

[34] J. Rohlena, L. Dong, S. Ralph, and J. Neuzil, Anticancer drugs targeting the mitochondrial electron transport chain. *Antioxid. Redox. Signal,* 15 (2011) 2951-74.

[35] M. Vanhuyse, J. Kluza, C. Tardy, G. Otero, C. Cuevas, C. Bailly, and A. Lansiaux, Lamellarin D: a novel pro-apoptotic agent from marine origin insensitive to P-glycoprotein-mediated drug efflux. *Cancer Lett.,* 221 (2005) 165-75.

[36] M. Brand, and D. Nicholls, Assessing mitochondrial dysfunction in cells. *Biochem. J.,* 435 (2011) 297-312.

[37] G. Solaini, G. Sgarbi, and A. Baracca, Oxidative phosphorylation in cancer cells. *Biochim. Biophys. Acta.,* 1807 (2011) 534-42.

[38] A. Lemarie, and S. Grimm, Mitochondrial respiratory chain complexes: apoptosis sensors mutated in cancer ? *Oncogene,* 30 (2011) 3985-4003.

[39] P. Pedersen, Warburg, me and Hexokinase 2: Multiple discoveries of key molecular events underlying one of cancers' most common phenotypes, the "Warburg Effect", i.e., elevated glycolysis in the presence of oxygen. *J. Bioenerg. Biomembr.,* 39 (2007) 211-22.

[40] S. Ralph, S. Rodriguez-Enriquez, J. Neuzil, E. Saavedra, and R. Moreno-Sanchez, The causes of cancer revisited: "mitochondrial malignancy " and ROS-induced oncogenic transformation - why mitochondria are targets for cancer therapy. *Mol. Aspects Med.,* 31 (2010) 145-70.

[41] C. Ballot, A. Martoriati, M. Jendoubi, S. Buche, P. Formstecher, L. Mortier, J. Kluza, and P. Marchetti, Another facet to the anticancer response to lamellarin D: induction of cellular senescence through inhibition of topoisomerase I and intracellular Ros production. *Mar. Drugs,* 12 (2014) 779-98.

[42] M. Tillhon, L. M. Guaman Ortiz, P. Lombardi, and A. I. Scovassi, Berberine: new perspectives for old remedies. *Biochem. Pharmacol.*, 84 (2012) 1260-7.

[43] P. R. Vuddanda, S. Chakraborty, and S. Singh, Berberine: a potential phytochemical with multispectrum therapeutic activities. *Expert Opin. Investig. Drugs,* 19 (2010) 1297-307.

[44] G. C. Pereira, A. F. Branco, J. A. Matos, S. L. Pereira, D. Parke, E. L. Perkins, T. L. Serafim, V. A. Sardao, M. S. Santos, A. J. Moreno, J. Holy, and P. J. Oliveira, Mitochondrially targeted effects of berberine [Natural Yellow 18, 5,6-dihydro-9,10-dimethoxybenzo(g)-1,3-benzodioxolo(5,6-a) quinolizinium] on K1735-M2 mouse melanoma cells: comparison with direct effects on isolated mitochondrial fractions. *J. Pharmacol. Exp. Ther.,* 323 (2007) 636-49.

[45] C. V. Diogo, N. G. Machado, I. A. Barbosa, T. L. Serafim, A. Burgeiro, and P. J. Oliveira, Berberine as a promising safe anti-cancer agent - is there a role for mitochondria ? *Curr. Drug Targets,* 12 (2011) 850-9.

[46] C. V. Pereira, N. G. Machado, and P. J. Oliveira, Mechanisms of berberine (natural yellow 18)-induced mitochondrial dysfunction: interaction with the adenine nucleotide translocator. *Toxicol. Sci.,* 105 (2008) 408-17.

[47] A. Chevrollier, D. Loiseau, P. Reynier, and G. Stepien, Adenine nucleotide translocase 2 is a key mitochondrial protein in cancer metabolism. *Biochim. Biophys. Acta.,* 1807 (2011) 562-7.

[48] O. Sharaf el dein, E. Mayola, J. Chopineau, and C. Brenner, The adenine nucleotide translocase 2, a mitochondrial target for anticancer biotherapy. *Curr. Drug Targets*, 12 (2011) 894-901.

[49] V. Kaminskyy, K. W. Lin, Y. Filyak, and R. Stoika, Differential effect of sanguinarine, chelerythrine and chelidonine on DNA damage and cell viability in primary mouse spleen cells and mouse leukemic cells. *Cell Biol. Int.,* 32 (2008) 271-7.

[50] A. Kemeny-Beke, J. Aradi, J. Damjanovich, Z. Beck, A. Facsko, A. Berta, and A. Bodnar, Apoptotic response of uveal melanoma cells upon treatment with chelidonine, sanguinarine and chelerythrine. *Cancer Lett.,* 237 (2006) 67-75.

[51] T. Serafim, J. Matos, V. Sardao, G. Pereira, A. Branco, S. Pereira, D. Parke, E. Perkins, A. Moreno, J. Holy, and P. Oliveira, Sanguinarine cytotoxicity on mouse melanoma K1735-M2 cells--nuclear vs. mitochondrial effects. *Biochem. Pharmacol.,* 76 (2008) 1459-75.

[52] W. Choi, G. Kim, W. Lee, and Y. Choi, Sanguinarine, a benzophenanthridine alkaloid, induces apoptosis in MDA-MB -231 human breast carcinoma cells through a reactive oxygen species -mediated mitochondrial pathway. *Chemotherapy,* 54 (2008) 279-87.

[53] A. Hussain, N. Al-Jomah, A. Siraj, P. Manogaran, K. Al-Hussein, J. Abubaker, L. Platanias, K. Al-Kuraya, and S. Uddin, Sanguinarine-dependent induction of apoptosis in primary effusion lymphoma cells. *Cancer Res.,* 67 (2007) 3888-97.

[54] M. Barreto, R. Pinto, J. Arrabaça, and M. Pavão, Inhibition of mouse liver respiration by Chelidonium majus isoquinoline alkaloids. *Toxicol. Lett.,* 146 (2003) 37-47.

[55] J. Chen, and J. Russo, Dysregulation of glucose transport, glycolysis, TCA cycle and glutaminolysis by oncogenes and tumor suppressors in cancer cells. *Biochim. Biophys. Acta.,* 1826 (2012) 370-84.

[56] R. A. Cairns, and T. W. Mak, Oncogenic isocitrate dehydrogenase mutations: mechanisms, models, and clinical opportunities. *Cancer Discov.*, 3 (2013) 730-41.

[57] C. Dang, Glutaminolysis: supplying carbon or nitrogen or both for cancer cells ?. *Cell Cycle,* 9 (2010) 3884-6.

[58] L. Yang, T. Moss, L. Mangala, J. Marini, H. Zhao, S. Wahlig, G. Armaiz-Pena, D. Jiang, A. Achreja, J. Win, R. Roopaimoole, C. Rodriguez-Aguayo, I. Mercado-Uribe, G. Lopez-Berestein, J. Liu, T. Tsukamoto, A. Sood, P. Ram, and D. Nagrath, Metabolic shifts toward glutamine regulate tumor growth, invasion and bioenergetics in ovarian cancer. *Mol. Syst. Biol.,* 10 (2014).

[59] J. Son, C. Lyssiotis, H. Ying, X. Wang, S. Hua, M. Ligorio, R. Perera, C. Ferrone, E. Mullarky, N. Shyh-Chang, Y. Kang, J. Fleming, N. Bardeesy, J. Asara, M. Haigis, R. DePinho, L. Cantley, and A. Kimmelman, Glutamine supports pancreatic cancer growth through a KRAS-regulated metabolic pathway. *Nature,* 496 (2013) 101-5.

[60] S. Gross, R. Cairns, M. Minden, E. Driggers, M. Bittinger, H. Jang, M. Sasaki, S. Jin, D. Schenkein, S. Su, L. Dang, V. Fantin, and T. Mak, Cancer-associated metabolite 2-hydroxyglutarate accumulates in acute myelogenous leukemia with isocitrate dehydrogenase 1 and 2 mutations. *J. Exp. Med.*, 207 (2010) 339-44.

[61] M. Bayet-Robert, S. Lim, C. Barthomeuf, and D. Morvan, Biochemical disorders induced by cytotoxic marine natural products in breast cancer cells as revealed by proton NMR spectroscopy -based metabolomics. *Biochem. Pharmacol.,* 80 (2010) 1170-9.

[62] V. M. Dirsch, S. O. Kirschke, M. Estermeier, B. Steffan, and A. M. Vollmar, Apoptosis signaling triggered by the marine alkaloid ascididemin is routed via caspase-2 and JNK to mitochondria. *Oncogene,* 23 (2004) 1586-93.

[63] A. G. Thomas, C. Rojas, C. Tanega, M. Shen, A. Simeonov, M. B. Boxer, D. S. Auld, D. V. Ferraris, T. Tsukamoto, and B. S. Slusher, Kinetic characterization of ebselen, chelerythrine and apomorphine as glutaminase inhibitors. *Biochem. Biophys. Res. Commun.,* 438 (2013) 243-8.

[64] Y. Zhang, and J. Yang, Altered energy metabolism in cancer: a unique opportunity for therapeutic intervention. *Cancer Biol. Ther.,* 14 (2013) 81-9.

[65] D. G. Hardie, AMP-activated/SNF1 protein kinases: conserved guardians of cellular energy. *Nat. Rev. Mol. Cell. Biol.*, 8 (2007) 774-85.

[66] S. Fogarty, and D. G. Hardie, Development of protein kinase activators: AMPK as a target in metabolic disorder s and cancer. *Biochim. Biophys. Acta.,* 1804 (2010) 581-91.

[67] C. Li, V. Liu, P. Chiu, K. Yao, H. Ngan, and D. Chan, Reduced expression of AMPK-β1 during tumor progression enhances the oncogenic capacity of advanced ovarian cancer. *Mol. Cancer,* 13 (2014).

[68] S. Hadad, L. Baker, P. Quinlan, K. Robertson, S. Bray, G. Thomson, D. Kellock, L. Jordan, C. Purdie, D. Hardie, S. Fleming, and A. Thompson, Histological evaluation of AMPK signalling in primary breast cancer. *BMC Cancer,* 9 (2009).

[69] K. A. Brown, N. U. Samarajeewa, and E. R. Simpson, Endocrine-related cancers and the role of AMPK. *Mol. Cell. Endocrinol.,* 366 (2013) 170-9.

[70] D. R. Alessi, K. Sakamoto, and J. R. Bayascas, LKB1-dependent signaling pathway s. *Annu. Rev. Biochem.,* 75 (2006) 137-63.

[71] M. Tiainen, K. Vaahtomeri, A. Ylikorkala, and T. P. Makela, Growth arrest by the LKB1 tumor suppressor: induction of p21(WAF1/CIP1). *Hum. Mol. Genet.,* 11 (2002) 1497-504.

[72] I. Kim, and Y. He, Targeting the AMP-Activated Protein Kinase for Cancer Prevention and Therapy. *Front. Oncol.*, 3 (2013).

[73] S. A. Hinke, G. A. Martens, Y. Cai, J. Finsi, H. Heimberg, D. Pipeleers, and M. Van de Casteele, Methyl succinate antagonises biguanide-induced AMPK-activation and death of pancreatic beta-cells through restoration of mitochondrial electron transfer. *Br. J. Pharmacol.,* 150 (2007) 1031-43.

[74] Y. Lee, W. Kim, K. Kim, M. Yoon, H. Cho, Y. Shen, J. Ye, C. Lee, W. Oh, C. Kim, C. Hohnen-Behrens, A. Gosby, E. Kraegen, D. James, and J. Kim, Berberine, a natural plant product, activates AMP-activated protein kinase with beneficial metabolic effects in diabetic and insulin -resistant states. *Diabetes,* 55 (2006) 2256-64.

[75] A. Cok, C. Plaisier, M. Salie, D. Oram, J. Chenge, and L. Louters, Berberine acutely activates the glucose transport activity of GLUT1. *Biochimie,* 93 (2011) 1187-92.

[76] L. Zhou, Y. Yang, X. Wang, S. Liu, W. Shang, G. Yuan, F. Li, J. Tang, M. Chen, and J. Chen, Berberine stimulates glucose transport through a mechanism distinct from insulin. *Metabolism,* 56 (2007) 405-12.

[77] J. Yin, Z. Gao, D. Liu, Z. Liu, and J. Ye, Berberine improves glucose metabolism through induction of glycolysis. *Am. J. Physiol. Endocrinol. Metab.*, 294 (2008) E148-56.

[78] S. Hawley, F. Ross, C. Chevtzoff, K. Green, A. Evans, S. Fogarty, M. Towler, L. Brown, O. Ogunbayo, A. Evans, and D. Hardie, Use of cells expressing gamma subunit variants to identify diverse mechanisms of AMPK activation. *Cell. Metab.*, 11 (2010) 554-65.

[79] N. Turner, J. Li, A. Gosby, S. To, Z. Cheng, H. Miyoshi, M. Taketo, G. Cooney, E. Kraegen, D. James, L. Hu, J. Li, and J. Ye, Berberine and its more biologically available derivative, dihydroberberine, inhibit mitochondrial respiratory complex I: a mechanism for the action of berberine to activate AMP-activated protein kinase and improve insulin action. *Diabete,* 57 (2008) 1414-8.

[80] J. J. Park, S. M. Seo, E. J. Kim, Y. J. Lee, Y. G. Ko, J. Ha, and M. Lee, Berberine inhibits human colon cancer cell migration via AMP-activated protein kinase-mediated downregulation of integrin beta1 signaling. *Biochem. Biophys. Res. Commun.,* 426 (2012) 461-7.

[81] H. S. Kim, M. J. Kim, E. J. Kim, Y. Yang, M. S. Lee, and J. S. Lim, Berberine-induced AMPK activation inhibits the metastatic potential of melanoma cells via reduction of ERK activity and COX-2 protein expression. *Biochem. Pharmacol.,* 83 (2012) 385-94.

[82] M. Tania, M. Khan, and J. Fu, Epithelial to mesenchymal transition inducing transcription factors and metastatic cancer. *Tumour Biol.*, (2014).

[83] H. Lahlou, and W. Muller, beta1 integrin as a molecular therapeutic target. *Int. J. Radiat. Biol.,* 83 (2011) 753-60.

[84] H. Kato, Z. Liao, J. Mitsios, H. Wang, E. Deryugina, J. Varner, J. Quigley, and S. Shattil, The primacy of β1 integrin activation in the metastatic cascade. *PLoS One,* 7 (2012) e46576.

[85] N. Cordes, and C.C. Park, beta1 integrin as a molecular therapeutic target. Int J Radiat Biol 83 (2007) 753-60.

[86] V. Nikoletopoulou, M. Markaki, K. Palikaras, and N. Tavernarakis, Crosstalk between apoptosis, necrosis and autophagy. *Biochim. Biophys. Acta.,* 1833 (2013) 3448-59.

[87] L. Galluzzi, S. A. Aaronson, J. Abrams, E. S. Alnemri, D. W. Andrews, E. H. Baehrecke, N. G. Bazan, M. V. Blagosklonny, K. Blomgren, C. Borner, D. E. Bredesen, C. Brenner, M. Castedo, J. A. Cidlowski, A. Ciechanover, G. M. Cohen, V. De Laurenzi, R. De Maria, M. Deshmukh, B. D. Dynlacht, W. S. El-Deiry, R. A. Flavell, S. Fulda, C. Garrido, P. Golstein, M. L. Gougeon, D. R. Green, H. Gronemeyer, G. Hajnoczky, J. M. Hardwick, M. O. Hengartner, H. Ichijo, M. Jaattela, O. Kepp, A. Kimchi, D. J. Klionsky, R. A. Knight, S. Kornbluth, S. Kumar, B. Levine, S. A. Lipton, E. Lugli, F. Madeo, W. Malomi, J. C. Marine, S. J. Martin, J. P. Medema, P. Mehlen, G. Melino, U. M. Moll, E. Morselli, S. Nagata, D. W. Nicholson, P. Nicotera, G. Nunez, M. Oren, J. Penninger, S. Pervaiz, M. E. Peter, M. Piacentini, J. H. Prehn, H. Puthalakath, G. A. Rabinovich, R. Rizzuto, C. M. Rodrigues, D. C. Rubinsztein, T. Rudel, L. Scorrano, H. U. Simon, H. Steller, J. Tschopp, Y. Tsujimoto, P. Vandenabeele, I. Vitale, K. H. Vousden, R. J. Youle, J. Yuan, B. Zhivotovsky, and G. Kroemer, Guidelines for the use and interpretation of assays for monitoring cell death in higher eukaryotes. *Cell. Death Differ.,* 16 (2009) 1093-107.

[88] K. Li, D. Wu, X. Chen, T. Zhang, L. Zhang, Y. Yi, Z. Miao, N. Jin, X. Bi, H. Wang, and J. Xu, Current and Emerging Biomarkers of Cell Death in Human Disease. 2014 (2014) 690103.

[89] L. Galluzzi, E. Morselli, O. Kepp, I. Vitale, A. Rigoni, E. Vacchelli, M. Michaud, H. Zischka, M. Castedo, and G. Kroemer, Mitochondrial gateways to cancer. *Mol. Aspects Med.,* 31 (2010) 1-20.

[90] S. Fulda, L. Galluzzi, and G. Kroemer, Targeting mitochondria for cancer therapy. *Nat. Rev. Drug Discov.,* 9 (2010) 447-464.

[91] D. A. Tennant, R. V. Duran, and E. Gottlieb, Targeting metabolic transformation for cancer therapy. *Nat. Rev. Cancer,* 10 (2010) 267-77.

[92] S. P. Mathupala, Y. H. Ko, and P. L. Pedersen, The pivotal roles of mitochondria in cancer: Warburg and beyond and encouraging prospects for effective therapies. *Biochim. Biophys. Acta.,* 1797 (2010) 1225-30.

[93] S. Kamarajugadda, L. Stemboroski, Q. Cai, N.E. Simpson, S. Nayak, M. Tan, and J. Lu, Glucose oxidation modulates anoikis and tumor metastasis. *Mol. Cell. Biol.*, 32 (2012) 1893-907.

[94] J. Y. Kim, S. G. Lee, J. Y. Chung, Y. J. Kim, J. E. Park, H. Koh, M. S. Han, Y. C. Park, Y. H. Yoo, and J. M. Kim, Ellipticine induces apoptosis in human endometrial cancer cells: the potential involvement of reactive oxygen species and mitogen -activated protein kinases. *Toxicology,* 289 (2011) 91-102.

[95] W. H. Chiu, S. J. Luo, C. L. Chen, J. H. Cheng, C. Y. Hsieh, C. Y. Wang, W. C. Huang, W. C. Su, and C. F. Lin, Vinca alkaloids cause aberrant ROS-mediated JNK activation, Mcl-1 downregulation, DNA damage, mitochondrial dysfunction, and apoptosis in lung adenocarcinoma cells. *Biochem. Pharmacol.*, 83 (2012) 1159-71.

[96] J. Poljaková, T. Eckschlager, J. Hrabeta, J. Hrebacková, S. Smutný, E. Frei, V. Martínek, R. Kizek, and M. Stiborová, The mechanism of cytotoxicity and DNA adduct formation by the anticancer drug ellipticine in human neuroblastoma cells. *Biochem. Pharmacol.*, 77 (2009) 1466-1479.

[97] E. Martinkova, M. Dontenwill, E. Frei, and M. Stiborova, Cytotoxicity of and DNA adduct formation by ellipticine in human U87MG glioblastoma cancer cells. *Neuro. Endocrinol. Lett.,* 30 (2009) 60-6.

[98] M. Stiborová, J. Poljaková, E. Martínková, L. Bořek-Dohalská, T. Eckschlager, R. Kizek, and E. Frei, Ellipticine cytotoxicity to cancer cell line s - a comparative study. *Interdiscip. Toxicol.*, 4 (2011) 98-105.

[99] H. Knölker, and K. Reddy, Isolation and synthesis of biologically active carbazole alkaloids. *Chem. Rev.,* 102 (2002) 4303-427.

[100] A. Schmidt, K. Reddy, and H. Knölker, Occurrence, biogenesis, and synthesis of biologically active carbazole alkaloids. *Chem. Rev.,* 112 (2012) 3193-328.

[101] S. Y. Fung, H. Yang, and P. Chen, Sequence effect of self-assembling peptides on the complexation and *In Vitro* delivery of the hydrophobic anticancer drug ellipticine. *PLoS ONE,* 3 (2008) e1956.

[102] S. Lu, H. Wang, Y. Sheng, M. Liu, and P. Chen, Molecular binding of self-assembling peptide EAK16-II with anticancer agent EPT and its implication in cancer cell inhibition. *J. Control Release,* 160 (2012) 33-40.

[103] M. Wu, Z. Ye, Y. Liu, B. Liu, and X. Zhao, Release of hydrophobic anticancer drug from a newly designed self-assembling peptide. *Mol. Biosyst.,* 7 (2011) 2040-7.

[104] L. Galluzzi, O. Kepp, S. Krautwald, G. Kroemer, and A. Linkermann, Molecular mechanisms of regulated necrosis. *Semin. Cell. Dev. Biol.,* (2014).

[105] L. Y. Zhang, Y. L. Wu, X. H. Gao, and F. Guo, Mitochondrial protein cyclophilin-D-mediated programmed necrosis attributes to berberine-induced cytotoxicity in cultured prostate cancer cells. *Biochem. Biophys. Res. Commun.,* (2014).

[106] A. Notte, L. Leclere, and C. Michiels, Autophagy as a mediator of chemotherapy - induced cell death in cancer. *Biochem. Pharmacol.*, 82 (2011) 427-34.

[107] M. B. Azad, Y. Chen, and S. B. Gibson, Regulation of autophagy by reactive oxygen species (ROS): implications for cancer progression and treatment. *Antioxid. Redox. Signal,* 11 (2009) 777-90.

[108] L. Li, G. Ishdorj, and S. B. Gibson, Reactive oxygen species regulation of autophagy in cancer: implications for cancer treatment. *Free Radic. Biol. Med.*, 53 (2012) 1399-410.

[109] K. Sharma, N. Le, M. Alotaibi, and D. Gewirtz, Cytotoxic autophagy in cancer therapy. *Int. J. Mol. Sci.,* 15 (2014) 10034-51.

[110] K. Gong, C. Chen, Y. Zhan, Y. Chen, Z. Huang, and W. Li, Autophagy-related gene 7 (ATG7) and reactive oxygen species /extracellular signal-regulated kinase regulate tetrandrine-induced autophagy in human hepatocellular carcinoma. *J. Biol. Chem.*, 287 (2012) 35576-88.

[111] C. Liu, K. Gong, X. Mao, and W. Li, Tetrandrine induces apoptosis by activating reactive oxygen species and repressing Akt activity in human hepatocellular carcinoma. *Int. J. Cancer*, 129 (2011) 1519-1531.

[112] P. Boya, R. A. Gonzalez-Polo, N. Casares, J. L. Perfettini, P. Dessen, N. Larochette, D. Metivier, D. Meley, S. Souquere, T. Yoshimori, G. Pierron, P. Codogno, and G. Kroemer, Inhibition of macroautophagy triggers apoptosis. *Mol. Cell. Biol.*, 25 (2005) 1025-40.

[113] G. L. Semenza, Mitochondrial autophagy: life and breath of the cell. *Autophagy*, 4 (2008) 534-6.

[114] H. Zhang, M. Bosch-Marce, L. A. Shimoda, Y. S. Tan, J. H. Baek, J. B. Wesley, F. J. Gonzalez, and G. L. Semenza, Mitochondrial autophagy is an HIF-1-dependent adaptive metabolic response to hypoxia. *J. Biol. Chem.*, 283 (2008) 10892-903.

[115] K. M. Rouschop, C. H. Ramaekers, M. B. Schaaf, T. G. Keulers, K. G. Savelkouls, P. Lambin, M. Koritzinsky, and B. G. Wouters, Autophagy is required during cycling hypoxia to lower production of reactive oxygen species. *Radiother. Oncol.*, 92 (2009) 411-6.

[116] P. Karna, S. Zughaier, V. Pannu, R. Simmons, S. Narayan, and R. Aneja, Induction of reactive oxygen species -mediated autophagy by a novel microtubule-modulating agent. *J. Biol. Chem.,* 285 (2010) 18737-48.

[117] C. Chien, K. Ma, H. Lee, P. Liu, Y. Li, Y. Huang, and S. Chueh, Dual effect of capsaicin on cell death in human osteosarcoma G292 cells. *Eur. J. Pharmaco.,* 718 (2013) 350-60.

[118] J. Yang, L. Wu, S. Tashino, S. Onodera, and T. Ikejima, Reactive oxygen species and nitric oxide regulate mitochondria -dependent apoptosis and autophagy in evodiamine-treated human cervix carcinoma HeLa cells. *Free Radic. Res.,* 42 (2008) 492-504.

[119] A. Liu, S. Wang, K. Chen, H. Kuei, Y. Shih, S. Hou, W. Chiu, S. Hsiao, and C. Shih, Evodiamine, a plant alkaloid, induces calcium /JNK-mediated autophagy and calcium/mitochondria -mediated apoptosis in human glioblastoma cells. *Chem. Biol. Interact.*, 205 (2013) 20-8.

[120] A. Rasul, B. Yu, L. Zhong, M. Khan, H. Yang, and T. Ma, Cytotoxic effect of evodiamine in SGC-7901 human gastric adenocarcinoma cells via simultaneous induction of apoptosis and autophagy. *Oncol. Rep.,* 27 (2012) 1481-7.

[121] N. Wang, Y. Feng, M. Zhu, C. Tsang, K. Man, Y. Tong, and S. Tsao, Berberine induces autophagic cell death and mitochondrial apoptosis in liver cancer cells: the cellular mechanism. *J. Cell. Biochem.,* 111 (2010) 1426-36.

[122] D. Ma, P. Tremblay, K. Mahngar, P. Akbari-Asl, J. Collins, T. Hudlicky, J. McNulty, and S. Pandey, A novel synthetic C-1 analogue of 7-deoxypancratistatin induces apoptosis in p53 positive and negative human colorectal cancer cells by targeting the mitochondria: enhancement of activity by tamoxifen. *Invest. New Drugs,* 30 (2012) 1012-27.

[123] D. Ma, P. Tremblay, K. Mahngar, J. Collins, T. Hudlicky, and S. Pandey, Selective cytotoxicity against human osteosarcoma cells by a novel synthetic C-1 analogue of 7-deoxypancratistatin is potentiated by curcumin. *PLoS One,* 6 (2011) e28780.

[124] C. Griffin, J. McNulty, and S. Pandey, Pancratistatin induces apoptosis and autophagy in metastatic prostate cancer cells. *Int. J. Oncol.*, 38 (2011) 1549-56.

[125] C. Griffin, A. Karnik, J. McNulty, and S. Pandey, Pancratistatin selectively targets cancer cell mitochondria and reduces growth of human colon tumor xenografts. *Mol. Cancer Ther.,* 10 (2011) 57-68.

[126] J. Hammerova, S. Uldrijan, E. Taborska, A. H. Vaculova, and I. Slaninova, Necroptosis modulated by autophagy is a predominant form of melanoma cell death induced by sanguilutine. *Biol. Chem.*, 393 (2012) 647-58.

[127] Y. S. Cho, and S. Y. Park, Harnessing of Programmed Necrosis for Fighting against Cancers. *Biomol. Ther. (Seoul),* 22 (2014) 167-75.

[128] S. Xiong, H. D. Pang, J. Fan, F. Ge, X. X. Yang, Q. Y. Liu, X. J. Liao, and S. H. Xu, In vitro and in vivo antineoplastic activity of a novel bromopyrrole and its potential mechanism of action. *Br. J. Pharmacol.,* 159 (2010) 909-18.

[129] H. T. Li, H. M. Wu, H. L. Chen, C. M. Liu, and C. Y. Chen, The pharmacological activities of (-)-anonaine. *Molecules,* 18 (2013) 8257-63.

[130] Z. F. Zhang, Y. Guo, J. B. Zhang, and X. H. Wei, Induction of apoptosis by chelerythrine chloride through mitochondrial pathway and Bcl-2 family proteins in human hepatoma SMMC-7721 cell. *Arch. Pharm. Res.*, 34 (2011) 791-800.

[131] J. Poljakova, T. Eckschlager, J. Hrabeta, J. Hrebackova, S. Smutny, E. Frei, V. Martinek, R. Kizek, and M. Stiborova, The mechanism of cytotoxicity and DNA adduct formation by the anticancer drug ellipticine in human neuroblastoma cells. *Biochem. Pharmacol.,* 77 (2009) 1466-79.

[132] Z. G. Yang, A. Q. Chen, and B. Liu, Antiproliferation and apoptosis induced by evodiamine in human colorectal carcinoma cells (COLO-205). *Chem. Biodivers.*, 6 (2009) 924-33.

[133] M. R. Cao, Q. Li, Z. L. Liu, H. H. Liu, W. Wang, X. L. Liao, Y. L. Pan, and J. W. Jiang, Harmine induces apoptosis in HepG2 cells via mitochondrial signaling pathway. *Hepatobiliary Pancreat. Dis. Int.*, 10 (2011) 599-604.

[134] R. Havelek, M. Seifrtova, K. Kralovec, L. Bruckova, L. Cahlikova, M. Dalecka, J. Vavrova, M. Rezacova, L. Opletal, and Z. Bilkova, The effect of Amaryllidaceae alkaloids haemanthamine and haemanthidine on cell cycle progression and apoptosis in p53 -negative human leukemic Jurkat cells. *Phytomedicine,* 21 (2014) 479-90.

[135] J. Kobayashi, and H. Morita, The Lycopodium alkaloids. *Alkaloids Chem. Biol.,* 61 (2005) 1-57.

[136] S. K. Mandal, R. Biswas, S. S. Bhattacharyya, S. Paul, S. Dutta, S. Pathak, and A. R. Khuda-Bukhsh, Lycopodine from Lycopodium clavatum extract inhibits proliferation of HeLa cells through induction of apoptosis via caspase-3 activation. *Eur. J. Pharmacol.,* 626 (2010) 115-22.

[137] Z. Jin, Amaryllidaceae and Sceletium alkaloids. *Nat. Prod. Rep.*, 26 (2009) 363-81.

[138] D. Lamoral-Theys, C. Decaestecker, V. Mathieu, J. Dubois, A. Kornienko, R. Kiss, A. Evidente, and L. Pottier, Lycorine and its derivatives for anticancer drug design. *Mini Rev. Med. Chem.,* 10 (2010) 41-50.

[139] J. McNulty, J. Nair, J. Bastida, S. Pandey, and C. Griffin, Structure-activity studies on the lycorine pharmacophore: A potent inducer of apoptosis in human leukemia cells. *Phytochemistry,* 70 (2009) 913-9.

[140] X. S. Liu, J. Jiang, X. Y. Jiao, Y. E. Wu, J. H. Lin, and Y. M. Cai, Lycorine induces apoptosis and down-regulation of Mcl-1 in human leukemia cells. *Cancer Lett.,* 274 (2009) 16-24.

[141] Y. Han, S. Zhang, J. Wu, K. Yu, Y. Zhang, L. Yin, and L. Bi, Matrine induces apoptosis of human multiple myeloma cells via activation of the mitochondrial pathway. *Leuk. Lymphoma,* 51 (2010) 1337-46.

[142] S. Carmely, M. Ilan, and Y. Kashman, 2-amino imidazole alkaloids from the marine sponge *Leucetta chagosensis*. *Tetrahedron,* 45 (1989) 2193-2200.

[143] D. V. LaBarbera, K. Modzelewska, A. I. Glazar, P. D. Gray, M. Kaur, T. Liu, D. Grossman, M. K. Harper, S. K. Kuwada, N. Moghal, and C. M. Ireland, The marine alkaloid naamidine A promotes caspase-dependent apoptosis in tumor cells. *Anticancer Drugs,* 20 (2009) 425-36.

[144] M. Liu, X. J. Luo, F. Liao, X. F. Lei, and W. G. Dong, Noscapine induces mitochondria -mediated apoptosis in gastric cancer cells in vitro and in vivo. *Cancer Chemother. Pharmacol.*, 67 (2011) 605-12.

[145] R. Aneja, T. Miyagi, P. Karna, T. Ezell, D. Shukla, M. Vij Gupta, C. Yates, S. R. Chinni, H. Zhau, L. W. Chung, and H. C. Joshi, A novel microtubule-modulating agent induces mitochondrially driven caspase-dependent apoptosis via mitotic checkpoint activation in human prostate cancer cells. *Eur. J. Cancer,* 46 (2010) 1668-78.

[146] A. McLachlan, N. Kekre, J. McNulty, and S. Pandey, Pancratistatin: a natural anti-cancer compound that targets mitochondria specifically in cancer cells to induce apoptosis. *Apoptosis,* 10 (2005) 619-30.

[147] C. Park, H. J. Shin, G. Y. Kim, T. K. Kwon, T. J. Nam, S. K. Kim, J. Cheong, I. W. Choi, and Y. H. Choi, Induction of apoptosis by streptochlorin isolated from Streptomyces sp. in human leukemic U937 cells. *Toxicol. In Vitro,* 22 (2008) 1573-81.

[148] Z. Li, X. Xu, Y. Huang, L. Ding, Z. Wang, G. Yu, D. Xu, W. Li, and D. Tong, Swainsonine activates mitochondria -mediated apoptotic pathway in human lung cancer A549 cells and retards the growth of lung cancer xenografts. *Int. J. Biol. Sci.*, 8 (2012) 394-405.

[149] R. Qin, H. Shen, Y. Cao, Y. Fang, H. Li, Q. Chen, and W. Xu, Tetrandrine induces mitochondria -mediated apoptosis in human gastric cancer BGC-823 cells. *PLoS One,* 8 (2013) e76486.

[150] S. H. Oh, and B. H. Lee, Induction of apoptosis in human hepatoblastoma cells by tetrandrine via caspase-dependent Bid cleavage and cytochrome c release. *Biochem. Pharmacol.,* 66 (2003) 725-31.

[151] S. M. Huang, P. C. Hsu, M. Y. Chen, W. S. Li, S. V. More, K. T. Lu, and Y. C. Wang, The novel indole compound SK228 induces apoptosis and FAK/Paxillin disruption in tumor cell line s and inhibits growth of tumor graft in the nude mouse. *Int. J. Cancer,* 131 (2012) 722-32.

[152] R. C. Scarpulla, R. B. Vega, and D. P. Kelly, Transcriptional integration of mitochondrial biogenesis. *Trends Endocrinol. Metab.*, 23 (2012) 459-66.
[153] T. Wenz, Regulation of mitochondrial biogenesis and PGC-1alpha under cellular stress. *Mitochondrion,* 13 (2013) 134-42.
[154] K. Elgass, J. Pakay, M. T. Ryan, and C. S. Palmer, Recent advances into the understanding of mitochondrial fission. *Biochim. Biophys. Acta.*, 1833 (2013) 150-61.
[155] M. Escobar-Henriques, and F. Anton, Mechanistic perspective of mitochondrial fusion: tubulation vs. fragmentation. *Biochim. Biophys. Acta.,* 1833 (2013) 162-75.
[156] X. Wang, and C. T. Moraes, Increases in mitochondrial biogenesis impair carcinogenesis at multiple levels. *Mol. Oncol.*, 5 (2011) 399-409.
[157] E. Samper, L. Morgado, J. C. Estrada, A. Bernad, A. Hubbard, S. Cadenas, and S. Melov, Increase in mitochondrial biogenesis, oxidative stress, and glycolysis in murine lymphomas. *Free Radic. Biol. Med.*, 46 (2009) 387-96.
[158] J. Zhao, J. Zhang, M. Yu, Y. Xie, Y. Huang, D. Wolff, P. Abel, and Y. Tu, Mitochondrial dynamics regulates migration and invasion of breast cancer cells. *Oncogene,* 32 (2013) 4814-24.
[159] V. Voccoli, and L. Colombaioni, Mitochondrial remodeling in differentiating neuroblasts. *Brain Res.,* 3 (2009) 15-19.
[160] M. Estève, M. Carré, and D. Braguer, Microtubules in apoptosis induction: are they necessary?. *Curr. Cancer Drug Targets,* 7 (2007) 713-29.
[161] M. Oropesa, M. de la Mata, J. Maraver, M. Cordero, D. Cotán, A. Rodríguez-Hernández, I. Domínguez-Moñino, M. de Miguel, P. Navas, and J. Sánchez-Alcázar, Apoptotic microtubule network organization and maintenance depend on high cellular ATP levels and energized mitochondria. *Apoptosis,* 16 (2011) 404-24.
[162] C. Cerquaglia, M. Diaco, G. Nucera, M. La Regina, M. Montalto, and R. Manna, Pharmacological and clinical basis of treatment of Familial Mediterranean Fever (FMF) with colchicine or analogues: an update. *Curr. Drug Targets Inflamm. Allergy,* 4 (2005) 117-24.
[163] E. Maldonado, J. Patnaik, M. Mullins, and J. Lemasters, Free tubulin modulates mitochondrial membrane potential in cancer cells. *Cancer Res.,* 70 (2010) 10192-201.
[164] T. Rostovtseva, K. Sheldon, E. Hassanzadeh, C. Monge, V. Saks, S. Bezrukov, and D. Sackett, Tubulin binding blocks mitochondrial voltage-dependent anion channel and regulates respiration. *Proc. Natl. Acad. Sci. USA,* 105 (2008) 18746-51.
[165] N. André, D. Braguer, G. Brasseur, A. Gonçalves, D. Lemesle-Meunier, S. Guise, M. Jordan, and C. Briand, Paclitaxel induces release of cytochrome c from mitochondria isolated from human neuroblastoma cells. *Cancer Res.,* 60 (2000) 5349-53.
[166] Y. V. Evtodienko, V. V. Teplova, S. S. Sidash, F. Ichas, and J. P. Mazat, Microtubule-active drugs suppress the closure of the permeability transition pore in tumour mitochondria. *FEBS Lett.,* 393 (1996) 86-8.
[167] M. Karbowski, J. H. Spodnik, M. Teranishi, M. Wozniak, Y. Nishizawa, J. Usukura, and T. Wakabayashi, Opposite effects of microtubule-stabilizing and microtubule-destabilizing drugs on biogenesis of mitochondria in mammalian cells. *J. Cell. Sci.,* 114 (2001) 281-91.
[168] D. Selimovic, H. Badura, A. El-Khattouti, M. Soell, B. Porzig, A. Spernger, F. Ghanjati, S. Santourlidis, Y. Haikel, and M. Hassan, Vinblastine-induced apoptosis of

melanoma cells is mediated by Ras homologous A protein (Rho A) via mitochondrial and non-mitochondrial-dependent mechanisms. *Apoptosis,* 18 (2013) 980-97.

[169] W. Chiu, S. Luo, C. Chen, J. Cheng, C. Hsieh, C. Wang, W. Huang, W. Su, and C. Lin, Vinca alkaloids cause aberrant ROS-mediated JNK activation, Mcl-1 downregulation, DNA damage, mitochondrial dysfunction, and apoptosis in lung adenocarcinoma cells. *Biochem. Pharmacol.,* 83 (2012) 1159-71.

[170] M. Le Grand, A. Rovini, V. Bourgarel-Rey, S. Honore, S. Bastonero, D. Braguer, and M. Carre, ROS-mediated EB1 phosphorylation through Akt/GSK3β pathway: implication in cancer cell response to microtubule-targeting agents. *Oncotarget,* 5 (2014) 3408-23.

[171] D. Newman, and G. Cragg, Marine-sourced anti-cancer and cancer pain control agents in clinical and late preclinical development. *Mar. Dugs,* 12 (2014) 255-78.

[172] G. D. Demetri, S. P. Chawla, M. von Mehren, P. Ritch, L. H. Baker, J. Y. Blay, K. R. Hande, M. L. Keohan, B. L. Samuels, S. Schuetze, C. Lebedinsky, Y. A. Elsayed, M. A. Izquierdo, J. Gomez, Y. C. Park, and A. Le Cesne, Efficacy and safety of trabectedin in patients with advanced or metastatic liposarcoma or leiomyosarcoma after failure of prior anthracyclines and ifosfamide: results of a randomized phase II study of two different schedules. *J. Clin. Oncol.*, 27 (2009) 4188-96.

[173] I. Ray-Coquard, Biology of ovarian cancer and trabectedin mechanism of action. *Future Oncol.*, 9 (2013) 11-7.

[174] T. Ueda, S. Kakunaga, M. Ando, K. Yonemori, H. Sugiura, K. Yamada, and A. Kawai, Phase I and pharmacokinetic study of trabectedin, a DNA minor groove binder, administered as a 24-h continuous infusion in Japanese patients with soft tissue sarcoma. *Invest. New Drugs,* (2014).

[175] M. D'Incalci, N. Badri, C. Galmarini, and P. Allavena, Trabectedin, a drug acting on both cancer cells and the tumour microenvironment. *Br. J. Cancer.*

[176] C. Galmarini, M. D'Incalci, and P. Allavena, Trabectedin and plitidepsin: drugs from the sea that strike the tumor microenvironment. *Mar. Drugs,* 12 (2014) 719-33.

[177] N. Penel, S. Cousin, A. Duhamel, and A. Kramar, Activity endpoints reported in soft tissue sarcoma phase II trials: quality of reported endpoints and correlation with overall survival. *Crit. Rev. Oncol. Hematol.,* 88 (2013) 309-17.

[178] X. Lin, M. Liu, C. Hu, and D. Liao, Targeting cellular proapoptotic molecules for developing anticancer agents from marine sources. *Curr. Drug Targets,* 11 (2010) 708-15.

[179] C. Gajate, F. An, and F. Mollinedo, Differential cytostatic and apoptotic effects of ecteinascidin-743 in cancer cells. Transcription-dependent cell cycle arrest and transcription -independent JNK and mitochondrial mediated apoptosis. *J. Biol. Chem.,* 277 (2002) 41580-9.

[180] H. Atmaca, E. Bozkurt, S. Uzunoglu, R. Uslu, and B. Karaca, A diverse induction of apoptosis by trabectedin in MCF-7(HER2−/ER+) and MDA-MB -453 (HER2+/ER−) breast cancer cells. *Toxicol. Lett.*, 221 (2013) 128-136.

[181] Y. C. Shen, D. L. Ou, C. Hsu, K. L. Lin, C. Y. Chang, C. Y. Lin, S. H. Liu, and A. L. Cheng, Activating oxidative phosphorylation by a pyruvate dehydrogenase kinase inhibitor overcomes sorafenib resistance of hepatocellular carcinoma. *Br. J. Cancer,* 108 (2013) 72-81.

[182] C. Yang, J. Sudderth, T. Dang, R. M. Bachoo, J. G. McDonald, and R. J. DeBerardinis, Glioblastoma cells require glutamate dehydrogenase to survive impairments of glucose metabolism or Akt signaling. *Cancer Res.,* 69 (2009) 7986-93.

[183] J. Son, C. A. Lyssiotis, H. Ying, X. Wang, S. Hua, M. Ligorio, R. M. Perera, C. R. Ferrone, E. Mullarky, N. Shyh-Chang, Y. Kang, J. B. Fleming, N. Bardeesy, J. M. Asara, M. C. Haigis, R. A. DePinho, L. C. Cantley, and A. C. Kimmelman, Glutamine supports pancreatic cancer growth through a KRAS-regulated metabolic pathway. *Nature,* 496 (2013) 101-5.

[184] M. Wu, A. Neilson, A. L. Swift, R. Moran, J. Tamagnine, D. Parslow, S. Armistead, K. Lemire, J. Orrell, J. Teich, S. Chomicz, and D. A. Ferrick, Multiparameter metabolic analysis reveals a close link between attenuated mitochondrial bioenergetic function and enhanced glycolysis dependency in human tumor cells. *Am. J. Physiol. Cell. Physiol.,* 292 (2007) C125-36.

[185] J. Kluza, P. Corazao-Rozas, Y. Touil, M. Jendoubi, C. Maire, P. Guerreschi, A. Jonneaux, C. Ballot, S. Balayssac, S. Valable, A. Corroyer-Dulmont, M. Bernaudin, M. Malet-Martino, E. M. de Lassalle, P. Maboudou, P. Formstecher, R. Polakowska, L. Mortier, and P. Marchetti, Inactivation of the HIF-1alpha/PDK3 signaling axis drives melanoma toward mitochondrial oxidative metabolism and potentiates the therapeutic activity of pro-oxidants. *Cancer Res.,* 72 (2012) 5035-47.

[186] C. W. Lu, S. C. Lin, K. F. Chen, Y. Y. Lai, and S. J. Tsai, Induction of pyruvate dehydrogenase kinase-3 by hypoxia -inducible factor -1 promotes metabolic switch and drug resistance. *J. Biol. Chem.,* 283 (2008) 28106-14.

[187] J. Kluza, M. Jendoubi, C. Ballot, A. Dammak, A. Jonneaux, T. Idziorek, S. Joha, V. Dauphin, M. Malet-Martino, S. Balayssac, P. Maboudou, G. Briand, P. Formstecher, B. Quesnel, and P. Marchetti, Exploiting mitochondrial dysfunction for effective elimination of imatinib-resistant leukemic cells. *PLoS One,* 6 (2011) e21924.

[188] G. Sutendra, P. Dromparis, A. Kinnaird, T. H. Stenson, A. Haromy, J. M. Parker, M. S. McMurtry, and E. D. Michelakis, Mitochondrial activation by inhibition of PDKII suppresses HIF1a signaling and angiogenesis in cancer. *Oncogene,* 32 (2013) 1638-50.

[189] Y. W. Choi, and I. K. Lim, Sensitization of metformin -cytotoxicity by dichloroacetate via reprogramming glucose metabolism in cancer cells. *Cancer Lett.,* 346 (2014) 300-8.

[190] L. H. Stockwin, S. X. Yu, S. Borgel, C. Hancock, T. L. Wolfe, L. R. Phillips, M. G. Hollingshead, and D. L. Newton, Sodium dichloroacetate selectively targets cells with defects in the mitochondrial ETC. *Int. J. Cancer,* 127 (2010) 2510-9.

[191] P. Corazao-Rozas, P. Guerreschi, M. Jendoubi, F. Andre, A. Jonneaux, C. Scalbert, G. Garcon, M. Malet-Martino, S. Balayssac, S. Rocchi, A. Savina, P. Formstecher, L. Mortier, J. Kluza, and P. Marchetti, Mitochondrial oxidative stress is the Achille's heel of melanoma cells resistant to Braf-mutant inhibitor. *Oncotarget,* 4 (2013) 1986-98.

[192] Y. Yamada, and H. Harashima, Mitochondrial drug delivery systems for macromolecule and their therapeutic application to mitochondrial diseases. *Adv. Drug Deliv. Rev.,* 60 (2008) 1439-62.

[193] M. Brandon, P. Baldi, and D. C. Wallace, Mitochondrial mutations in cancer. *Oncogene,* 25 (2006) 4647-62.

[194] U. Dier, D. H. Shin, L. P. Hemachandra, L. M. Uusitalo, and N. Hempel, Bioenergetic analysis of ovarian cancer cell line s: profiling of histological subtypes and identification of a mitochondria -defective cell line. *PLoS One,* 9 (2014) e98479.

[195] J. M. Cuezva, M. Krajewska, M. L. de Heredia, S. Krajewski, G. Santamaría, H. Kim, J. M. Zapata, H. Marusawa, M. Chamorro, and J. C. Reed, The Bioenergetic Signature of Cancer. *Cancer Res.*, 62 (2002) 6674-6681.

[196] M. Millard, J. D. Gallagher, B. Z. Olenyuk, and N. Neamati, A selective mitochondrial-targeted chlorambucil with remarkable cytotoxicity in breast and pancreatic cancer s. *J. Med. Chem.*, 56 (2013) 9170-9.

[197] A. Sturzu, S. Sheikh, H. Echner, T. Nagele, M. Deeg, B. Amin, C. Schwentner, M. Horger, U. Ernemann, and S. Heckl, Rhodamine-marked bombesin: a novel means for prostate cancer fluorescence imaging. *Invest. New Drugs*, 32 (2014) 37-46.

[198] C. M. Porteous, A. Logan, C. Evans, E. C. Ledgerwood, D. K. Menon, F. Aigbirhio, R. A. Smith, and M. P. Murphy, Rapid uptake of lipophilic triphenylphosphonium cations by mitochondria in vivo following intravenous injection: implications for mitochondria-specific therapies and probes. *Biochim. Biophys. Acta.*, 1800 (2010) 1009-17.

[199] N. Sassi, L. Biasutto, A. Mattarei, M. Carraro, V. Giorgio, A. Citta, P. Bernardi, S. Garbisa, I. Szabo, C. Paradisi, and M. Zoratti, Cytotoxicity of a mitochondriotropic quercetin derivative: mechanisms. *Biochim. Biophys. Acta.*, 1817 (2012) 1095-106.

[200] N. Sassi, A. Mattarei, M. Azzolini, P. Bernardi, I. Szabo, C. Paradisi, M. Zoratti, and L. Biasutto, Mitochondria -targeted resveratrol derivatives act as cytotoxic pro-oxidants. *Curr. Pharm. Des.*, 20 (2014) 172-9.

[201] L. Biasutto, A. Mattarei, E. Marotta, A. Bradaschia, N. Sassi, S. Garbisa, M. Zoratti, and C. Paradisi, Development of mitochondria -targeted derivatives of resveratrol. *Bioorg. Med. Chem. Lett.*, 18 (2008) 5594-7.

[202] S. Chalmers, S. T. Caldwell, C. Quin, T. A. Prime, A. M. James, A. G. Cairns, M. P. Murphy, J. G. McCarron, and R. C. Hartley, Selective uncoupling of individual mitochondria within a cell using a mitochondria-targeted photoactivated protonophore. *J. Am. Chem. Soc.*, 134 (2012) 758-61.

[203] S. J. McQuaker, C. L. Quinlan, S. T. Caldwell, M. D. Brand, and R. C. Hartley, A prototypical small-molecule modulator uncouples mitochondria in response to endogenous hydrogen peroxide production. *Chembiochem.*, 14 (2013) 993-1000.

[204] B. Chernyak, Y. Antonenko, E. Galimov, L. Domnina, V. Dugina, R. Zvyagilskaya, O. Ivanova, D. Izyumov, K. Lyamzaev, A. Pustovidko, T. Rokitskaya, A. Rogov, I. Severina, R. Simonyan, M. Skulachev, V. Tashlitsky, E. Titova, T. Trendeleva, and G. Shagieva, Novel mitochondria -targeted compounds composed of natural constituents: conjugates of plant alkaloids berberine and palmatine with plastoquinone. *Biochemistry (Mosc)*, 77 (2012) 983-95.

[205] K. Lyamzaev, A. Pustovidko, R. Simonyan, T. Rokitskaya, L. Domnina, O. Ivanova, I. Severina, N. Sumbatyan, G. Korshunova, V. Tashlitsky, V. Roginsky, Y. Antonenko, M. Skulachev, B. Chernyak, and V. Skulachev, Novel mitochondria -targeted antioxidants: plastoquinone conjugated with cationic plant alkaloids berberine and palmatine. *Pharm. Res.*, 28 (2011) 2883-95.

[206] A. Pustovidko, T. Rokitskaya, I. Severina, R. Simonyan, T. Trendeleva, K. Lyamzaev, Y. Antonenko, A. Rogov, R. Zvyagilskaya, V. Skulachev, and B. Chernyak, Derivatives of the cationic plant alkaloids berberine and palmatine amplify protonophorous activity of fatty acids in model membranes and mitochondria. *Mitochondrion,* 13 (2013) 520-5.

[207] V. Dedukhova, E. Mokhova, A. Starkov, and Y. Leikin, Carboxyatractylate inhibits the potentiating effect of lipophylic cation TPP+ on uncoupling activity of fatty acid. *Biochem. Mol. Biol. Int.,* 30 (1993) 1161-7.

In: Alkaloids
Editor: Eduardo Sobarzo-Sánchez

ISBN: 978-1-63482-074-5

Chapter 2

BOLDINE AS ANTIOXIDANT: FROM FARM TO PHARMACY

***Seyed Fazel Nabavi*[1]*, Maria Daglia*[2]*, Eduardo Sobarzo-Sánchez*[3] *and Seyed Mohammad Nabavi*[1*]**

[1]Applied Biotechnology Research Center,
Baqiyatallah University of Medical Sciences, Tehran, Iran
[2]Department of Drug Sciences, Medicinal Chemistry and Pharmaceutical Technology Section, University of Pavia, Italy
[3]Department of Pharmaceutical Technology, Faculty of Pharmacy,
University of Santiago de Compostela, Santiago de Compostela, Spain

ABSTRACT

Oxidative stress plays a causative and important role in the initiation and progression of various diseases such as cancer, cardiovascular and neurodegenerative pathologies. Over the last decades, much attention has been focused on diet as a rich source of natural antioxidants such as flavonoids, carotenoids, and alkaloids. Recent studies have demonstrated that different alkaloids, derived from plants, microorganisms, and marine organisms, are effective antioxidants both under in vitro and in vivo conditions. In addition, alkaloids are known as highly scavengers of reactive oxygen and nitrogen species (such as hydroxyl radical and peroxynitrite) and from this way, possess a wide range of health -promoting effects. Boldine is one of the most important antioxidant alkaloids, widely distributed in the Chilean medicinal plant, commonly known as boldo. There are several scientific studies about the in vitro antioxidant effects of boldine as well as in vivo protective role of this alkaloid against oxidative stress. According to its chemical structure, the phenol group bound to the aporphine ring and the benzylic amine function play important roles in the antioxidant and free radical scavenging activities of boldine. Furthermore, boldine inhibitory activity against lipid peroxidation is due to the presence of hydroxyl moieties in its chemical skeleton. The aim of the present chapter is to examine the antioxidant activity of boldine and its mechanism of action in order to explain its potential health-promoting effects. As a first step, the definitions and important aspects regarding reactive species, oxidative stress and potential role of oxidative stress have been discussed as background for next steps of this chapter. Subsequently, we provide information regarding boldo (the main source of boldine), boldine, its chemistry and toxicity. Finally we critically reviewed the available literature data on antioxidant and free radical scavenging activities of boldine and its potential mechanisms of actions.

Keywords: alkaloid, antioxidant, boldo, boldine, free radical, oxidative stress

INTRODUCTION

The term free radical refers to the group of reactive molecules having one or more unpaired electron (s) [1]. These molecules are known both as essential mediators of critical processes in the human body (i.e., neurotransmission and inflammatory responses) and as toxic byproducts of oxygen metabolisms [2]. In the aerobic respiration, reactive oxygen species are generated through oxygen reduction as well as electron transport processes [3]. During this reduction, most of oxygen is bonded to hydrogen which is involved in the oxidative phosphorylation [4]. However, a minor part of oxygen is partially reduced and leads to production of free radicals and/or other non-radical reactive species [5]. There are similar processes for reactive nitrogen species production [6, 7]. The produced radical forms of reactive species include superoxide, peroxyl, alloxyl, hydroxyl, nitric oxide and nitrogen dioxide [8, 9]. In addition, non-radical reactive species include singlet oxygen, hydrogen peroxide, hypochlorous acid, ozone, nitrous acid, nitrosyl, nitroxyl, peroxynitrite, etc. [10, 11]. Reactive species can easily react with different biomolecules such as proteins, lipids, carbohydrates, DNA and RNA and from this way lead to oxidative stress through subtraction of a hydrogen atom from these molecules [12]. This process leads to oxidation of hydrogen and/or electron donating molecules and therefore, reactive species are known as (pro)-oxidants [13].

The term oxidative stress refers to the imbalance between oxidant and antioxidant levels, which leads to oxidative injuries and cell death [14]. Oxidants are common products of aerobic metabolisms of living systems [15]. However, excessive production of oxidants under physiological conditions leads to oxidative stress and from this way causes a plethora of detrimental abnormalities in body tissues [16-18]. Superoxide anions, hydrogen peroxide and the hydroxyl radicals are oxidants which are generated by one, two and three electrons losses during the reduction of oxygen to water [19]. In addition, singlet oxygen is the first state of molecular oxygen which is known as highly reactive oxygen species. It easily reacts with singlet molecules [20]. Alkyl or peroxyl radicals are two common oxidants in lipids which are generated from oxygen radicals [21]. Furthermore, peroxynitrite is a well-known non-radical oxidative species which is generated by the reaction of nitric oxide with superoxide under physiological conditions [22]. In addition, different types of radiation such as X-radiation, ultraviolet radiation, and ultrasound produce reactive oxygen species [23, 24].

The term antioxidant refers to a substance which can significantly prevent, stop and/or delay the oxidation of the oxidizable materials as well as cellular organelles via prevention and/or inhibition of chain reactions [25, 26]. In living systems, there is a multiple range of antioxidants such as different antioxidant enzymes (such as catalase, superoxide dismutase, glutathione peroxidases, etc.) as well as non-enzymatic antioxidants (such as glutathione, L-ascorbic acid, α-tocopherol, etc.) [27-29]. However, insufficient antioxidant defense systems enhance the susceptibility to oxidative stress and lead to oxidative injuries and cell death [27]. Considering the numerous adverse effects of synthetic antioxidants, much attention has been paid to natural antioxidants such as phytochemicals [30].

Phytochemicals are bioactive natural antioxidants which occur in different fruits and vegetables [26]. These plant-derived antioxidants are defined as chemicals with health - promoting effects which derive from plants, but cannot be considered human essential nutrients [31]. Due to the complexity of some plant-derived antioxidants, there is limited information about the health-promoting effects of these compounds [32]. However, major groups of natural antioxidants are carotenoids, flavonoids, alkaloids, etc. [33].

The term alkaloid refers to the group of natural products which contains biological amine and cyclic compounds including basic nitrogen in their rings [34]. Alkaloids are reported in different organisms such as plants, bacteria, fungi, marine organisms and animals [35]. Nowadays, much attention has been paid to alkaloids due to their wide range of health - promoting effects including anti-inflammatory, antioxidant, neuroprotective effects, etc. [36-38]. There is a plethora of literature data on the antioxidant activities of alkaloids both under in vitro and in vivo conditions [39-41]. In the present paper, we critically reviewed the available literature data on the antioxidant activities of boldine, the major constituent of medicinal plant boldo.

BOLDO AS A MEDICINAL PLANT

Boldo (*Peumus boldus* Molina) is a member of Monimiaceae family and is known as one of the most important medicinal plants with a wide range of traditional uses in Chile [42]. Over the past two decades, much attention has been paid to the pharmacological activities of boldo due to its high alkaloid content and potent protective and beneficial effect s against digestive and hepato-biliary diseases [43]. This plant is the only species in the genus *Peumus* which is endemic to central and southern Chile [43]. Boldo tree is often growing to 6-12 m; however, there are some boldo trees reaching up to 20 m. The dark green leaves of boldo are perennial and leathery [44]. The leaf of boldo has been used as herbal tea in Latin America due to its bitter flavor and camphor-like aroma [45]. Its tasty and nutritious fruits are edible spheres (diameter 2 cm) which appear between December and February [44]. Dried fruits of boldo have been used as beads to make bracelets and necklaces [44].

BOLDO IN TRADITIONAL MEDICINE

Boldo is one of the most important medicinal plants, which have multiple therapeutic effects in Chilean folk medicine [42]. It has been used by Araucanian Mapuche as a tonic herb [46]. Boldo stimulates hepatic functions and bile flow [47, 48]. It is also used for prevention and/or mitigation of digestive and hepatobiliary disorders, gallstones and gallbladder pain [43]. In traditional medicine of Chile, boldo is used in the form of infusions, tinctures as well as extracts. The combination of boldo with other medicinal plants such as Barberry (Berberis thunbergii DC.) and/or White Fringetree (*Chionanthus virginicus* L.) possesses beneficial effect s on gallstones [46]. Boldo has also been used to treat rheumatisms

[43]. In addition, treatises dealing with medicinal plants showed that boldo has a wide range of therapeutic effects in the treatment of headache, earache, nasal congestion, dropsy, menstrual pain, syphilis and gonorrhea, dyspepsia, flatulence, rheumatism and nervous weakness [43]. The leaves of boldo have been used as antiseptic, anti-nociceptive and diuretic agent [49]. It has also been used to treat urinary tract infection s and intestinal parasites [49]. Furthermore, many years ago boldo has been used as antimalarial agent. Currently, boldo has been used in homeopathy [50].

Cultivation of Boldo

Cultivation of boldo is recognized as one of the interesting possibility of improving agricultural income for Chilean farmers due to its multiple culinary and pharmaceutical uses [51]. Boldo is a perennial tree that grows in full sun climates. This plant grows better in loamy well drained soils and full sun [51]. However, it can grow in the semi shaded environments. This plant is commonly propagated by seeds and also half-ripe cutting [51]. In addition, boldo dislikes moisture and needs sandy and acid soil to develop to its best potential [52].

Boldine, One of the Main Constituents of Boldo

Boldine ($C_{19}H_{21}NO_4$ or 2,9-dihydroxy-1, 10-dimethoxyaporphine) (figure 1) is an aporphine alkaloid which is widely distributed in leaves and bark of boldo tree [53]. Boldine is a solid compound with pale yellow color and bitter taste [53]. It is soluble in dilute mineral acids, ammonia, acetic acid and ethyl alcohol. The level of boldine in boldo leaves is more than 0.14% [54]. However, it has been identified in negligible amount also in *Nectandra glandiflora* Nees & Mart., *Dehaasia kurzii* King ex Hook. f., *Lindera angustifolia* W.C. Cheng, *Litsea rotundifolia* Hemsl., as well as *Actinodaphne pruinosa* Nees [55-59].

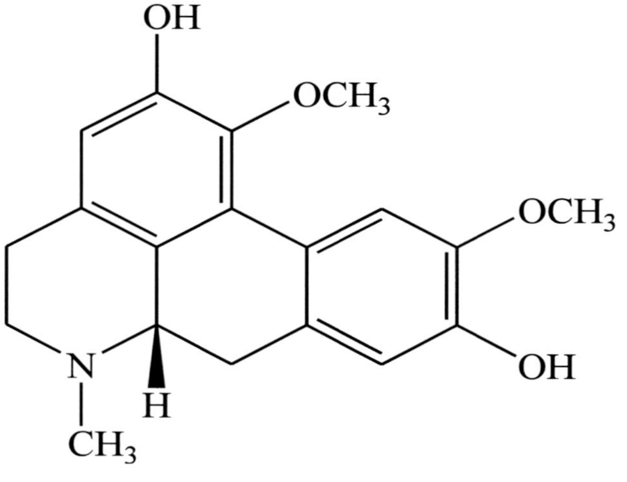

Figure 1. Chemical structure of boldine.

It is well known that the phenol groups (which are bonded to the aporphine ring) and the benzylic amine function are responsible for the antioxidant and free radical scavenging activities of boldine [60]. In addition, the presence of hydroxyl moieties in the chemical skeleton of boldine is responsible for its lipid peroxidation inhibitory activity [61]. The level of boldine in the bark of boldo is more than 6% [62]. Paper electrophoresis, voltammetric method, thin-layer chromatography and gas chromatography are common analytical methods for boldine determination [63].

TOXICITY OF BOLDINE

Considering the wide range of pharmacological effects of boldo, there are few data on the toxicity and adverse effects of boldo and its active constituent, boldine, to date [64]. Boldine does not cause death in the mammalians even at high doses. According to animal studies, orally administration of boldine at 15 g kills a 12 kg dog [43]. Therefore, it can be concluded that 500 and 1000 mg/kg of body weight are maximum non-fatal doses of boldine for mice and guinea pigs [65]. There are no scientific reports about genotoxicity and/or carcinogenicity of boldine [66]. In addition, oral administration of boldine-rich extract at 3000 mg/kg of body weight causes no toxicity in rats [67].

Furthermore, boldine is known as non-mutagenic compound according to the SOS chromotest [68]. It has also been reported that intraperitoneal administration of boldine causes no genotoxicity in mouse bone marrow cell [69]. Despite to the safety of boldo extract and boldine, there is a toxic volatile constituent in the essential oil of boldo. It has been reported that ascaridole, 1,8-cineole and *p*-cymene are the main volatile constituents of boldo among which ascaridole is a highly toxic compound [70].

BOLDINE AS ANTIOXIDANT

Due to the deleterious effects of reactive oxygen and nitrogen species and oxidative stress in different human diseases, recently, much attention has been paid to antioxidant compounds [71]. However, there are numerous scientific reports regarding the wide range of adverse effects of some synthetic antioxidants such as butylated hydroxytoluene and butylated hydroxyanisole [72, 73]. Therefore, the search for new natural antioxidants, with high efficacy and low adverse effects, has been developed over the past decade. [74]. Moreover, over the last two decades, much attention has been paid to the antioxidant and free radical scavenging potential of boldine, which is the main alkaloidal constituent of Chilean medicinal plant, boldo, and its related compounds [75]. Hirschmann et al. [76] showed that hot water extract of boldo leaves has scavenger activity against 1,1-diphenyl-2-picryl-hydrazyl radical and inhibits lipid peroxidation. These in vitro properties were ascribed to the high content of bioactive constituents such as catechin and boldine [76]. In addition to this, Quezada et al. [54] showed that crude extract and alkaloid and flavonoid fractions of boldo possess antioxidant activity under in vitro conditions. They found that the reducing power of boldo crude extract and fractions is comparable to other natural and synthetic antioxidants [54]. Furthermore, Fernandez et al. [77] evaluated the inhibitory effects of boldo infusion against cisplatin-induced lipid peroxidation in liver of experimental mice. They found that pretreatment with boldo leaf infusion significantly reduces lipid peroxidation in liver tissues [77]. Fale et al. also reported that water extract of boldo inhibits acetylcholinesterase activity and possesses a potent antioxidant activity [78]. In addition, Lara-Fernandez et al. [79] demonstrated that boldo extract scavenges 2,2 diphenyl-1-picrylhydrazyl radicals. They found that there is low correlation between antioxidant activity of boldo and its boldine content [79]. Recently, Klimaczewski et al. [75] found that leaf extract of boldo and its active alkaloid, boldine, inhibit Fe^{2+}-citrate-induced mitochondrial oxidative stress in rat liver tissues.

Speisky et al. [61] reported that boldine decreases lipid peroxidation and prevents lysozyme inactivation in red cell plasma membrane s auto-oxidation induced by 2,2'-azobis(2-amidinopropane). Valenzuela et al. [80] evaluated the inhibitory role of boldine against the spontaneous and metal-induced sardine oil oxidation. They found that boldine shows a potent protective effect against oil oxidation which is comparable with standard antioxidants, used as positive control [80]. Another study by Cederbaum et al. [81] reported that boldine inhibits both iron -dependent and iron-independent liver microsomal lipid peroxidation. In accordance with these results, Ubeda et al. [82] reported that boldine and other benzylisoquinoline alkaloids inhibit microsomal lipid peroxidation in a dose-dependent manner. Kringstein and Cederbaum [83] also found that boldine inhibits human liver microsomal lipid peroxidation and inactivates human cytochrome P4502E1. Hu et al. [84] also reported that boldine and its dimethoxy analogue, glaucine, prevent 12-*O*-tetradecanoylphorbol-13-acetate-induced downregulation of gap junction in a dose-dependent manner. They also found that this protective effect is associated with the inhibition of intracellular peroxides, translocation of protein kinase C, and phosphorylation of connexin 43 [84]. Moreover, boldine possesses cytoprotective activity against tert-butyl hydroperoxide-induced oxidative stress and lytic damage in isolated rat hepatocytes through modification of the level of non-enzymatic antioxidants such as thiobarbituric acid reactive substances and reduced glutathione [85]. According to another study reported by Jimenez et al. [86], boldine protects erythrocyte against hemolytic damage induced by 2,2'-azobis-(2-amidinopropane) through its antioxidant effects. In addition, Kubinova et al. [87] showed that boldine possesses chemoprotective effects through inhibition of the activities of CYP1A-dependent 7-ethoxyresorufin-*O*-deethylase and CYP3A-dependent testosterone 6β-hydroxylase in the Hepa-1 cells microsomes. They also found that boldine mitigates the abnormality in the activity of glutathione *S*-transferase [87]. It has also been reported that boldine mitigates brain mitochondrial dysfunction induced by catecholamine oxidation and also reduces the PC12 cells death induced by dopamine through its antioxidant and free radical scavenging activities [88]. In addition, Lau et al. [89] demonstrated that boldine treatment causes endothelial protective activity in hypertensive rats through aortic superoxide inhibition and peroxynitrite generation as well as down-regulation of p47phox protein expression in rat's aortas. Lau et al. [89] also reported that oral administration of boldine at 20 mg/kg/day for one week improves endothelium -dependent relaxation in the aortas of diabetic mice through suppression of oxidative stress and angiotensin II -mediated BMP4-dependent oxidative stress [89]. They also reported that the incubation of mouse aortic endothelial cells with boldine (1 μM/L) restores the level of phosphorylation of eNOS [89]. In addition to this, Lau et al. [90] examined the protective effects of boldine against hyperglycemia -induced oxidative stress and endothelial function in the rat aortas through the reduction of reactive oxygen species production and normalization of NOX2 and p47phox protein expression. Konrath et al. [91]

examined the antioxidant and neuroprotective effects of boldine against oxygen -glucose deprivation exposure in the hippocampal slices of rat brain. They concluded that boldine possesses both potent antioxidant and prooxidant activities (at high concentrations) [91]. Moreover, Santanam et al. [92] examined the inhibitory effect of boldine on the oxidative stress under in vitro and in vivo conditions. They found that boldine reduces in vitro LDL oxidation and also mitigates the lesion formation in atherosclerotic mice [92]. Jang et al. [93] showed that 8 weeks treatment with boldine mitigate oxidative stress in diabetic rats through the decrease of the levels of malondialdehyde and carbonyls as well as the activities of mitochondrial manganese superoxide dismutase in different tissues of rats. They also found that boldine treatment increases the activity of glutathione peroxidase and decreases the production of superoxide anions, hydrogen peroxide s, hydroxyl and nitric oxide radicals in different tissues of diabetic rats [93].

Structure Activity Relationship

Rìos et al. in 1990 reported that phenolic benzylisoquinoline alkaloids inhibit lipid peroxidation and also possess superoxide scavenging activity and from this way interfere in the peroxidative processes [94]. Thereafter, a plethora of studies has been performed on the antioxidant and free radical scavenging activities of different isoquinoline alkaloids and related derivatives [95, 96]. Martìnez et al. [97] showed that different benzylisoquinoline alkaloids (i.e., benzylisoquinolines, phtalideisoquinolines, aporphines, protoberberines, and benzophenanthridines) inhibit Fe^{2+}/cysteine -induced lipid peroxidation in the microsomal fractions of rat liver. They found that boldine possesses potent lipid peroxidation inhibitory actions and from this way, they concluded that the presence of hydroxyl group in the C1 moiety of aporphine structures plays an important role in the lipid peroxidation inhibitory action of aporphine derivatives [97]. According to these results, the antioxidant activity of bulbocapnine is greater than anonaine (which has the same chemical structure of bulbocapnine, and IC_{50} values are 12.5 μM for bulbocapnine and 27 μM for anonaine) due to the presence of a hydroxyl group in the C11 position [64]. It is also reported that the dihydroxy-containing apomorphine derivate showed the best antioxidant activity (IC_{50} is 3.3 μM) which is due to the presence of catechol with hydroxyl moieties at C10 and C11 positions [64]. *In vitro* studies on glaucine (*O*-dimethylated boldine) showed that although the presence of hydroxyl groups is not essential for the antioxidant activity of aporphine analogues, it can increase their antioxidant activity [64]. Casselst et al. [98] showed that among aporphine derivatives, apomorphine is the most active antioxidant in in vitro model of auto-oxidation of brain homogenates. They also found that *O*-methylation in boldine and glaucine induces a marginal decrease in their antioxidant activity [98]. According to these results, Ubeda et al. [82] reported that glaucine has slightly lower activity than boldine (similar compound) in the Fe^{3+}–EDTA–H_2O_2-induced deoxyribose degradation model. In another study reported by Loghin et al. [99], glaucine was shown to possess only half of the antioxidant activity of its similar compound, boldine, both in Fe^{2+}/H_2O_2/ascorbic acid - induced lipid peroxidation in the liver microsomes and in AAPH-induced lysozyme inactivation models. The same author also reported that boldine halogenation with bromo or iodine at the C-3 position causes no protective effects on lysozyme inactivation, but affects the inhibitory effects of boldine in the microsomes. This result was ascribed to the lipophilicity on the chain-breaking lipid peroxidation process within the membranes [99]. Another study reported by Martìnez et al. showed that in the absence of phenolic moieties, aporphines are easily oxidized into the forms of dehydro- and oxo- aporphines [97]. Cassels et al. [98] also reported that in the aporphine dervatives without phenolic moieties, the benzylic C–6a–H bond is the initial site of free radical attack and also instable free radicals are stabilized by contribution of nitrogen lone pair through electron donation. They also reported that N-methylglaucinium (analogue of glaucine with protonated nitrogen) is unable to inhibit lysozyme inactivation [98]. In the hypoxanthine–xanthine oxidase model, *N*-methylboldinium showed only half of free radical scavenging activity of boldine. Therefore, it can be concluded that the presence of the nitrogen lone pair plays an important role in the antioxidant activity of both phenolic and non-phenolic aporphines derivatives [100]. In addition, a structure–activity study showed that the piperidine ring does not seem to be essential for the antioxidant activity of some of aporphine derivatives such as boldine [100]. Teng et al. [101] also reported that *N*-allylsecoboldine (a phenanthrenic derivative of boldine in which nitrogen atom acts as side chain in its chemical structure) inhibits the lipid peroxidation in rat brain homogenates and erythrocyte membranes and also shows scavenger activity against the hydroxyl and peroxyl radicals as well as boldine. Furthermore, it has been reported that boldine possesses lower free radical scavenging activity than secoboldine, *N*-methylsecoboldine and *N*-methylsecoboldinium in the hypoxanthine–xanthine oxidase model [100]. It can be concluded that the better antioxidant activity of these phenolic phenanthrenic compounds is due to the presence of the third benzylic ring which provides more capacity for phenoxy free radical scavenging activity. However, boldine possesses a better antioxidant activity than non-phenolic phenanthrene such as *O*-dimethylated-*N*-methyl secoboldinium, which is due to the presence of free hydroxyl moieties in the phenolic rings of boldine [100]. Martìnez et al. [97] found that although reticuline and laudanosine possess a weak antioxidant activity, laudanosoline shows potent antioxidant activity which is due to the effects of *O*-methylation on the antioxidant activity of benzylisoquinoline alkaloids. In addition, it can be concluded that the presence of catechol moiety can influence the antioxidant activity of both aporphine and benzylisoquinoline alkaloids [97]. However, despite the presence of more catechol moiety in the chemical structure of laudanosoline versus apormorphine, it possesses lower antioxidant activity due to the greater ability of aporphines to trap free radicals trapping associated with increasing of the spin delocalization of phenoxy radicals across the biphenyl system [97].

Conclusion and Future Prospects

In the present chapter, we showed that boldine is the main alkaloidal constituent of the Chilean medicinal plant, boldo. Boldine shows potent antioxidant activity and free radical scavenging effects both under in vitro and in vivo conditions. We also showed that the antioxidant activity of boldine is due to the presence of phenol groups in the aporphine ring of its chemical structure. In view of the multiple applications of boldo in the traditional medicine, it can be concluded that boldine causes no adverse effects and therefore it could be used as antioxidant for mitigation of oxidative stress. Nevertheless, there is no clinical study on the beneficial effect s of boldo and its alkaloidal constituent boldine till now, and so it is very difficult to make a clear decision about the effective doses of this compound in humans. In view of its safety, we suggested that future clinical trials should be performed on the efficacy of boldine for treatment of oxidative-related diseases. In addition, we recommend that future studies should focus on:

- Finding of best analytical methods for boldine separation and purification.
- Increasing the bioavailability of boldine by employing of new drug delivery systems.
- Finding the molecular mechanisms of health -promoting effects of boldine.
- Finding the most effective doses of boldine for clinical trials.

REFERENCES

[1] Halliwell, B., Murcia, M. A., Chirico, S., Aruoma, O. I. (1995). Free radical s and antioxidants in food and in vivo: what they do and how they work. *Crit. Rev. Food Sci. Nutr.*, 35(1-2), 7-20.

[2] McCord, J. M. (2000). The evolution of free radicals and oxidative stress. *Am. J. Med.*, 108(8), 652-659.

[3] Hendry, G. A. (1993). Oxygen, free radical processes and seed longevity. *Seed Sci. Res.*, 3(03), 141-153.

[4] Matés, J. M., Sanchez-Jimenez, F. M. (2000). Role of reactive oxygen species in apoptosis: implications for cancer therapy. *Int. J. Biochem. cell Biol.*, 32(2), 157-170.

[5] Halliwell, B. (1989). Free radical s, reactive oxygen species and human disease: a critical evaluation with special reference to atherosclerosis. *Brit. J. Exp. Pathology*, 70(6), 737.

[6] Dedon, P. C., Tannenbaum, S. R. (2004). Reactive nitrogen species in the chemical biology of inflammation. *Arch. Biochem. Biophysic.*, 423(1), 12-22.

[7] Darley-Usmar, V., Halliwell, B. (1996). Blood radicals: reactive nitrogen species, reactive oxygen species, transition metal ions, and the vascular system. *Pharm. Res.*, 13(5), 649-662.

[8] Halliwell, B. (2005). *Free radical s and other reactive species in disease*. John Wiley & Sons, Ltd.

[9] Beckman, J. S., Koppenol, W.H. (1996). Nitric oxide, superoxide, and peroxynitrite: the good, the bad, and the ugly. *Am. J. Physiol.-Cell Physiol.*, 40(5), C1424.

[10] Halliwell, B. A. R. R. Y. (2001). Free radical reactions in human disease. *Oxidative Stress Dis.*, 7, 1-16.

[11] Pryor, W. A., Houk, K.N., Foote, C. S., Fukuto, J.M., Ignarro, L.J., Squadrito, G.L., Davies, K.J. (2006). Free radical biology and medicine: it's a gas, man !. American Journal of Physiology-Regulatory, *Integr. Comp. Physiol.*, 291(3), R491-R511.

[12] Apel, K., Hirt, H. (2004). Reactive oxygen species: metabolism, oxidative stress, and signal transduction. *Annu. Rev. Plant Biol.*, 55, 373-399.

[13] Winterbourn, C. C. (2008). Reconciling the chemistry and biology of reactive oxygen species. *Nature Chem. Biol.*, 4(5), 278-286.

[14] Klaunig, J. E., Kamendulis, L. M. (2004). The role of oxidative stress in carcinogenesis. *Annu. Rev. Pharmacol. Toxicol.*, 44, 239-267.

[15] Finkel, T., Holbrook, N.J. (2000). Oxidants, oxidative stress and the biology of ageing. *Nature*, 408(6809), 239-247.

[16] Finkel, T. (2003). Oxidant signals and oxidative stress. *Curr. Opin. Cell Biol.*, 15(2), 247-254.

[17] Sayre, L. M., Smith, M. A., Perry, G. (2001). Chemistry and biochemistry of oxidative stress in neurodegenerative disease. *Curr. Med. Chem.*, 8(7), 721-738.

[18] Halliwell, B. (2006). Oxidative stress and neurodegeneration: where are we now?. *J. Neurochem.*, 97(6), 1634-1658.

[19] Raha, S., Robinson, B.H. (2000). Mitochondria, oxygen free radicals, disease and ageing. *Trend Biochem. Sci.*, 25(10), 502-508.

[20] Lee, J., Koo, N., Min, D.B. (2004). Reactive oxygen species, aging, and antioxidative nutraceuticals. *Compr. Rev. Food Sci. Food Safety*, 3(1), 21-33.

[21] Patel, R. P., Boersma, B. J., Crawford, J. H., Hogg, N., Kirk, M., Kalyanaraman, B., Parks, D.A., Barnes, S., Darley-Usmar, V. (2001). Antioxidant mechanisms of isoflavones in lipid systems: paradoxical effects of peroxyl radical scavenging. *Free Radic. Biol. Med.*, 31(12), 1570-1581.

[22] Virag, L., Szabo, E., Gergely, P., Szabo, C. (2003). Peroxynitrite-induced cytotoxicity: mechanism and opportunities for intervention. *Toxicol. Lett.*, 140, 113-124.

[23] Rugo, R. E., Secretan, M. B., Schiestl, R. H. (2009). X radiation causes a persistent induction of reactive oxygen species and a delayed reinduction of TP53 in normal human diploid fibroblasts. *Radiat. Res.*, 158(2), 210-219.

[24] A-H-Mackerness, S., John, C. F., Jordan, B., Thomas, B. (2001). Early signaling components in ultraviolet-B responses: distinct roles for different reactive oxygen species and nitric oxide. *FEBS lett.*, 489(2), 237-242.

[25] Kagiya, G., Ogawa, R., Tabuchi, Y., Feril Jr, L. B., Nozaki, T., Fukuda, S., Yamamoto, K., Kondo, T. (2006). Expression of heme oxygenase -1 due to intracellular reactive oxygen species induced by ultrasound. *Ultrason. Sonochem.*, 13(5), 388-396.

[26] Nabavi, S. F., Nabavi, S. M., N Setzer, W., Nabavi, S. A., Nabavi, S.A., Ebrahimzadeh, M.A. (2013). Antioxidant and antihemolytic activity of lipid-soluble bioactive substances in avocado fruits. *Fruits*, 68(03), 185-193.

[27] Nabavi, S. M., Nabavi, S. F., Eslami, S., Moghaddam, A.H. (2012). *In vivo* protective effects of quercetin against sodium fluoride-induced oxidative stress in the hepatic tissue. *Food Chem.*, 132(2), 931-935.

[28] Herrera, E., Barbas, C. (2001). Vitamin E: action, metabolism and perspectives. *J. Physiol. Biochem.*, 57(1), 43-56.

[29] Padayatty, S. J., Katz, A., Wang, Y., Eck, P., Kwon, O., Lee, J. H., Chen S, Corpe, C., Dutta, A., Dutta, S.K., Levine, M. (2003). Vitamin C as an antioxidant: evaluation of its role in disease prevention. *J. Am. Coll. Nutr.*, 22(1), 18-35.

[30] Nabavi, S. F., Nabavi, S. M., Habtemariam, S., Moghaddam, A. H., Sureda, A., Jafari, M., Latifi, A. M. (2013). Hepatoprotective effect of gallic acid isolated from *Peltiphyllum peltatum* against sodium fluoride-induced oxidative stress. *Ind. Crop Prod.*, 44, 50-55.

[31] Martin, C., Zhang, Y., Tonelli, C., Petroni, K. (2013). Plants, diet, and health. *Ann. Rev. Plant Biol.*, 64, 19-46.

[32] Martin, C., Butelli, E., Petroni, K., Tonelli, C. (2011). How can research on plants contribute to promoting human health ?. *Plant Cell Online*, 23(5), 1685-1699.

[33] Chabert, P., Auger, C., Pincemail, J., Schini-Kerth, V. B. (2014). Overview of Plant-Derived Antioxidants. *Syst. Biol. Free Radic. Antiox.*, 4005-4022.

[34] Gueven, K. C., Percot, A., Sezik, E. (2010). Alkaloids in marine algae. *Mar. Drugs*, 8(2), 269-284.

[35] Cushnie, T. P., Cushnie, B., Lamb, A. J. (2014). Alkaloids: An overview of their antibacterial, antibiotic -enhancing and antivirulence activities. *Int. J Antimicrob. Agents.* Doi: 10.1016/j.ijantimicag.2014.06.001.

[36] Jung, H. A., Min, B. S., Yokozawa, T., Lee, J. H., Kim, Y. S., Choi, J.S. (2009). Anti-Alzheimer and antioxidant activities of Coptidis Rhizoma alkaloids. *Biol. Pharm. Bull.*, 32(8), 1433-1438.

[37] Chonpathompikunlert, P., Wattanathorn, J., Muchimapura, S. (2010). Piperine, the main alkaloid of Thai black pepper, protects against neurodegeneration and cognitive impairment in animal model of cognitive deficit like condition of Alzheimer's disease. *Food Chem. Toxicol.*, 48(3), 798-802.

[38] Ivanovska, N., Philipov, S. (1996). Study on the anti-inflammatory action of *Berberis vulgaris* root extract, alkaloid fractions and pure alkaloids. *Int. J. Immunopharmacol.*, 18(10), 553-561.

[39] Maiza-Benabdesselam, F., Khentache, S., Bougoffa, K., Chibane, M., Adach, S., Chapeleur, Y., Max, H., Laurain-Mattar, D. (2007). Antioxidant activities of alkaloid extracts of two Algerian species of Fumaria: Fumaria capreolata and Fumaria bastardii. *Rec. Nat. Prod.*, 1(2-3), 28-35.

[40] Cao, Z. F., Zhu, X. Q. (1997). Antioxidant action of tetrandrine: An alkaloid from the roots of Radix stephania tetrandra, S Moore. *J. Sci. Food Agric.*, 73(1), 106-110.

[41] El-Desouky, S. K., Kim, K. H., Ryu, S. Y., Eweas, A. F., Gamal-Eldeen, A. M., Kim, Y.K. (2007). A new pyrrole alkaloid isolated fromArum palaestinum Boiss. and its biological activities. *Arch. Pharm. Res.*, 30(8), 927-931.

[42] Bittner, M., Aguilera, M. A., Hernández, V., Arbert, C., Becerra, J., Casanueva, M.E. (2009). Fungistatic activity of essential oils extracted from Peumus boldus Mol., Laureliopsis philippiana (Looser) Schodde and Laurelia sempervirens (Ruiz & Pav.) Tul.(Chilean monimiaceae). *Chilean J. Agric. Res.*, 69(1), 30-37.

[43] Speisky, H., Cassels, B. K. (1994). Boldo and boldine: an emerging case of natural drug development. *Pharmacol. Res.*, 29(1), 1-12.

[44] Chester, S. (2010). *A Wildlife Guide to Chile: Continental Chile, Chilean Antarctica, Easter Island, Juan Fernandez Archipelago*. Princeton University Press.

[45] Silva, E., Jopia, M., Edwards, A. M., Lemp, E., Lissi, E. (2002). Protective Effect of Boldo and Tea Infusions on the Visible Light–mediated Pro-oxidant Effects of Vitamin B2, Riboflavin. *Photochem. Photobiol.*, 75(6), 585-590.

[46] Torri, M. C. (2010). Medicinal plants used in Mapuche traditional medicine in Araucania, Chile: linking sociocultural and religious values with local heath practices. *Complement. Health. Pract. Rev.*, 15(3), 132-148.

[47] Kringstein, P., Cederbaum, A.I. (1995). Boldine prevents human liver microsomal lipid peroxidation and inactivation of cytochrome P4502E1. *Free Radic. Biol. Med.*, 18(3), 559-563.

[48] Capasso, F., Gaginella, T. S., Grandolini, G., Izzo, A. A. (2003). *Plants, Liver and Biliary System. In Phytotherapy* (pp. 295-308). Springer Berlin Heidelberg.

[49] Cartaxo, S. L., de Almeida Souza, M. M., de Albuquerque, U.P. (2010). Medicinal plants with bioprospecting potential used in semi-arid northeastern Brazil. *J. Ethnopharmacol.*, 131(2), 326-342.

[50] Milliken, W. (1997). Traditional anti-malarial medicine in Roraima, Brazil. *Economic Bot.*, 51(3), 212-237.

[51] Vogel, H., Gonzalez, B., & Razmilic, I. (2011). Boldo (*Peumus boldus*) cultivated under different light conditions, soil humidity and plantation density. *Ind. Crop. Prod.*, 34(2), 1310-1312.

[52] Susana, F., Marisol, B., Rosemarie, W., Marcelo, B., Edgar, P., Luis, I., Claudia, T., Gonzalez, W. (2011). Characterization and propagation of some medicinal plants in the central-south region of Chile. *Ind. Crop. Prod.*, 34(2), 1313-1321.

[53] Cassels, B. K. (1999). *11. Recent developments in the chemistry and pharmacology of boldo and boldine.* Chemistry, Biological and Pharmacological Properties of Medicinal Plants from the Americas, 197.

[54] Quezada, N., Asencio, M., Valle, J.D., Aguilera, J.M., Gómez, B. (2004). Antioxidant activity of crude extract, alkaloid fraction, and flavonoid fraction from Boldo (*Peumus boldus* Molina) leaves. *J. Food Sci.*, 69(5), C371-C376.

[55] Moreno, P.R.H., Von Poser, G. L., Rates, S.M.K., Yoshida, M., Gottlieb, O. R., Souza, I. A., Henriques, A.T. (1993). Chemical Constituents and Antitumour Acitivity of Nectandra Grandiflora. *Pharm. Biol.*, 31(3), 189-192.

[56] Hasan, C. M., Rahman, P., Karim, A., Jabbar, A., Gray, A.I., Waterman, P.G. (1987). Boldine from the stem bark of *Dehaasia kurzii*. *Fitoterapia*, 58, 430.

[57] Zhao, Q. Z., Zhao, Y. M., Wang, K.J. (2005). Alkaloids from the root of Lindera angustifolia. *Acta Pharm. Sinica*, 40(10), 931-934.

[58] XiaoHong, Y., XiaoYi, W., HaiHui, X., MeiFang, L., FengXian, Z. (2000). Aporphine alkaloids of *Litsea rotundifolia* and *L. rotundifolia* var. oblongifolia. *J.Trop. Subtrop. Bot.*, 8(4), 324-328.

[59] Rachmatiah, T., Mukhtar, M. R., Nafiah, M.A., Hanafi, M., Kosela, S., Morita, H., Litaudon, M., Awang, K., Omar, H., Hadi, A. H. A. (2009). (+)-N-(2-Hydroxypropyl) lindcarpine: A New Cytotoxic Aporphine Isolated from Actinodaphne pruinosa Nees. *Molecules*, 14(8), 2850-2856.

[60] Hidalgo, M. E., Alarcón, M. G., Ojeda, J. R., Fernández, E.C., Sobarzo-Sánchez, E. M., De la Fuent, J.R. (2010). Spectroscopic and photochemical properties of some annulated boldine derivatives. *J. Braz. Chem. Soc.*, 21(12), 2205-2210.

[61] Speisky, H., Cassels, B.K., Lissi, E.A., Videla, L.A. (1991). Antioxidant properties of the alkaloid boldine in systems undergoing lipid peroxidation and enzyme inactivation. *Biochem. Pharmacol.*, 41(11), 1575-1581.

[62] Urzua, A., Acuna, P. (1983). Alkaloids from the bark of Peumus boldus. *Fitoterapia*, 54, 175-177.

[63] Jiménez, I., Speisky, H. (2000). Biological disposition of boldine: in vitro and in vivo studies. *Phytother. Res.*, 14(4), 254-260.

[64] O'Brien, P., Carrasco-Pozo, C., Speisky, H. (2006). Boldine and its antioxidant or health -promoting properties. *Chem. Biol. Interact.*, 159(1), 1-17.

[65] Kreitmair, H. (1952). Pharmakologische wirkung des alkaloids aus Peumus boldus Molina. *Pharmazie*, 7, 507-511.

[66] Speisky, H., Ellahuefie, M., Cofr, S., Nfifiez-Vergara, L. J., & De la Fuente, N. (1993). Evaluation of the lethality and genotoxicity of boldine in mice. Murat Res.

[67] Magistretti, M. J. (1980). Remarks on the pharmacological examination of plant extracts. *Fitoterapia*.

[68] Moreno, P.R.H., Andrade, H.H.R., Henriques, A.T., Henriques, J.A.P. (1991). Genotoxicity of the boldine aporphine alkaloid in prokaryotic and eukaryotic organisms. *Mutat. Res. Genetic Toxicol.*, 260(2), 145-152.

[69] Moreno, P.R.H., Andrade, H.H.R., Henriques, A.T., Henriques, J.A.P. (1991). Genotoxicity of the boldine aporphine alkaloid in prokaryotic and eukaryotic organisms. *Mutat. Res./Genetic Toxicol.*, 260(2), 145-152.

[70] Tisserand, R. (1995). Balacs, T. *Essential oil safety,* Churchill Livingstone, London.

[71] Reuter, S., Gupta, S.C., Chaturvedi, M.M., Aggarwal, B.B. (2010). Oxidative stress, inflammation, and cancer: how are they linked?. *Free Radic. Biol. Med.*, 49(11), 1603-1616.

[72] Branen, A. L. (1975). Toxicology and biochemistry of butylated hydroxyanisole and butylated hydroxytoluene. *J. Am. Oil Chem. Soc.*, 52(2), 59-63.

[73] Williams, G. M., Iatropoulos, M.J., Whysner, J. (1999). Safety assessment of butylated hydroxyanisole and butylated hydroxytoluene as antioxidant food additive s. *Food Chem. Toxicol.*, 37(9), 1027-1038.

[74] Nabavi, S. M., Habtemariam, S., Nabavi, S. F., Sureda, A., Daglia, M., Moghaddam, A.H., Amani, M.A. (2013). Protective effect of gallic acid isolated from Peltiphyllum peltatum against sodium fluoride-induced oxidative stress in rat's kidney. *Mol. Cell. Biochem.*, 372(1-2), 233-239.

[75] Klimaczewski, C.V., Saraiva, R.D.A., Roos, D.H., Boligon, A., Athayde, M.L., Kamdem, J. P., Barbosa, N.V., Rocha, J.B.T. (2014). Antioxidant activity of *Peumus boldus* extract and alkaloid boldine against damage induced by Fe (II)–citrate in rat liver mitochondria *in vitro*. *Ind. Crop. Prod.*, 54, 240-247.

[76] Schmeda-Hirschmann, G., Rodriguez, J.A., Theoduloz, C., Astudillo, S.L., Feresin, G.E., Tapia, A. (2003). Free-radical scavengers and antioxidants from Peumus boldus Mol.(" Boldo "). *Free Rad. Res.*, 37(4), 447-452.

[77] Fernández, J., Lagos, P., Rivera, P., Zamorano-Ponce, E. (2009). Effect of boldo (Peumus boldus Molina) infusion on lipoperoxidation induced by cisplatin in mice liver. *Phytother. Res.*, 23(7), 1024-1027.

[78] Falé, P. L., Amaral, F., Amorim Madeira, P.J., Sousa Silva, M., Florencio, M.H., Frazao, F.N., Serralheiro, M.L.M. (2012). Acetylcholinesterase inhibition, antioxidant activity and toxicity of *Peumus boldus* water extracts on HeLa and Caco-2 cell line s. *Food Chem. Toxicol.*, 50(8), 2656-2662.

[79] Lorena, L. F., Heliodoro, D.L.G.T., Jorge, E.W.P., Ruth, B., Raul, R.H., & Cristobal, N.A. (2013). Separation conditions and evaluation of antioxidant properties of boldo (Peumus boldus) extracts. J. Med. Plant Res., 7(15), 911-917.

[80] Valenzuela, A., Nieto, S., Cassels, B.K., Speisky, H. (1991). Inhibitory effect of boldine on fish oil oxidation. *J. Am. Oil Chem. Soc.*, 68(12), 935-937.

[81] Cederbaum, A. I., Ukielka, E. K., Speiskyf, H. (1992). Inhibition of rat liver microsomal lipid peroxidation by boldine. *Biochem. Pharmacol.*, 44(9), 1765-1772.

[82] Ubeda, A., Montesinos, C., Paya, M., Alcaraz, M.J. (1993). Iron-reducing and free-radical-scavenging properties of apomorphine and some related benzylisoquinolines. *Free Radic. Biol. Med.*, 15(2), 159-167.

[83] Kringstein, P., Cederbaum, A. I. (1995). Boldine prevents human liver microsomal lipid peroxidation and inactivation of cytochrome P4502E1. *Free Radic. Biol. Med.*, 18(3), 559-563.

[84] Hu, J., Speisky, H., Cotgreave, I. A. (1995). The inhibitory effects of boldine, glaucine, and probucol on TPA -induced down regulation of gap junction function: relationships to intracellular peroxides, protein kinase C translocation, and connexin 43 phosphorylation. *Biochem. Pharmacol.*, 50(10), 1635-1643.

[85] Bannach, R., Valenzuela, A.L.F.O.N.S.O., Cassels, B.K., Nunez-Vergara, L.J., Speisky, H.E.R.N.A.N. (1996). Cytoprotective and antioxidant effects of boldine on tert-butyl hydroperoxide-induced damage to isolated hepatocytes. *Cell Biol. Toxicol.*, 12(2), 89-100.

[86] Jiménez, I., Garrido, A., Bannach, R., Gotteland, M., Speisky, H. (2000). Protective effects of boldine against free radical-induced erythrocyte lysis. *Phytother. Res.*, 14(5), 339-343.

[87] Kubinova, R., Machala, M., Minksova, K., Neca, J., Suchy, V. (2001). Chemoprotective activity of boldine: modulation of drug-metabolizing enzymes. Pharmazie, 56(3), 242-243.

[88] Youn, Y. C., Kwon, O.S., Han, E. S., Song, J.H., Shin, Y.K., Lee, C.S. (2002). Protective effect of boldine on dopamine -induced membrane permeability transition in brain mitochondria and viability loss in PC12 cells. *Biochem Pharmacol.*, 63(3), 495-505.

[89] Lau, Y. S., Machha, A., Achike, F.I., Murugan, D., Mustafa, M.R. (2012). The aporphine alkaloid boldine improves endothelial function in spontaneously hypertensive rats. *Exp. Biol. Med.*, 237(1), 93-98.

[90] Lau, Y. S., Tian, X. Y., Huang, Y., Murugan, D., Achike, F.I., Mustafa, M.R. (2013). Boldine protects endothelial function in hyperglycemia -induced oxidative stress through an antioxidant mechanism. *Biochem. Pharmacol.*, 85(3), 367-375.

[91] Konrath, E. L., Santin, K., Nassif, M., Latini, A., Henriques, A., Salbego, C. (2008). Antioxidant and pro-oxidant properties of boldine on hippocampal slices exposed to oxygen –glucose deprivation *in vitro. Neurotoxicology*, 29(6), 1136-1140.

[92] Santanam, N., Penumetcha, M., Speisky, H., Parthasarathy, S. (2004). A novel alkaloid antioxidant, Boldine and synthetic antioxidant, reduced form of RU486, inhibit the oxidation of LDL in-vitro and atherosclerosis in vivo in $LDLR^{-/-}$ mice. *Atherosclerosis*, 173(2), 203-210.

[93] Jang, Y. Y., Song, J. H., Shin, Y. K., Han, E.S., Lee, C.S. (2000). Protective effect of boldine on oxidative mitochondrial damage in streptozotocin-induced diabetic rats. *Pharmacol. Res.*, 42(4), 361-371.

[94] Ríos, J. L., Cholbi, R., Huguet, A.I., Mora, A., Manez, S., Paya, M., Alcaraz, J. (1990). Effects of benzylisoquinolines on lipid peroxidation and superoxide scavenging. *Planta Med.*, 56(06), 645-646.

[95] Shirwaikar, A., Shirwaikar, A., Rajendran, K., Punitha, I.S.R. (2006). In vitro antioxidant studies on the benzyl tetra isoquinoline alkaloid berberine. *Biol. Pharm. Bull.*, 29(9), 1906-1910.

[96] Rackova, L., Oblozinsky, M., Kostalova, D., Kettmann, V., Bezakova, L. (2007). Free radical scavenging activity and lipoxygenase inhibition of Mahonia aquifolium extract and isoquinoline alkaloids. *J. Inflamm.*, 4(1), 15.

[97] Martinez, L. A., Rios, J., Paya, M., Alcaraz, M.J. (1992). Inhibition of nonenzymic lipid peroxidation by benzylisoquinoline alkaloids. *Free Radic. Biol. Med.*, 12(4), 287-292.

[98] Casselst, B.K., Asencio, M., Conget, P., Speisky, H., Videla, L.A., Lissi, E.A. (1995). Structure-antioxidative activity relationships in benzylisoquinoline alkaloids. *Pharmacol. Res.*, 31(2), 103-107.

[99] Loghin, F., Chagraoui, A., Asencio, M., Comoy, E., Speisky, H., Cassels, B.K., Protais, P. (2003). Effects of some antioxidative aporphine derivatives on striatal dopaminergic transmission and on MPTP-induced striatal dopamine depletion in B6CBA mice. *Eur. J. Pharm. Sci.*, 18(2), 133-140.

[100] Milian, L., Estelles, R., Abarca, B., Ballesteros, R., Sanz, M. J., Blázquez, M. A. (2004). Reactive oxygen species (ROS) generation inhibited by aporphine and phenanthrene alkaloids semi-synthesized from natural boldine. *Chem. Pharm. Bull.*, 52(6), 696-699.

[101] Teng, C. M., Hsiao, G., Ko, F.N., Lin, D.T., Lee, S.S. (1996). *N*-Allylsecoboldine as a novel antioxidant against peroxidative damage. *Eur. J. Pharmacol.*, 303(1), 129-139.

In: Alkaloids
Editor: Eduardo Sobarzo-Sánchez

ISBN: 978-1-63482-074-5

Chapter 3

BIOSYNTHESIS OF AMARYLLIDACEAE ALKALOIDS: A BIOCHEMICAL OUTLOOK

***Aparna Singh*[1] *and Isabel Desgagné-Penix*[1,2,*]**
[1]Université du Québec à Trois-Rivières, Canada
[2]Groupe de Recherche en Biologie Végétale, Department of Chemistry, Biochemistry and Physics, Trois-Rivières, Québec, Canada

ABSTRACT

Amaryllidaceae alkaloids (AAs) are a diverse group of biologically active specialized metabolites produced mainly in Amaryllidaceae plant family. Several AAs possess potent pharmaceutical properties making them interesting target for drug development. For example, the AA galanthanmine, an acetylcholinesterase inhibitor, is used to treat neurodegenerative disorder including Alzeihmer's disease. Also, AAs such as lycorine possess antimicrobial activity, whereas others such as crinine and narciclasine are potentially anticancer agents. Ironically, more is understood about the effects of alkaloids on humans than on their biosynthesis or their roles in plants. Understanding the biochemical genetics underpinning AA biosynthesis could enable improved production of these important pharmaceutical both in plants and in other systems.

Keywords: Amaryllidaceae alkaloid; plant secondary metabolism; alkaloid biosynthesis; natural products; systems biology; galanthamine

INTRODUCTION

Alkaloids are low-molecular weight, nitrogenous specialized metabolites occuring in approximately 20% of plant species. Many of the about 12,000 structurally elucidated alkaloids show potent biological activity [1, 2]. Particularly, the pharmacological properties of Amaryllidaceae alkaloids (AAs) have been exploited for centuries, and modern medecine

[*] E-mail: Isabel.Desgagne-Penix@uqtr.ca.

continues to rely on plant-derived compounds such as acetylcholinesterase inhibitor galanthamine used to treat neurodegenerative disorder including Alzeihmer's disease [3-7]. AAs such as lycorine possess antimicrobial activity, whereas others such as crinine and narciclasine are potentially anticancer agents.

Ironically, more is understood about the effects of alkaloids on humans than on the roles of these compounds in the plants that produce them. Although not considered essential for normal growth and development, AAs likely play key roles in plant defense as several AAs possess potent pharmalogical activities (Table 1), which in turn may be a good indication of their biological function. For example, the potency of crinine as a hallucinogenic or haemanthamine as an analgesic suggest their role as herbivores deterrents. Similarly, the anti-microbial properties of lycorine and cytotoxicity of narciclasine suggest that they may confer protection against pathogens.

The bulbous monocot Amaryllidaceae (*Amaryllis, Narcissus, Galanthus,* etc.), comprising over 1100 species, are among the top 20 of the most widely used plant families as traditional source of medicines [6, 7]. For example, in India, *Crinum bulbispermum* has been used for treating rheumatic disorder whereas in Nigeria, bulbs of *Crinum jagus* and *C. glaucum* has been traditionally recommended for various mental illness. The presence of AAs such as crinine and vittatine in *Crinum* bulbs (Table 1), with potency to inhibit acetylcholinesterase has been documented, extending their therapeutic usage towards Alzheimer's disease [6].

In Egypt, mixture from roots of *Amaryllis belladonna* and *Clivia miniata* were used to treat and prevent infection of wounds such as insect and snake bites. The anti-microbial activities are achieved by virtue of certain types of AAs namely lycorine, homolycorine, haemanthamine and tazettine present in the root mixture [10, 15, 19].

Following on the commercial success of galanthamine as a prescription drug in the treatment of Alzheimer's disease, much work has focus on the discovery and production of derivatives as drug candidates. For example, the improved ability of galanthamine derivatives to cross blood -brain barrier coupled with stronger analgesic and analeptic properties make them interesting pharmaceutical compounds for the treatment of numerous neurological conditions such as paralysis and schizophrenia [20-23].

Given the interesting biological properties of most AAs, Amaryllidaceae plants provide a diverse and accessible platform for phytochemical-based drug discovery. For example, lycorine- and crinine-types of AAs have shown much promise as remarkably potent and selective anticancer agents [24].

As it is the cases for many valuable plant natural products, the commercial development of most AAs is restricted by their limited availabilities due to their low concentrations in plants. Currently, among the ~500 known AAs, only galanthamine is used clinically and is majoritarily obtained from *Leucojum* and *Narcissus* plant sources [5, 7]. Alhtough, several complexe chemical syntheses and semi-syntheses have been developed successfully, they are not commercially viable for the pharmaceutical industry [3].

Compared with the large knowledge for the potential pharmaceutical activities of AAs, the biochemistry of AA biosynthesis remains elusive. However, an impressive array of biosynthetic enzymes must function together to yield the vast diversity of AAs. Understanding the biochemical genetics underpinning AA biosynthesis could enable improved production of these important pharmaceutical and related molecules both in plants and in other systems.

Table 1. Examples of important AA, plant source and biological activity

AA	Group	Plant source	Biological activity	Ref.
Galanthamine	*para-ortho'*	*Narcissus pseudonarcissus* *Leucojum aestivum* *Galanthus niveali* *Galanthus elewesii* *Pancratium maritimum*	acetylcholinesterase inhibitor, treatment for Alzheimer's disease	[4, 8-10]
Narciclasine	*para-para'*	*Narcissus spp.* *Leucojum aestivum* *Lycoris radiata,* *Pancratium maritimum* *Hymenocallis expansa*	cytotoxic	[3, 4]
Montanine	*papa-para'*	*Hippeastrum vittatum*	anxiolytic, antidepressant,	[4, 11]
Pretazettine	*para-para'*	*Narcissus spp.*	antiviral, cytotoxic, analgesic	[12]
Crinine	*para-para'*	*Crinum spp.* *Galanthus reginae–olgae* *Brunsvigia gregaria* *Pancratinum maritimum* *Boophone disticha*	cytotoxic, hallucinogenic	[3, 4, 9, 13]
Haemanthamine	*para-para'*	*Narcissus pseudonarcissus* *Narcissus primigenius* *Pancratinum maritimum* *Clivia miniata*	cytotoxic, analgesic, antimalarial, anti-retroviral, fungicide hypotensive	[4, 13-15]
Vittatine	*para-para'*	*Narcissus pseudonarcissus* *Crinum bulbispermum*	analgesic, anti-microbial, cytotoxic	[4, 13, 16]
Galanthine	*ortho-para'*	*Galanthus elewesii* *Lycoris sunguinea*	cytotoxic	[16, 17]
Narcissidine	*ortho-para'*	*Leucojium autumnale*	cytotoxic	[4]
Lycorine	*ortho-para'*	*Narcissus spp.* *Leucojum aestivum* *Leucojium autumnale* *Galanthus reginae–olgae* *Galanthus elewesii* *Hippeastrum equestre* *Lycoris radiata* *Lycoris sunguinea* *Pancratinum maritimum*	antiviral, analgesic, anti-microbial, fungicide, antimalarial, anti-inflammatory, cytotoxic	[3, 4, 9, 16, 18]
Homolycorine	*ortho-para'*	*Narcissus primigenius* *Narcissus pseudonarcissus*	cytotoxic, antiviral antimalarial	[3, 4]

Biosynthesis

In contrast to the extensive literature on the biological effects of AAs, information on their biochemical pathways and molecular genetics is incomplete. The majority of the research done on AA biosynthesis is supported by early feeding experiments using radiolabeled precursors. These studies led to the biochemical elucidation of the initial steps in AA biosynthesis [25-31]. Despite their vast structural diversity, AAs share a common biosynthetic origin, norbelladine, which is formed through the condensation of amino acid s derivatives; 3,4-dihydroxybenzaldehyde (3,4-DHBA also named protocatechuic aldehyde) and tyramine. The resulting norbelladine is central to the biosynthesis of many structural types of AAs (Figure 1).

Currently, there is no consensus about the names or number of classes of AAs however it may be influenced by the alkaloids present in the particular species. As the number of AA discovered increases, new core structures are discovered. Evidente and Kornienko, recognized twelve distinct ring types [3] of AAs whereas Jin [6] distinguished eighteen subgroups and this expansion is mainly due to the continued isolation of AAs with rare skeleton types. Bastida et al. recognized nine subgroups but mostly focussed on *Narcissus spp*. [4].

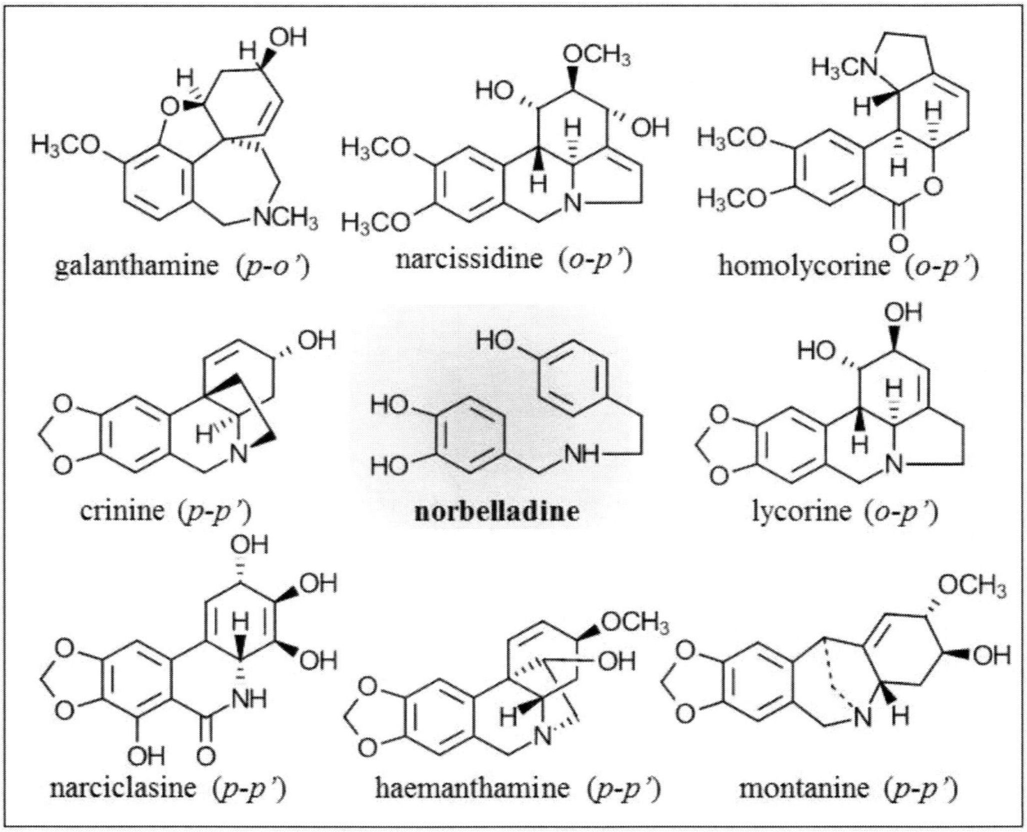

Figure 1. Examples of diverse chemical structures of AA types derived from simple skeletal structure nobelladine (shaded gray).

A classification based on the biosynthetic origins of the different AAs as three groups (*para-ortho', para-para', ortho-para'*) and nine AA-types (Table 2) is depicted in this chapter in order to discussed the biosynthesis of AAs (Figure 2).

Initial Biosynthetic Reactions

Despite the enormous variety of plant specialized metabolites, the number of corresponding basic biosynthetic pathways is restricted and distinct.

Table 2. Proposed classification based on biosynthetic origin

	Group	AA-type
1		Norbelladine
2	*para-ortho'*	Galanthamine
3	*para-para'*	Crinine
4	*para-para'*	Narciclasine
5	*para-para'*	Haemanthamine
6	*para-para'*	Montanine
7	*ortho-para'*	Homolycorine
8	*ortho-para'*	Lycorine
9	*ortho-para'*	Narcissidine

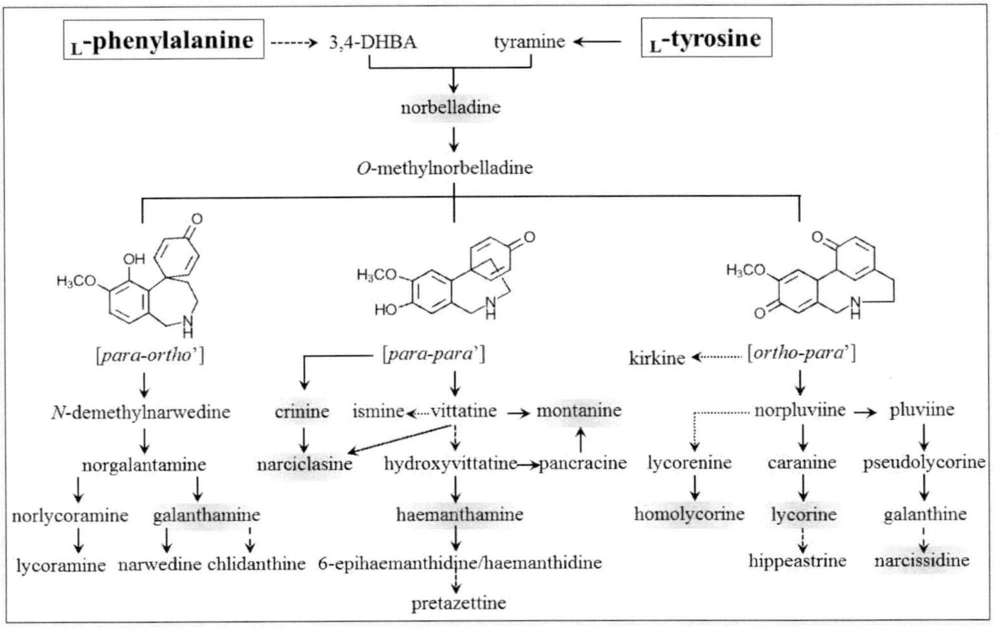

Figure 2. Proposed AA biosynthetic pathways. Three major groups of AA (*para-ortho', para-para', ortho-para'*) are represented along with the nine proposed AA types: norbelladine-, galanthamine-, crinine-, narciclasine-, haemanthamine-, montanine-, homolycorine-, lycorine-, and narcissidine-type (shaded gray). Broken arrow indicates more than one step.

The precursor substrates and enzymes necessary for the first committed steps often appear to have been recruited from primary metabolic pathways, such as glycolysis, the Krebs cycle, the pentose phosphate pathway and the shikimate pathway [32]. For example, the aromatic amino acid s L-phenylalanine and L-tyrosine, produced by the shikimate pathway, are precursors for a wide spectrum of natural products including phenylpropanoids, flavonoids, lignins, coumarins, cyanogenic glycosides, glucosinolates, and alkaloids [33].

AA biosynthesis starts with the formation of two precursors units: 1) the C_6-C_1 unit from L-phenylalanine named 3,4-DHBA and 2) the C_6-C_2-N unit from L-tyrosine called tyramine (Figure 3). The sequence of reactions leading to the conversion of L-phe into 3,4-DHBA are not known but studies suggested that 3,4-DHBA originates from phenylpropanoid C_6-C_3 compounds as in the formation of the fragrant phenolic compound, vanillin, in *Vanilla planifolia* plants [34-38]. Interestingly, vanillin is a metabolic intermediate involved in the biosynthesis of capsaicinoids which are benzylamine alkaloids responsible of the pungent flavor in chili pepper *Capsicum spp*. [39]. Altogether, these results suggest that initial reactions and biosynthetic enzymes of the phenylpropanoid pathway are commun in the synthesis of AA precursor 3,4-DHBA.

The first reaction of phenylpropanoid biosynthesis is catalyzed by the L-phenylalanine ammonia -lyase (PAL), a key regulatory enzyme promoting the deamination of L-phe to generate *trans*-cinnamic acid (Figure 3A). PAL reaction not only controls carbon flux into this pathway but also links specialized metabolism to primary metabolism leading to the production of a variety of phenolic compounds including cell wall structural polymer lignin [38]. This enzyme is among the most studied in plant specialized metabolism and many *PAL* genes have been cloned and characterized from various species including Amaryllidaceae species [40, 41]. Recently, *Lycoris radiata PAL1* gene was isolated, characterized and a correlation between *LrPAL1* expression and content of galanthamine was reported supporting its role in AA biosynthesis [40]. The hydroxylation of the benzyl ring of *trans*-cinnamic acid is catalyzed by the cytochrome P450-dependent monooxygenase (CYP) cinnamate-4-hydroxylase (C4H) which belongs to the CYP73 subfamily, and results in the formation of *p*-coumaric acid (Figure 3A) [34, 37, 42]. CYPs participate in a variety of biochemical pathways and the number of *CYP* genes in plant genomes is estimated to be up to 1%, implying that plants are huge sources for various CYP-dependent reactions [43]. CYP catalyze a variety of monooxygenation/hydroxylation reactions along with some unusal reactions such as methyledioxy bridge formation, phenol coupling reactions and oxidative rearrangement of carbon skeletons. cDNA sequences encoding *C4H* have been identified in numerous plants and were shown to be highly conserved across species [42, 44]. Additionally, *C4H* sequences were present in the transcriptome database of the Amaryllidaceae *L. radiata* [41].

From this stage, as it is the case for vanillin biosynthesis, two separate routes are proposed as to how p-coumaric acid is converted to 3,4-DHBA: the oxidative 'ferulate' and the non-oxidative 'benzoate' pathways (Figure 3A). The ferulate pathway, proposed by Zenk in 1965 [34], suggested that the aromatic ring of p-coumaric acid undergoes hydroxylation by the coumarate-3-hydroxylase (C3H), giving rise to caffeic acid. The later undergo chain shortening (loss of two carbon atoms) to form the precursor 3,4-DHBA, via a mechanism similar to the β-oxidation sequence for the catabolic degradation of fatty acids, [4, 28, 31]. The other route, the benzoate pathway, proposes a non-oxidative chain shortening of p-coumaric acid first, followed by the hydroxylation of the aromatic ring to yield 3,4-DHBA.

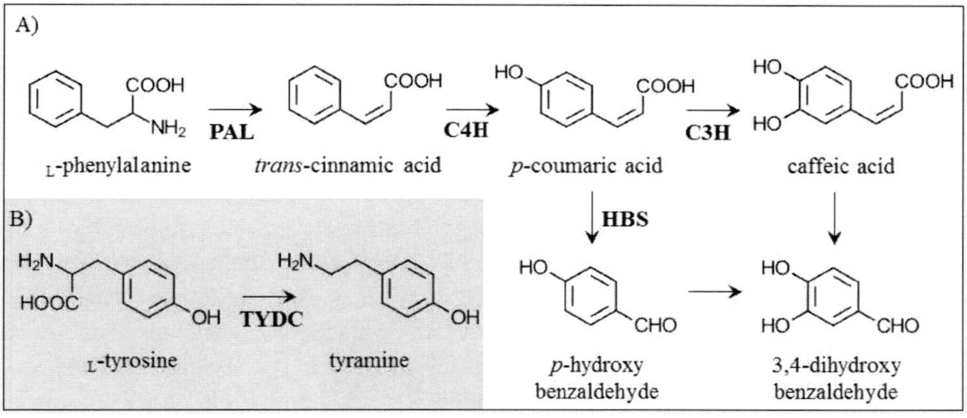

Figure 3. Proposed pathways to AA precursors. A) 3,4-dihydroxybenzaldehyde (3,4-DHBA) biosynthesis depicting the two possible routes from *p*-coumaric acid to form 3,4-DHBA: the oxidative 'ferulate' and the non-oxidative 'benzoate' pathways B) Tyramine biosynthesis. Arrows without labeling reflect chemical reactions that have not been enzymatically characterized. Enzymes that have been cloned, characterized and identified are labeled in black bold. Enzyme abbreviations: PAL, phenylalanine ammonia -lyase; C4H, cinnamate 4-hydroxylase; C3H, coumarate 3-hydroxylase; HBS, 4-hydroxybenzaldehyde synthase; TYDC, tyrosine decarboxylase.

Recently, an unusual cysteine protease named 4-hydroxybenzaldehyde synthase (HBS) was shown to catalyze the conversion of p-coumaric acid to p-hydroxybenzaldehyde. However, the role of HBS in 3-4-DHBA and vanillin biosynthesis remains under investigation [36, 45].

The other precursor of AA biosynthesis, the C_6-C_2-N unit, originates from the decarboxylation of L-tyr to tyramine by the L-tyrosine decarboxylase (TYDC) [4, 5, 28, 31]. TYDC belongs to the DOPA decarboxylase family of the pyridoxal phosphate (PLP)-dependent aspartate aminotransferase superfamily. In addition to controlling the transition from primary to specialized metabolism, TYDC ensures an adequate supply of tyramine for the synthesis of various alkaloids including the well-known benzylisoquinoline alkaloids (BIAs) such as the narcotic analgesics morphine and codeine [1, 2, 46-48]. Several molecular clones for TYDC have been isolated from multiple organisms (*e.g.,* microorganisms, fungi, plants, animals) [46]. Plant TYDCs share great sequence similarity to each others within and across species [46, 49]. Transcriptome analysis of of *L. aurea* [41] have confirmed the presence of *TYDC* sequences in Amaryllidaceae.

Although, the enzymatic reactions leading to the formation of precursors are still hypothetical, biochemical evidences from biotransformation studies with labeled precursors [25-31] coupled with molecular evidences from recent transcriptome analyses [40, 41] have confirmed the presence of metabolite intermediates and transcripts for *PAL, C4H, C3H* and *TYDC* in Amaryllidaceae species supporting, in parts, the current proposed pathways.

Norbelladine-Type Alkaloids

The first committed step in AA biosynthesis in plants starts with the coupling of the two precursors, 3,4-DHBA and tyramine, defining the entry point of primary metabolites into AA biosynthetic pathway. The condensation of the aldehyde (3,4-DHBA) and the amine (tyramine), called Pictet-Spengler condensation, results in a Schiff's base intermediate which following reduction yields norbelladine (Figure 4) [50, 51]. Pictet-Spengler reactions are widely used in plant alkaloid biosynthesis to yield either a β–carboline or a tetrahydroquinoline product from the condensation of an aldehyde and an aromatic amine [52]. For example, the first committed step in the formation of BIAs is the combination of two L-tyr derivatives, namely 4-hydroxyphehylacetaldehyde and dopamine, by a Pictet-Spenglerase, the norcoclaurine synthase (NCS), to produce the trihydroxyisoquinoline alkaloid norcoclaurine [53-59]. NCS catalyzes a two-step reaction mechanism; a condensation followed by an intramolecular cyclization [53, 57]. The electron -donating oxygen of the hydroxyl group of dopamine (missing in tyramine) is essential for the reaction to proceed and may provide a mechanistic explanation for the absence of cyclization in the condensation step of AA biosynthesis. To date, two different protein families (pathogenesis - related PR10-Betv1 and 2-oxoglutarate-dependant dioxygenase) have been reported to catalyze this reaction *in vitro*. However the PR10/Betv1-NCS was shown to be the major one involved *in vivo* in BIA biosynthesis of *Papaver somniferum* and *Thalictrum flavum* plants [54]. Another example of a Pictet-Spengler condensation is the combination of tryptamine and secologanin to form strictosidine (β-carboline product), the general precursor to terpene indole alkaloids [60, 61]. In *Rauvolfia serpentina*, this reaction is catalyzed by strictosidine synthase, a member of the six-bladed four-stranded β-propeller fold protein family [61]. These examples show that the initial Pictet-Spengler reactions in alkaloid biosynthesis can be catalyzed by members of very different protein families and suggests that the type of enzyme recruited for the non-cyclizing condensation reaction in AA biosynthesis may belong to one of these families or to an additional class of enzymes.

In Amaryllidaceae plants, an *O*-methylation of norbelladine on the '3,4-DHBA' aromatic ring takes place prior to intramolecular oxidative coupling which results in the formation of different skeletons of AAs. Thus, the key biosynthetic intermediate of AAs is 4'-*O*-norbelladine (Figure 4).

Plant *O*-methylation reactions are common transformation in the biosynthesis of alkaloids and are most often catalyzed by *S*-adenosyl-L-methionine (SAM)-dependent methyltransferases (MTs) [62-72]. Thus, norbelladine must be 4'-*O*-methylated to form 4'-*O*-methylbelladine, a central intermediate from which multiple biosynthetic pathways lead to various structural types of AAs (Figures 1-2).

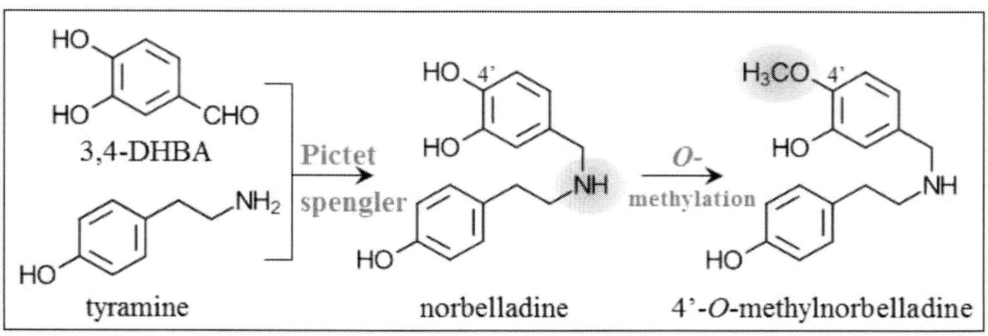

Figure 4. Biosynthesis of 4'-*O*-methylnorbelladine. Functional groups shaded gray highlights chemical conversion.

Mann et al. isolated a catechol-*O*-methyltransferase from the Amaryllidaceae *Nerine bowdenii* bulbs that catalyzed the highly regiospecific methylation of norbelladine at the *para* position at a yield of more than 90% but it also methylated a wide range of catechol substrates [73]. Additionally, protein extracts from leaves of *L. vernum* showed specific activity with regiospecificities favoring the 4' position of norbelladine over the 3' position in a 18:1 ratio [28] and transcript sequences for *OMTs* were detected in the transcriptome of *L. radiata* [41].

Phenol Coupling Reaction and Formation of Diverse AA-Type Backbones

Many specialized metabolites are produced by the coupling of two or more phenolic rings. The reactions involve hydrogen abstraction from a phenol giving the radical followed by the delocalization of the unpaired electron via resonance forms in which the free electron is dispersed to positions *ortho* and *para* to the original oxygen function. These phenol-derived radicals are quenched by coupling with other radicals and the coupling of two of these resonance structures, in various combinations, gives a range of dimeric systems. Thus, C-C bonds involving positions *ortho* or *para* to the original phenols may be formed. The reactive dienone systems formed as intermediates may in some cases be rearranged and further modified extending the range or structures ultimately derived from the basic reaction.

A crucial step in AA biosynthesis is the cyclization of 4'-*O*-methylbelladine by three different ways of intramolecular C-C oxidative phenol coupling named *para-ortho'*(*p-o'*), *para-para'* (*p-p'*) and *ortho-para'* (*o-p'*) and these alternative phenol coupling reactions generate three backbone structures (Figure 2). The different AA subgroups are referred to as *p-o'*, *p-p'* and *o-p'* are derived from these different backbones (Figure 1; Tables 1-2). For instance, antitumor AAs of the narciclasine-type are derived by a *p-p'* phenol coupling step whereas galanthamine originates from the *p-o'* coupling (Figure 2) [3-5].

Recently, several enzymes discovered to be responsible for the formation of intramolecular C-C phenol -couples in various alkaloids pathways, were found to belong to the plant CYP superfamily [43]. For example, the intramolecular C-C phenol-coupling catalyzed by CYP80G2 participates in the formation of aporphine-type alkaloid in *Coptis japonica* [74]. In the formation of morphinan alkaloids from *P.somniferum*, the enzyme salutaridine synthase, a member of the CYP719B1, catalyzed a pivotal bridge-forming reaction through the C-C phenol coupling of (*R*)-reticuline into salutaridine, the *p-o'*-coupled of morphine [75].

Altogether, these examples indicate that specific CYP are involved in the C-C phenol coupling reaction in alkaloid biosynthesis. Specifically, members of two CYP families, CYP80 and CYP719, have been identified. A recent comparative analysis of plant CYP sequences has shown that only the CYP80 family occurs in monocot plant species however no members of the Amaryllidaceae family were represented in this study [44]. This suggests that the phenol coupling steps in AA biosynthesis is likely to involve CYP and may be of the CYP80 family. Interestingly, AAs from all three groups of phenol coupling can co-occur in a single plant, whereas some species and cultivars contains only one group of AA [3, 76] suggesting that multiples genes and/or different isoforms for the phenol coupling enzyme exist. Biotransformation of deuterated-4'-*O*-methylnorbelladine in tissue cultures of *L. aestivum* showed incorporation in all three groups of AAs further supporting the three modes of intramolecular oxidative phenol coupling for cyclization in this Amaryllidaceae.

The three core skeletons obtained from the phenol coupling steps form the basis of further alkaloid diversity. A complex network of enzymatic reactions exists to produce a spectrum of compounds that differs between species, varieties and cultivars and even between the different tissues and vegetative phases of the same plant. These biochemical modifications are achieved by a multitude of enzymes catalyzing various types of reactions, such as C-C and C-O bond formations, *O*- and *N*-methylations, demethylations, hydroxylations, oxidations and reductions. The various products obtained from these reactions yields the several hundred of structurally related AAs known to date (Table 1, Figures 1-2) [3-6, 13].

Alkaloids Derived from the *Para-Ortho'*-Coupling

The *p-o'* phenol coupling reaction of 4'-*O*-methylnorbelladine leads to the formation of galanthamine-type of AAs (Figure 2). Feeding experiments using ^{13}C-labelled and deuterium-labelled *O*-methylnorbelladine to *L. aestivum* plants showed that the biosynthesis of galanthanmine involves the phenol oxidative coupling of *O*-methylnorbelladine to a dienone intermediate, which undergoes spontaneous closure of the ether bridge to yield *N*-demethylnarwedine (Figure 5) [4, 28-30]. The stereo-specific reduction of *N*-demethylnarwedine yields norgalanthanmine which is *N*-methylated to form galanthamine (Figure 5) [28-30].

Feeding studies also showed that narwedine is not the direct precursor of galanthanmine, and possibly exist in equilibrium with galanthamine [28], a reaction catalyzed by a hypothetically reversible oxido-reductase. Norgalanthamnine can also be further reduced to yield norlycoramine which is *N*-methylated into lycoramine (Figure 5). Chlidanthine might arise from galanthamine by *O*-demethylation, a reaction confirmed with ^{3}H-labeled incorporation into chlidanthine [77, 78].

Recently, opium poppy (*P. somniferum*) has emerged as a model system to investigate alkaloid metabolism. Great progress towards the elucidation of BIA metabolism including the well-known morphinan alkaloids has been made at the biochemical and molecular levels and a restricted number of enzyme families have been implicated.

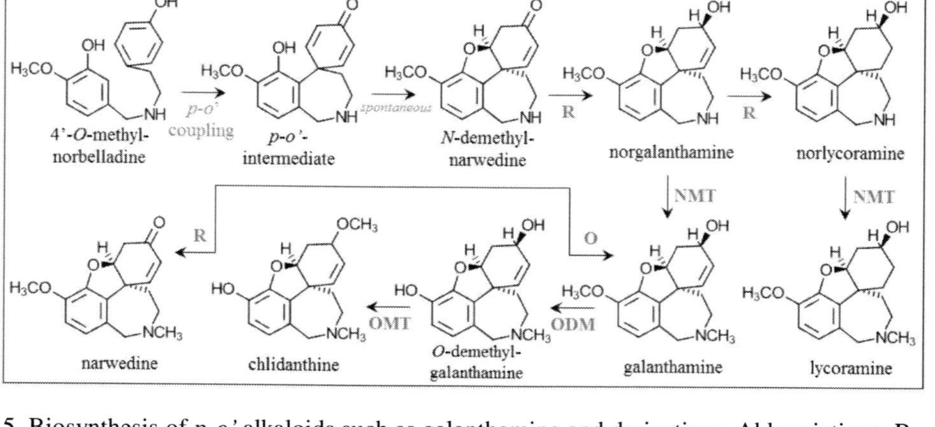

Figure 5. Biosynthesis of *p-o'* alkaloids such as galanthamine and derivatives. Abbreviations: R, reduction; NMT, *N*-methylation; ODM, *O*-demethylation; OMT, *O*-methylation.

Enzymes of the lyase, oxidoreductase and transferase categories have been characterized from several BIA -producing plants and reviewed [2, 47, 48]. The analogy between morphine and galanthamine biosyntheses is quite stricking and is highlighted in Figure 6. The similarities between reactions suggested that AA biosynthetic enzymes may belong to similar gene families. As described above, the Pictet-Spengler condensation of precursor 3,4-DHBA and tyramine, into norbelladine ressembles the condensation of 4-HPAA and dopamine to form norcoclaurine (Figure 6). Similarly, the 4'-*O*-methylation catalyzed by the 3'-hydroxy-*N*-methylcoclaurine-4'-*O*-methyltransferase (4'OMT) and the phenol coupling reaction catalyzed by SalSyn yielding salutaridine mirror the *O*-methylation and phenol coupling reaction in galanthamine biosynthesis (Figure 6). Additionally, the reduction reaction of *N*-demethylnarwedine is similar to the reduction of salutaridine (Figure 6). Reduction reactions are commun in alkaloid metabolism and two major families have been reported to be involved in morphinan alkaloid biosynthesis, the short chain dehydrogenases/reductases (SDRs) and the aldo-keto reductases (AKRs) [2, 47, 48]. For example, the stereo-specific reduction of salutaridine to form salutaridinol is catalyzed by the cytosolic NADPH-dependent salutaridine reductase (SalR) which belongs to the SDR family whereas the reduction of codeinone or morphinone by the NADPH-dependent codeinone reductase (CoR), member of the AKRs, yields codeine and morphine respectively [79-81]. Recently, the last step in noscapine biosynthesis, a phtalideisoquinoline alkaloid from *P. somniferum* with anti-cancer properties, as shown to be catalyzed by a reductase member of the SDRs [82]. From this step, norgalanthanmine is *N*-methylated to galanthamine whereas the *N*-methylation reaction in BIA biosynthesis occurs at an earlier step, i.e., prior to 4'-*O*-methylation, and is catalyzed by the coclaurine *N*-methyltransferase (Figure 6) [1, 2, 47, 48]. In some Amaryllidaceae plants such as in *Harmanthus multiflorus*, galanthamine is *O*-demethylated to form *O*-demethylgalanthamine. This reaction is analogous to the antepenultimate and final steps in morphine biosynthesis catalyzed by the 2-oxoglutarate/Fe(II)-dependent dioxygenases, thebaine 6-*O*-demethylase (T6ODM) and codeine *O*-demethylase (CODM) (Figure 6) [83-85]. Since the comparison of morphine and galanthamine biosynthetic pathways showed great similarities, it may be a strategy to identify interesting biosynthetic gene candidates for the elucidation of AA biosynthesis.

Alkaloids Derived from the *para-para'*-Subgroup

Great structural diversity arises from the *p-p'* phenol coupling of 4'-*O*-methylnorbelladine including crinine-, haemanthamine-, narciclasine- and montanine-type of AAs (Figure 2). Experiments using labelled precursors suggested that the oxidation of *p-p'* core skeleton promoting the formation of methylenedioxy bridge leads to the formation of two possible intermediates, crinine and vittatine (Figure 7) [4].

Different biosynthetic routes from the vittatine intermediate are possible. For example, vittatine (and crinine) may be converted into narciclasine via several oxidation reactions whereas one *O*-methylation of vittatine yields montanine (Figure 7). Hydroxylation of vitattine yields 11-hydroxyvittatine which is subsequently reduced into pancracine and *O*-methylated to form montanine (Figure 7). Alternatively, *O*-methylation of 11-hydroxyvittatine gives haemanthamine which can be hydroxylated and *N*-methylated to yield pretazettine (Figure 7).

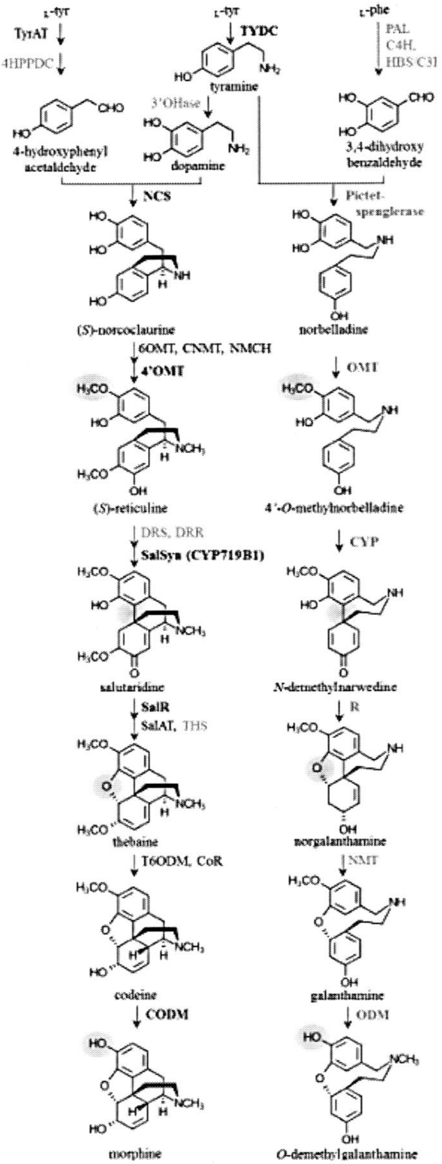

Figure 6. Analogy of reactions between the BIA (left) and AA (right) biosynthetic pathways. Broken arrows represent more than one reaction. Enzymes for which corresponding genes have been isolated from *P. somniferum* are shown in black. Enzymes for which corresponding genes have not been isolated or proposed enzymes are shown in gray. Functional groups shaded gray highlights analogous chemical conversions. Abbreviations: TyrAT, tyrosine aminotransferase; 4HPPDC, 4-hydroxyphenylpuruvate decarboxylase; TYDC, tyrosine/DOPA decarboxylase; 3OHase, tyrosine/tyramine 3-hydroxylase; PAL, phenylalanine ammonia -lyase; C4H, cinnamate hydroxylase; HBS, hydroxybenzyl synthase; C3H, coumarate hydroxylase; NCS, norcoclaurine synthase; 6OMT, norcoclaurine 6-*O*-methyltransferase; CNMT, coclaurine *N*-methyltransferase; NMCH, *N*-methylcoclaurine 3'-hydroxylase; 4'OMT, 3'-hydroxyl -*N*-methylcoclaurine 4'-*O*methyltransferase; SalSyn, salutaridine synthase; SalR, salutaridine reductase; SalAT, salutaridinol 7-*O*-acetyltransferase; T6ODM, thebaine 6-O-demethylase; CoR, codeinone reductase; CODM, codeine *O*-demethylase; OMT, *O*-methyltransferase; CYP, cytochrome P450; R, reductase; NMT, *N*-methyltransferase; ODM, *O*-demethylase.

Enzymes from the methyltransferase families (OMTs and NMTs) are likely to catalyze the methylation reactions.

The formation of C-C or C-O bonds to establish the various AA backbone sturucture is catalyzed by oxidoreductases such as CYPs and FAD -dependent enzymes. For example, the methyledioxy bridge formation of the *p-p'* intermediate leading to the formation of crinine and vittatine may be catalyzed by enzymes of the CYP superfamily (Figure 7). For example, members of the CYP719A subfamily were reported to catalyze methylenedioxy bridge formation in the biosynthesis of alkaloids such as berberine and sanguinarine including *C. japonica* CYP719A1, *E. californica* CYP719A2 and CYP719A3, and *Argemone mexicana* CYP719A13 and CYP719A14 [86-88]. This suggests that members of the CYP719A subfamily are likely to catalyze oxidation reactions in the biosynthesis of AAs. However, other enzyme family are also possible including other CYP families or FAD-dependent oxidoreductases such as berberine bridge enzyme (BBE), (*S*)-tetrahydroprotoberberine oxidase (STOX) and dihydrobenzophenanthridine oxidase (DBOX) which are known to be involved in alkaloid metabolism [2, 89, 90].

Alkaloids Derived from the *ortho-para'*-Subgroup

The *o-p'* phenolic coupling of 4'-*O*-methylnorbelladine results in the formation of the key branchpoint intermediate norpluviine, the precursor of lycorine-, narcissidine- and homolycorine-type of AAs. Feeding of isotopically labeled precursors to different cultivars of *Narcissus* including 'King Alfred', Sempre Avanti', 'Deanna Durbin' and 'Texas' revealed that norpluviine may be processed by three possible route to form the different AA-types [4]. In the first route, nororpluviine is *O*-methylated to pluviine which is hydroxylated to 9-*O*-methylpseudolycorine and then *O*-methylated to galanthine (Figure 8). Galantine may be further converted via oxidation -reduction reactions to form narcissidine. In another route, norpluviine is converted to lycorine via caranine with steps involing oxidation (methylenedioxy bridge formation) and hydroxylation of the benzyl ring (Figure 8). Alternatively, norpluviine may undergoe hydroxylation (benzylic oxidation at position 6), ring opening, rotation and closure, followed by subsequent *O*- and *N*-methylations to form lycorenine, which after oxidation yields homolycorine (Figure 8). The *O*- and *N*-methylation reactions involved in these pathways are likely to be catalyzed by similar enzyme families (OMTs, NMTs) than those discussed earlier (Figure 6). Similarly, the reduction reactions may be catatalyzed by enzyme of the SDR and the AKR families.

C-C or C-O bond formation is catalyzed by oxidoreductases such as CYPs and FAD -dependent enzymes. For example, the hydroxylation of norpluviine leads to the formation of a C-O bond which disturbed the electronic state of the hydroxynorpluviine rings subsequently promoting ring opening, rotation and rearrangement of the core structure (Figure 8). The hydroxylation reactions in alkaloid biosynthesis are catalyzed by CYP enzymes. Characterized members of the CYP80 family including berbamunine synthase (*Berberis stolonifera* CYP80A1) and (*S*)-*N*-methylcolcaurine 3'-hydroxylase (*Eschscholzia californica* CYP80B1, *P. somniferum* CYP80B3 and *C. japonica* CYP80G2) [2, 74]. In addition to hydroxylation, members of the CYP80 family catalyze intermolecular C-O and intramolecular C-C phenol coupling reaction. Recently, members of another CYP family were reported to be involved in hydroxylation reactions of alkaloid biosynthesis.

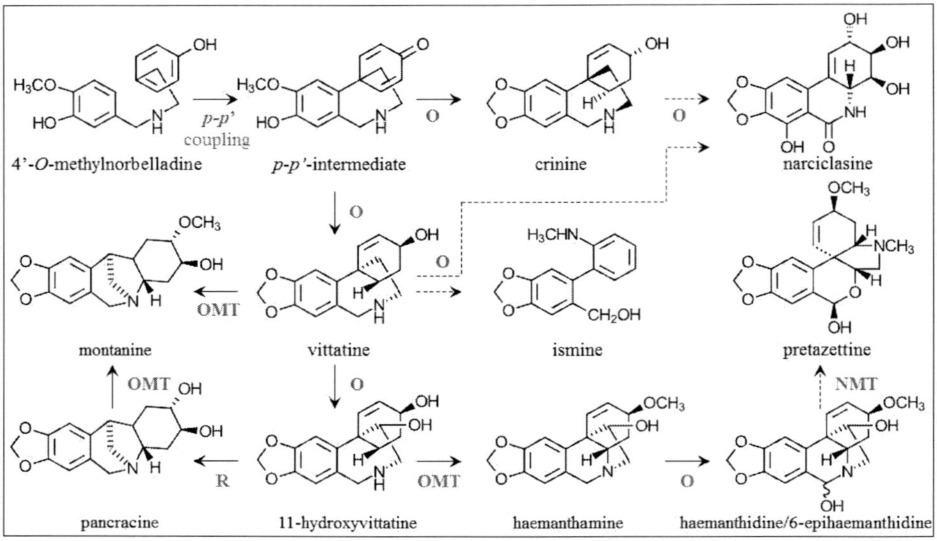

Figure 7. Biosynthesis of *p-p'* alkaloids such as crinine, haemanthamine, narciclasine and montanine. Broken arrow represents more than one biochemical reaction. Abbreviations: R, reduction; O, oxidation; NMT, *N*-methylation; OMT, *O*-methylation.

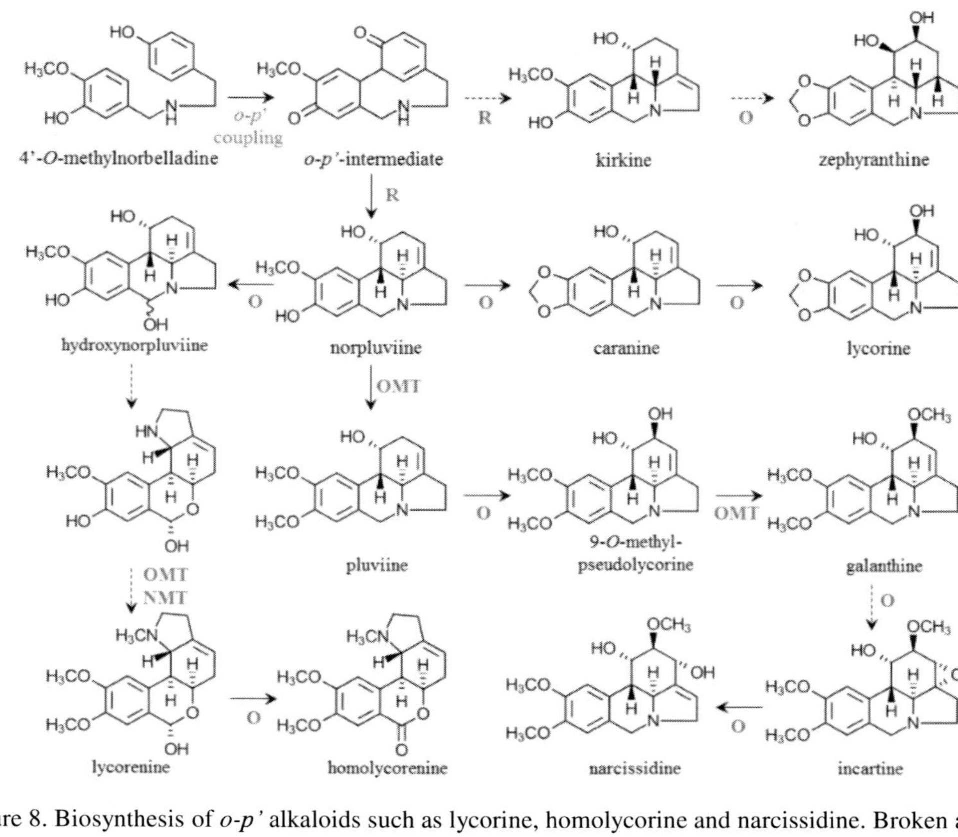

Figure 8. Biosynthesis of *o-p'* alkaloids such as lycorine, homolycorine and narcissidine. Broken arrow represents more than one biochemical reaction. Abbreviations: R, reduction; O, oxidation; NMT, *N*-methylation; OMT, *O*-methylation.

Specifically, the characterization of the *C. japonica* CYP82N2v2 and *P. somniferum* CYP82 showed hydroxylase activity in the last steps of sanguinarine biosynthesis [91, 92]. The methylenedioxy bridge formation from norpluviine to caranine may also be catalyzed by CYP enzymes specifically the CYP719A subfamily as discussed earlier (Figures 7-8). Altogether, these results showed that members of the CYP80 and CYP82 are involved in hydroxylation reactions whereas CYP719A members promoted methylenebridge formation in alkaloid biosynthetic pathways. This further supports the involvement of members of these CYP families in the AA biosynthesis. Finally, despite the vast structural diversity of AAs, a restricted number of enzyme families seems to be involved in AA biosynthesis. Additionally, some enzymes may exhibit a relatively broad substrate range whereas others are may be highly substrate specific further participating in the formation of a pletora of AAs.

METABOLIC ENGINEERING AND SYNTHETIC BIOLOGY

The commercial demand to increase certain AA in plants is very difficult to achieve by conventional breeding techniques and nowadays technologies may help to reach these goals. Until recently, non-model organisms such as Amaryllidaceae plants were recalcitrant to modern molecular biology approaches for gene and pathway discovery due in part to the lack of transformation protocols. However, rapid progress in the development of integrative approaches such as genomics, transcriptomics, proteomics, metabolomics and fluxomics, has provided essential information for the understanding of many complex biological processes at different levels [93]. Comparative and integrative analyses of such 'omics' databases has facilitated the determination of gene function and pinpoint rate limiting steps and factors in the biosynthetic pathway of natural products in non-model species [93, 94]. Correlations between plant transcriptome, proteome and metabolome have been successfully used for the identification of novel genes involved in alkaloid biosynthesis [69, 70, 79, 95-97].

Each Amaryllidaceae species studied to date displays specific alkaloid composition (AA types and levels), often with a few dominant AAs and a larger number at lower concentrations [3, 4, 76]. Although the molecular origin of this biochemical diversity has not yet been defined, these AA compositions likely result from differences in the expression level and substrate specificity of the various biosynthetic enzymes. The development and subsequent integration of 'omics' databases for Amaryllidaceae plants displaying different AA composition would allow for the identification of candidate genes involved in AA biosynthesis. For example, searches by sequence similarity to *P. somniferum*, of orthologous genes involved in synthesis of the precursors 3,4-DHBA and tyramine, e.g., *PAL, C4H, C3H, HBS* and *TYDC,* can be identified. Recently, the transcriptome of the Amaryllidaceae *L. aurea* was sequenced, assembled and putative genes involved in AAs biosynthesis (*PAL, TYDC, OMT, NMT, CYP*) were identified based on sequence analysis [41] however none of the biosynthetic enzymes involved in these pathways have been isolated or functionally characterized. Those could be targeted for silencing or over-expression study in plants to confirm their role(s) in AA biosynthesis and to alter AA composition. Additionally, information obtained would help in designing rational plant breeding methods to raise new cultivars with optimized alkaloid production beneficial for pharmaceutical industries and other research applications.

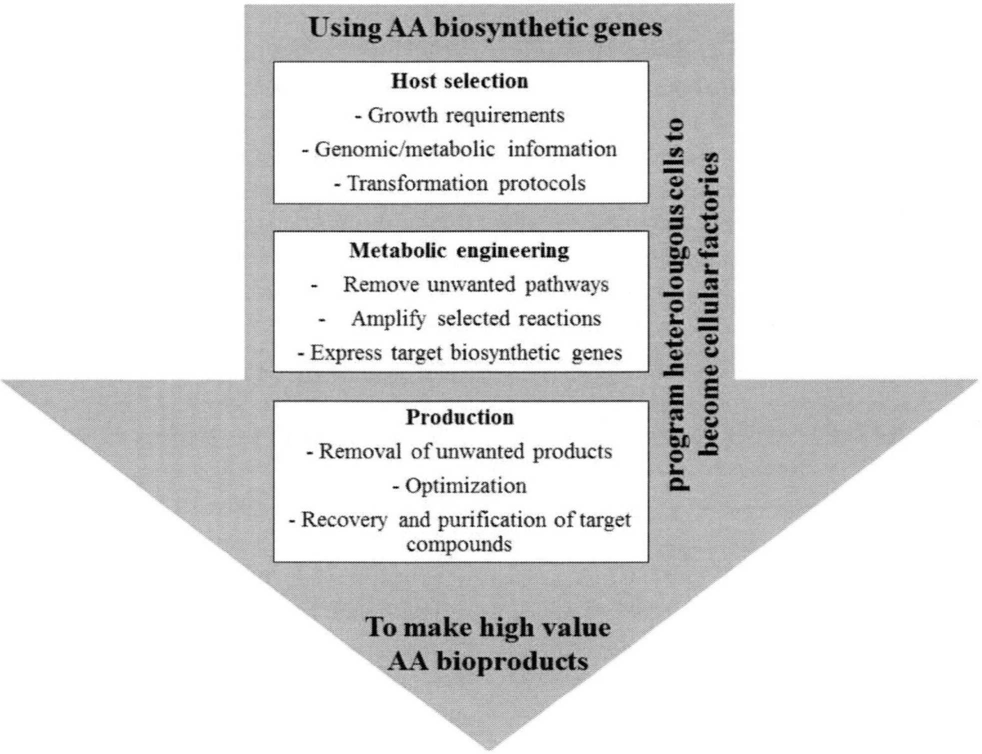

Figure 9. Representation of the strategy for the metabolic engineering of microorganism for AA drug production.

For example, acetylcholinesterase inhibitor galanthamine and anticancerous agent narciclasine and crinine, could be produced commercially by such newly designed plant breeding programs.

Currently, large-scale production of AA-based pharmaceuticals is limited to extraction of alkaloids that accumulate in plants. However the synthesis of AAs in microbial hosts could bypass such limitations and transform industrial production of AAs (Figure 9). Thus, a precise knowledge of AA biosynthetic networks is mandatory to the emerging field of synthetic biology, which aims to reconstruct plant pathways in microorganisms with the ultimate goal of creating alternative systems for the production of valuable bioproducts. To date, several strategies have been developed to engineered microbial hosts for the production of alkaloids specifically related to the BIAs [98-102]. Recently, the introduction of ten BIA biosynthetic genes in yeast led to the production of dihydrosanguinarine and its oxidized derivative sanguinarine from (*R,S*)-norlaudanosoline further demonstrating the feasibility of production of high-value alkaloids in microbial systems [102].

Conclusion

Over the past several years, extensive phytochemical and pharmacological analyses have reported the numerous biological activities of Amaryllidaceae alkaloids and numerous AAs display interesting and valuable pharmacological capabilities. Earlier biochemical tracer studies have led to the current AA biosynthetic proposed pathways however the molecular identity of the biosynthetic enzymes remains unknown. The technological advances brought as part of the post-genomics era have revolutionized the study of alkaloid metabolism and discoveries made over the past years have relied largely on 'omics' tools including transcriptome libraries, proteomic analyses and targeted metabolomics methods. Integration of these 'omics' resources such as comparative analyses of metabolite and transcript data from Amaryllidaceae species will lead to the identification and isolation of numerous enzymes involved in AA biosynthesis. A deeper understanding of the molecular mechanisms involved in AA biosynthesis will support breeding efforts to produced cultivars of Amaryllidaceae species with enhanced AA production. In addition, it will pave the way for the successful metabolic engineering of microbial systems for the production of valuable AAs.

REFERENCES

[1] Facchini, P. J., Alkaloid Biosynthesis In Plants: Biochemistry, Cell Biology, Molecular Regulation, and Metabolic Engineering Applications. *Annu. Rev. Plant Physiol. Plant Mol. Biol.,* 2001. 52: p. 29-66.

[2] Ziegler, J. and P. J. Facchini, Alkaloid biosynthesis: metabolism and trafficking. *Annu. Rev. Plant Biol.*, 2008. 59: p. 735-69.

[3] Kornienko, A. and A. Evidente, Chemistry, biology, and medicinal potential of narciclasine and its congeners. *Chem. Rev.,* 2008. 108(6): p. 1982-2014.

[4] Bastida, J., et al., Chap3 Chemical and biological aspects of Amaryllidaceae alkaloids, E. D. Muñoz-Torrero, et al., Editors. 2011, Recent Advances in Pharmaceutical Sciences, *Recent Advances in Pharmaceutical Sciences,* 2011: 65-100 ISBN: 978-81-7895-528-5.

[5] Takos, A. M. and F. Rook, Towards a molecular understanding of the biosynthesis of amaryllidaceae alkaloids in support of their expanding medical use. *Int. J. Mol. Sci.,* 2013. 14(6): p. 11713-41.

[6] Jin, Z., Amaryllidaceae and Sceletium alkaloids. *Nat. Prod. Rep.,* 2013. 30(6): p. 849-68.

[7] Singh, A. and I. Desgagne-Penix, Biosynthesis of the Amaryllidaceae alkaloids. *Plant Science Today,* 2014. 1(2): p. 114-120.

[8] Heinrich, M. and H. Lee Teoh, Galanthamine from snowdrop--the development of a modern drug against Alzheimer's disease from local Caucasian knowledge. *J. Ethnopharmacol.*, 2004. 92(2-3): p. 147-62.

[9] Conforti, F., et al., Quantitative determination of Amaryllidaceae alkaloids from Galanthus reginae-olgae subsp. vernalis and in vitro activities relevant for neurodegenerative diseases. *Pharm. Biol.*, 2010. 48(1): p. 2-9.

[10] Lubbe, A., et al., Seasonal accumulation of major alkaloids in organs of pharmaceutical crop Narcissus Carlton. *Phytochemistry,* 2013. 88: p. 43-53.

[11] da Silva, A. F., et al., Anxiolytic-, antidepressant - and anticonvulsant -like effects of the alkaloid montanine isolated from Hippeastrum vittatum. *Pharmacol. Biochem. Behav.,* 2006. 85(1): p. 148-54.

[12] Furusawa, E. and S. Furusawa, Effect of pretazettine and viva-natural, a dietary seaweed extract, on spontaneous AKR leukemia in comparison with standard drugs. *Oncology,* 1988. 45(3): p. 180-6.

[13] Nair, J. J. and J. van Staden, Pharmacological and toxicological insights to the South Africa n Amaryllidaceae. *Food Chem. Toxicol.*, 2013. 62C: p. 262-275.

[14] Lubbe, A., R. Verpoorte, and Y. H. Choi, Effects of fungicides on galanthamine and metabolite profiles in Narcissus bulbs. *Plant Physiol. Biochem.,* 2012. 58: p. 116-23.

[15] Sener, B., I. Orhan, and J. Satayavivad, Antimalarial activity screening of some alkaloids and the plant extracts from Amaryllidaceae. *Phytother. Res.,* 2003. 17(10): p. 1220-3.

[16] Pettit, G. R., et al., Antineoplastic agents, 99. Amaryllis belladonna. *J. Nat. Prod.,* 1984. 47(5): p. 796-801.

[17] Weniger, B., et al., Cytotoxic activity of Amaryllidaceae alkaloids. *Planta Med.,* 1995. 61(1): p. 77-9.

[18] Hua, D. H., S. Saha, and D. J. Takemoto, Anticancer activities of 2,5,8,9-substituted 6-oxo-1,2,3,4,5,6-hexahydrophenanthridines on multi-drug-resistant phenotype cells. *Anticancer Res.,* 1997. 17(4A): p. 2435-41.

[19] Lamoral-Theys, D., et al., Lycorine and its derivatives for anticancer drug design. *Mini Rev. Med. Chem.,* 2010. 10(1): p. 41-50.

[20] Koola, M. M., et al., Potential role of the combination of galantamine and memantine to improve cognition in schizophrenia. *Schizophr. Res.,* 2014.

[21] Konrath, E. L., et al., Alkaloids as a source of potential anticholinesterase inhibitors for the treatment of Alzheimer's disease. *J. Pharm. Pharmacol.,* 2013. 65(12): p. 1701-25.

[22] Birks, J. and D. Craig, Galantamine for vascular cognitive impairment. *Cochrane Database Syst. Rev.*, 2013. 4: p. CD004746.

[23] Furst, R. and I. Zundorf, Plant-Derived Anti-Inflammatory Compounds: Hopes and Disappointments regarding the Translation of Preclinical Knowledge into Clinical Progress. *Mediators Inflamm.,* 2014. 2014: p. 146832.

[24] Nair, J. J., et al., Cytotoxic agents of the crinane series of amaryllidaceae alkaloids. *Nat. Prod. Commun.*, 2012. 7(12): p. 1677-88.

[25] Barton, D. H. R. and T. Cohen, In Festschrift Arthur Stoll p.117 Birkhauser, Basel., 1957.

[26] Barton, D. H. R., et al., Phenol oxidation and biosynthesis. Part VI. The biogenesis of Amaryllidaceae alkaloids. *J. Chem. Soc.,* 1963: p. 4545-4558.

[27] Battersby, A. R., H. M. Fales, and W. C. Wildman, Biosynthesis in the Amaryllidaceae. Tyrosine and norbelladine as precursors of haemanthamine. *J. Amer. Chem. Soc.,* 1961. 83: p. 4098-4099.

[28] Eichhorn, J., et al., Biosynthesis of the Amaryllidaceae alkaloid galanthamine. *Phytochemistry,* 1998. 49(4): p. 1037-1047.

[29] El Tahchy, A., et al., New method for the study of Amaryllidaceae alkaloid biosynthesis using biotransformation of deuterium-labeled precursor in tissue cultures. *Acta. Biochim. Pol.*, 2010. 57(1): p. 75-82.

[30] El Tahchy, A., et al., Kinetic study of the rearrangement of deuterium-labeled 4'-O-methylnorbelladine in Leucojum aestivum shoot cultures by mass spectrometry. Influence of precursor feeding on amaryllidaceae alkaloid accumulation. *J. Nat. Prod.,* 2011. 74(11): p. 2356-61.

[31] Grisebach, H., Comparative biosynthetic pathways in higher plants. *Pure and Applied chemistry,* 1973. 34(3-4): p. 487-513.

[32] Chu, H. Y., E. Wegel, and A. Osbourn, From hormones to secondary metabolism: the emergence of metabolic gene clusters in plants. *Plant J.,* 2011. 66(1): p. 66-79.

[33] Wink, M., Biochemistry of plant secondary metabolism. *Annual plant reviews,* 2010. 40.

[34] Zenk, M., Biosynthese von vanillin in Vanilla planifolia Andr. *Z. Pflanzen Physiol.,* 1965. 53: p. 404-414.

[35] Walton, N. J., M. J. Mayer, and A. Narbad, Vanillin. *Phytochemistry,* 2003. 63(5): p. 505-515.

[36] Podstolski, A., et al., Unusual 4-hydroxybenzaldehyde synthase activity from tissue cultures of the vanilla orchid Vanilla planifolia. *Phytochemistry,* 2002. 61(6): p. 611-620.

[37] Negishi, O., K. Sugiura, and Y. Negishi, Biosynthesis of vanillin via ferulic acid in Vanilla planifolia. *J. Agric. Food Chem.,* 2009. 57(21): p. 9956-61.

[38] Havkin-Frenkel, D. and F. Belanger, Application of Metabolic Engineering to Vanillin Biosynthetic Pathways in Vanilla Planifolia, in Applications of Plant Metabolic Engineering, R. Verpoorte, A. W. Alfermann, and T. S. Johnson, Editors. 2007, Springer Netherlands. p. 175-196.

[39] Aza-Gonzalez, C., H. G. Nunez-Palenius, and N. Ochoa-Alejo, Molecular biology of capsaicinoid biosynthesis in chili pepper (Capsicum spp.). *Plant Cell. Rep.,* 2011. 30(5): p. 695-706.

[40] Jiang, Y., et al., Molecular and analysis of a phenylalanine ammonia -lyase gene (LrPAL2) from Lycoris radiata. *Mol. Biol. Rep.,* 2013. 40(3): p. 2293-300.

[41] Wang, R., et al., De novo sequence assembly and characterization of Lycoris aurea transcriptome using GS FLX titanium platform of 454 pyrosequencing. *PLoS One,* 2013. 8(4): p. e60449.

[42] Ehlting, J., et al., Cytochromes P450 in phenolic metabolism. *Phytochemistry Reviews,* 2006. 5(2-3): p. 239-270.

[43] Mizutani, M. and F. Sato, Unusual P450 reactions in plant secondary metabolism. *Arch. Biochem. Biophys.,* 2011. 507(1): p. 194-203.

[44] Nelson, D. and D. Werck-Reichhart, A P450-centric view of plant evolution. *Plant J.,* 2011. 66(1): p. 194-211.

[45] Sircar, D. and A. Mitra, Accumulation of p-hydroxybenzoic acid in hairy roots of Daucus carota 2: confirming biosynthetic steps through feeding of inhibitors and precursors. *J. Plant Physiol.,* 2009. 166(13): p. 1370-80.

[46] Facchini, P. J., K. L. Huber-Allanach, and L. W. Tari, Plant aromatic L-amino acid decarboxylases: evolution, biochemistry, regulation, and metabolic engineering applications. *Phytochemistry,* 2000. 54(2): p. 121-38.

[47] Hagel, J. M. and P. J. Facchini, Benzylisoquinoline Alkaloid Metabolism: A Century of Discovery and a Brave New World. *Plant and Cell Physiology,* 2013. 54(5): p. 647-672.

[48] Desgagné-Penix, I. and P. J. Facchini, Benzylisoquinoline Alkaloid Biosynthesis, in Plant Metabolism and Biotechnology. 2011, John Wiley & Sons, Ltd. p. 241-261.

[49] Torrens-Spence, M. P., et al., Biochemical evaluation of a parsley tyrosine decarboxylase results in a novel 4-hydroxyphenylacetaldehyde synthase enzyme. *Biochemical and Biophysical Research Communications,* 2012. 418(2): p. 211-216.

[50] Dewick, P. M., Medicinal natural products: a biosynthetic apporach. 3rd ed. 2009: John Wiley & Sons Ltd. 539.

[51] Ghosal, S., A. Shanthy, and S. K. Singh, Isocraugsodine, an n-arylidenephenethylamine from Crinum asiaticum and its e-z isomerism. *phytochemistry,* 1988. 27(6): p. 1849-1852.

[52] Maresh, J. J., et al., Strictosidine synthase: mechanism of a Pictet-Spengler catalyzing enzyme. *J. Am. Chem. Soc.,* 2008. 130(2): p. 710-23.

[53] Luk, L. Y., et al., Mechanistic studies on norcoclaurine synthase of benzylisoquinoline alkaloid biosynthesis: an enzymatic Pictet-Spengler reaction. *Biochemistry,* 2007. 46(35): p. 10153-61.

[54] Lee, E. J. and P. Facchini, Norcoclaurine synthase is a member of the pathogenesis - related 10/Bet v1 protein family. *Plant Cell,* 2010. 22(10): p. 3489-503.

[55] Minami, H., et al., Functional analysis of norcoclaurine synthase in Coptis japonica. *J. Biol. Chem.*, 2007. 282(9): p. 6274-82.

[56] Ilari, A., et al., Structural basis of enzymatic (S)-norcoclaurine biosynthesis. *J. Biol. Chem.*, 2009. 284(2): p. 897-904.

[57] Bonamore, A., et al., Norcoclaurine synthase: mechanism of an enantioselective pictet-spengler catalyzing enzyme. *Molecules,* 2010. 15(4): p. 2070-8.

[58] Pasquo, A., et al., Cloning, expression, crystallization and preliminary X-ray data analysis of norcoclaurine synthase from Thalictrum flavum. *Acta. Crystallogr. Sect. F. Struct. Biol. Cryst. Commun.*, 2008. 64(Pt 4): p. 281-3.

[59] Samanani, N., D. K. Liscombe, and P. J. Facchini, Molecular cloning and characterization of norcoclaurine synthase, an enzyme catalyzing the first committed step in benzylisoquinoline alkaloid biosynthesis. *Plant J.*, 2004. 40(2): p. 302-13.

[60] Kutchan, T. M., Strictosidine: From alkaloid to enzyme to gene. *Phytochemistry,* 1993. 32(3): p. 493-506.

[61] Stockigt, J., et al., 3D-Structure and function of strictosidine synthase--the key enzyme of monoterpenoid indole alkaloid biosynthesis. *Plant Physiol. Biochem.,* 2008. 46(3): p. 340-55.

[62] Liscombe, D. K., G. V. Louie, and J. P. Noel, Architectures, mechanisms and molecular evolution of natural product methyltransferases. *Nat. Prod. Rep.*, 2012. 29(10): p. 1238-50.

[63] Lam, K. C., et al., Structure, function, and evolution of plant O-methyltransferases. *Genome,* 2007. 50(11): p. 1001-13.

[64] Rueffer, M., N. Nagakura, and M. H. Zenk, Partial Purification and Properties of S-Adenosylmethionine: (R), (S)-Norlaudanosoline-6-O-Methyltransferase from Argemone platyceras Cell Cultures. *Planta Med.,* 1983. 49(11): p. 131-7.

[65] Frick, S. and T. M. Kutchan, Molecular cloning and functional expression of O-methyltransferases common to isoquinoline alkaloid and phenylpropanoid biosynthesis. *Plant J.,* 1999. 17(4): p. 329-39.

[66] Morishige, T., et al., Molecular characterization of the S-adenosyl-L-methionine:3'-hydroxy-N-methylcoclaurine 4'-O-methyltransferase involved in isoquinoline alkaloid biosynthesis in Coptis japonica. *J. Biol. Chem.,* 2000. 275(30): p. 23398-405.

[67] Morishige, T., et al., Molecular cloning of columbamine O-methyltransferase from cultured Coptis japonica cells. *Eur. J. Biochem.*, 2002. 269(22): p. 5659-67.

[68] Ounaroon, A., et al., (R,S)-Reticuline 7-O-methyltransferase and (R,S)-norcoclaurine 6-O-methyltransferase of Papaver somniferum - cDNA cloning and characterization of methyl transfer enzymes of alkaloid biosynthesis in opium poppy. *Plant J.*, 2003. 36(6): p. 808-19.

[69] Ziegler, J., et al., Comparative macroarray analysis of morphine containing Papaver somniferum and eight morphine free Papaver species identifies an O-methyltransferase involved in benzylisoquinoline biosynthesis. *Planta,* 2005. 222(3): p. 458-71.

[70] Desgagne-Penix, I., et al., Integration of deep transcript and targeted metabolite profiles for eight cultivars of opium poppy. *Plant Mol. Biol.,* 2012. 79(3): p. 295-313.

[71] Pienkny, S., et al., Functional characterization of a novel benzylisoquinoline O-methyltransferase suggests its involvement in papaverine biosynthesis in opium poppy (Papaver somniferum L). *Plant J.*, 2009. 60(1): p. 56-67.

[72] Nomura, T. and T. M. Kutchan, Three new O-methyltransferases are sufficient for all O-methylation reactions of ipecac alkaloid biosynthesis in root culture of Psychotria ipecacuanha. *J. Biol. Chem.,* 2010. 285(10): p. 7722-38.

[73] Mann, J. D., H. M. Fales, and S. H. Mudd, Alkaloids And Plant Metabolism. VI. O-Methylation In Vitro Of Norbelladine, A Precursor Of Amaryllidaceae Alkaloids. *J. Biol. Chem.*, 1963. 238: p. 3820-3.

[74] Ikezawa, N., K. Iwasa, and F. Sato, Molecular cloning and characterization of CYP80G2, a cytochrome P450 that catalyzes an intramolecular C-C phenol coupling of (S)-reticuline in magnoflorine biosynthesis, from cultured Coptis japonica cells. *J. Biol. Chem.*, 2008. 283(14): p. 8810-21.

[75] Gesell, A., et al., CYP719B1 is salutaridine synthase, the C-C phenol -coupling enzyme of morphine biosynthesis in opium poppy. *J. Biol. Chem.*, 2009. 284(36): p. 24432-42.

[76] Berkov, S., et al., Evolution of alkaloid biosynthesis in the genus Narcissus. *Phytochemistry,* 2014. 99: p. 95-106.

[77] Bhandarkar, J. G. and G. W. Kirby, Structure and biosynthesis of chlidanthine. *Journal of the Chemical Society C: Organic,* 1970(9): p. 1224-1227.

[78] Abdallah, O. M., A. A. Ali, and H. Itokawa, 2-O-acetyl chlidanthine; An alkaloid from Haemanthus multiflorus. *Phytochemistry,* 1989. 28(11): p. 3248-3249.

[79] Ziegler, J., et al., Comparative transcript and alkaloid profiling in Papaver species identifies a short chain dehydrogenase/reductase involved in morphine biosynthesis. *Plant J.,* 2006. 48(2): p. 177-92.

[80] Unterlinner, B., R. Lenz, and T. M. Kutchan, Molecular cloning and functional expression of codeinone reductase: the penultimate enzyme in morphine biosynthesis in the opium poppy Papaver somniferum. *Plant J.,* 1999. 18(5): p. 465-75.

[81] Lenz, R. and M. H. Zenk, Purification and properties of codeinone reductase (NADPH) from Papaver somniferum cell culture s and differentiated plants. *Eur. J. Biochem.,* 1995. 233(1): p. 132-9.

[82] Chen, X. and P. J. Facchini, Short-chain dehydrogenase/reductase catalyzing the final step of noscapine biosynthesis is localized to laticifers in opium poppy. *Plant J.*, 2014. 77(2): p. 173-84.

[83] Hagel, J. M. and P. J. Facchini, Dioxygenases catalyze the O-demethylation steps of morphine biosynthesis in opium poppy. *Nat. Chem. Biol.,* 2010. 6(4): p. 273-5.

[84] Hagel, J. M. and P. J. Facchini, Biochemistry and occurrence of o-demethylation in plant metabolism. *Front. Physiol.*, 2010. 1: p. 14.

[85] Farrow, S. C. and P. J. Facchini, Dioxygenases catalyze o-demethylation and o,o-demethylenation with widespread roles in benzylisoquinoline alkaloid metabolism in opium poppy. *J. Biol. Chem.*, 2013. 288(40): p. 28997-9012.

[86] Ikezawa, N., et al., Molecular cloning and characterization of CYP719, a methylenedioxy bridge-forming enzyme that belongs to a novel P450 family, from cultured Coptis japonica cells. *J. Biol. Chem.*, 2003. 278(40): p. 38557-65.

[87] Ikezawa, N., K. Iwasa, and F. Sato, Molecular cloning and characterization of methylenedioxy bridge-forming enzymes involved in stylopine biosynthesis in Eschscholzia californica. *FEBS J.*, 2007. 274(4): p. 1019-35.

[88] Diaz Chavez, M. L., et al., Characterization of two methylenedioxy bridge-forming cytochrome P450-dependent enzymes of alkaloid formation in the Mexican prickly poppy Argemone mexicana. *Arch. Biochem. Biophys.*, 2011. 507(1): p. 186-93.

[89] Hagel, J. M., et al., Characterization of a Flavoprotein Oxidase from Opium Poppy Catalyzing the Final Steps in Sanguinarine and Papaverine Biosynthesis. *Journal of Biological Chemistry*, 2012. 287(51): p. 42972-42983.

[90] Gesell, A., et al., Heterologous expression of two FAD -dependent oxidases with (S)-tetrahydroprotoberberine oxidase activity from Arge mone mexicana and Berberis wilsoniae in insect cells. *Planta*, 2011. 233(6): p. 1185-97.

[91] Takemura, T., et al., Molecular cloning and characterization of a cytochrome P450 in sanguinarine biosynthesis from Eschscholzia californica cells. *Phytochemistry*, 2013. 91: p. 100-8.

[92] Beaudoin, G. A. W. and P. J. Facchini, Isolation and characterization of a cDNA encoding (S)-cis-N-methylstylopine 14-hydroxylase from opium poppy, a key enzyme in sanguinarine biosynthesis. *Biochemical and Biophysical Research Communications*, 2013. 431(3): p. 597-603.

[93] Schilmiller, A. L., E. Pichersky, and R. L. Last, Taming the hydra of specialized metabolism: how systems biology and comparative approaches are revolutionizing plant biochemistry. *Curr. Opin. Plant Biol.*, 2012. 15(3): p. 338-44.

[94] Xiao, M., et al., Transcriptome analysis based on next-generation sequencing of non-model plants producing specialized metabolites of biotechnological interest. *Journal of Biotechnology*, 2013. 166(3): p. 122-134.

[95] Desgagne-Penix, I., et al., Integration of deep transcriptome and proteome analyses reveals the components of alkaloid metabolism in opium poppy cell culture s. *BMC Plant Biol.*, 2010. 10: p. 252.

[96] Desgagne-Penix, I. and P. J. Facchini, Systematic silencing of benzylisoquinoline alkaloid biosynthetic genes reveals the major route to papaverine in opium poppy. *Plant J.*, 2012. 72(2): p. 331-44.

[97] Liscombe, D. K., et al., Targeted metabolite and transcript profiling for elucidating enzyme function: isolation of novel N-methyltransferases from three benzylisoquinoline alkaloid-producing species. *Plant J.*, 2009. 60(4): p. 729-43.

[98] Minami, H., et al., Microbial production of plant benzylisoquinoline alkaloids. *Proc. Natl. Acad. Sci. U S A*, 2008. 105(21): p. 7393-8.

[99] Hawkins, K. M. and C. D. Smolke, Production of benzylisoquinoline alkaloids in Saccharomyces cerevisiae. *Nat. Chem. Biol.*, 2008. 4(9): p. 564-73.

[100] Nakagawa, A., et al., Bench-top fermentative production of plant benzylisoquinoline alkaloids using a bacterial platform. *Bioeng. Bugs,* 2012. 3(1): p. 49-53.
[101] Facchini, P. J., et al., Synthetic biosystems for the production of high-value plant metabolites. *Trends Biotechnol.,* 2012. 30(3): p. 127-31.
[102] Fossati, E., et al., Reconstitution of a 10-gene pathway for synthesis of the plant alkaloid dihydrosanguinarine in Saccharomyces cerevisiae. *Nat. Commun.,* 2014. 5: p. 3283.

In: Alkaloids
Editor: Eduardo Sobarzo-Sánchez
ISBN: 978-1-63482-074-5

Chapter 4

MARINE BROMINATED ALKALOIDS: ISOLATION, BIOSYNTHESIS, BIOLOGICAL ACTIVITY AND SYNTHESIS

Saeed Ahmad and Faiz Ahmed Khan**[*]**
Department of Chemistry, Indian Institute of
Technology Hyderabad, Yeddumailaram, India

ABSTRACT

A large area of earth surface is covered by ocean and it has a diverse range of flora and fauna. A variety of secondary metabolites are produced by them for their specific purposes, which includes protection from predator, reproduction etc. Produced secondary metabolites comprise from simple molecules to complex molecules in terms of skeletal architectures. A large number of marine natural products have been isolated, in which a considerable percentage belongs to organohalogen compounds. In the category of organohalogen compounds, a major portion (approximate 50%) contributed by organobromine compounds and most of them are alkaloids. These compounds ranging from pyrrole, pyridine, pyrimidine, quinoline, isoquinoline to indole, steroid, guanidine, styrene and pyridoacridine. The isolated marine brominated alkaloids exhibited a wide range of biological activity profile like cytotoxic, antiviral, antifungal, antibacterial etc. and some of them are very important from pharmaceutical aspects. In recent past, many pharmaceutically important natural products have been isolated from the marine bryozoans and marketed as drugs and some are in clinical trials. Still in the search of new chemical entities of biological importance with novel mode of actions, the isolation of new marine natural products is continuously growing with a speedy rate. In this chapter, the importance and the key aspects of marine brominated alkaloids have been discussed.

Keywords: antibacterial, biological activity, cytotoxic, marine brominated alkaloids, marine natural products, nematocidal, organohalogens, pharmaceuticals, tyrosine derived alkaloids

---

[*] Corresponding author: Faiz Ahmed Khan. Department of Chemistry, Indian Institute of Technology Hyderabad, Ordnance Factory Estate, Yeddumailaram-502205, India. E-mail: faiz@iith.ac.in, tel: (+91) 40 2301 6084.

INTRODUCTION

To cure various disorders, ailments and diseases, mankind is using natural medicinal ingredients from ages. Still these natural medicinal ingredients are the only sources to treat diseases in some parts of the world. According to a report, more than 50% drugs available in the market are either directly extracted from natural sources or synthesized on natural products platform [1]. Marine organisms often produce halogenated secondary metabolites unlike their terrestrial counterparts. These halogenated secondary metabolites are generally alkaloids [2]. The majority of marine halogenated natural products are brominated and are abundant in marine environment, while chlorinated marine natural products are less abundant even though bromide ions concentration is much less than chloride ions concentration in ocean water (bromide 65 mg/L; chloride 19,000 mg/L).

The reason behind this is the easy oxidation of bromide ion into bromine or bromonium ion or biological equivalent [3]. There are many halogenated natural products known in literature that have been reviewed [4-7]. However, to the best of our knowledge, none of the review exclusively dedicated for the marine brominated alkaloids is reported in literature.

In this chapter, we have discussed about marine brominated alkaloids. We highlighted and emphasised the following points in this chapter (1) Why the attention towards marine brominated alkaloids is important (2) Specificity of marine brominated alkaloids in terms of their biological activities. (3) Biosynthesis of brominated alkaloids. Why the bromination of tribrominated alkaloids is not studied so far (4) Synthetic challenges involved in the synthesis of marine brominated alkaloids.

1. SYNOXAZOLIDINONES A-C

Tadesse and coworkers isolated a new class of alkaloids synoxazolidinones A (**1**) and B (**2**) from ascidian *Synoicum pulmonaria* through bioassay guided fractionation (Figure 1) [8]. They determined the structures by extensive spectroscopic techniques, especially NMR, and assigned the molecules as a tyrosine derived dibrominated alkaloids possessing 4-oxazolidinone ring at center and guanidine part at the terminal end. This type of combination is rarely seen in nature. Synoxazolidinones A (**1**) and B (**2**) were evaluated for antibacterial as well as antifungal activity. Synoxazolidinone A (**1**) exhibited antibacterial activity against the Gram-positive bacterium *Corynebacterium glutamicum* with the MIC value of 6.25 μg/mL and against the fungi *Saccharomyces cerevisiae* with the MIC value of 12.5 μg/mL. While its counterpart synoxazolidinone B (**2**) exhibited lower activities. Based on the observations, Tadesse and coworkers suggested that might be chlorine atom is essential for biological activities. In their continuous studies of ascidian *Synoicum pulmonaria*, they isolated a new bicyclic alkaloid synoxazolidinone C (**3**) of synoxazolidinones family and it has an additional pyrrolidine ring (Figure 1) [9]. The configuration of all stereocenters of synoxazolidinones A (**1**) and C (**3**) were assigned by vibrational chiroptical spectroscopy [10]. Synoxazolidinone C (**3**) exhibited activity against both Gram-positive bacteria (*Staphylococcus aureus* and methicillin-resistant *S. aureus* at the concentration of 10 μg/mL) and Gram-negative bacteria (*Escherichia coli* and *Enterococcus faecalis* at the concentration of 30 μg/mL and 20 μg/mL, respectively).

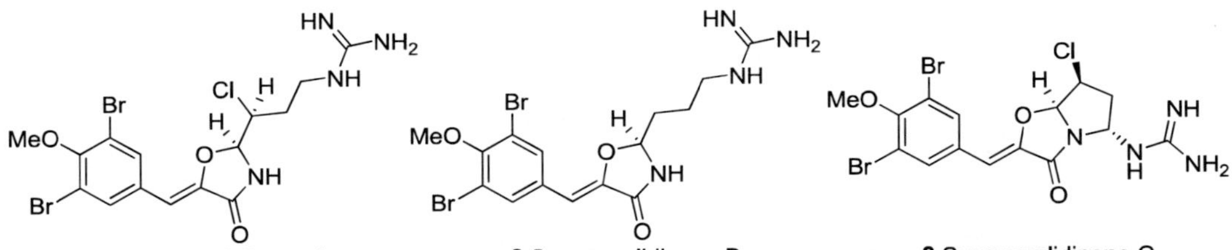

1 Synoxazolidinone A **2** Synoxazolidinone B **3** Synoxazolidinone C

Figure 1. Structures of synoxazolidinones A-C.

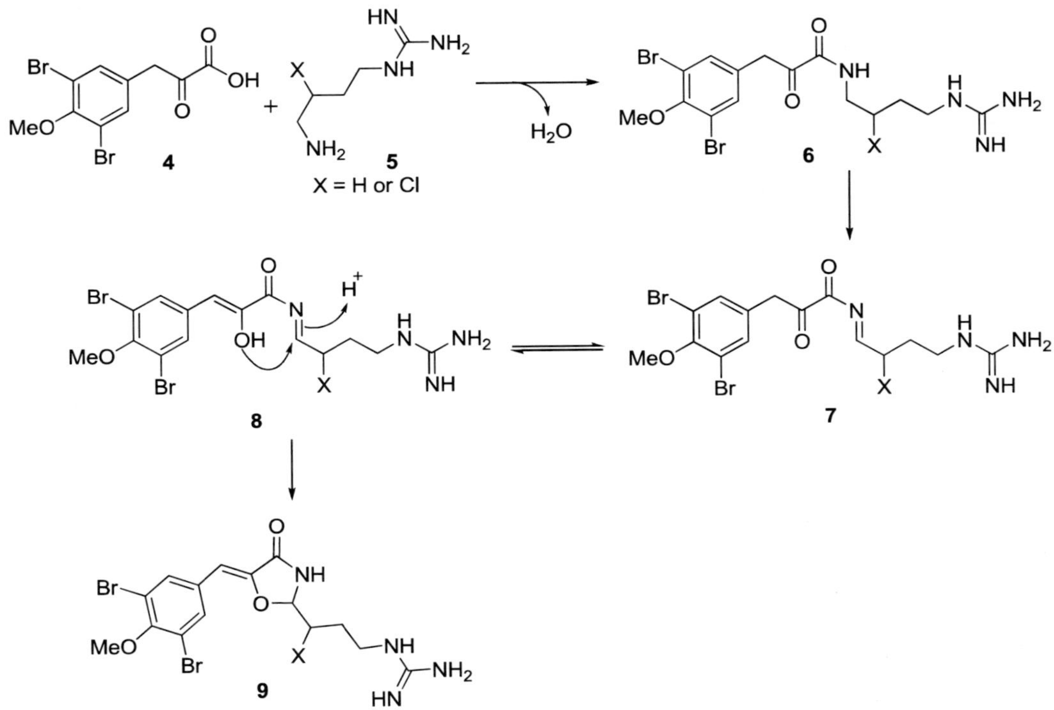

Scheme 1. Postulated biosynthesis of synoxazolidinones A (**1**) and B (**2**).

Synoxazolidinone C (**3**) also exhibited cytotoxic activity against the human cancer cell line s A2058 (melanoma), MCF-7 (breast adenocarcinoma) and HT-29 (colon carcinoma) with an IC_{50} value of 30.5 μM. It was also found that at the same level of concentration normal lung fibroblast cells (MRC-5) were killed.

The authors suggested that the increase in cytotoxic activity is due to additional pyrrolidine ring present in synoxazolidinone C (**3**).

The biosynthesis of synoxazolidinones A (**1**) and B (**2**) was postulated by Tadesse and coworkers and it is depicted in Scheme 1. It was assumed that brominated tyrosine derived metabolite **4** [11, 12] and an arginine /agmatine derivative **5** [13] react with each other and form a dipeptide **6**, which further cyclize to form oxazolidinone ring. The biosynthesis of synoxazolidinone C (**3**) has not been studied and it was assumed that synoxazolidinone C (**3**) might be derived from synoxazolidinone A (**1**) [9].

First total syntheses of synoxazolidinones A and B was reported by Shymanska, et al. [14] and it is depicted in Scheme 2.

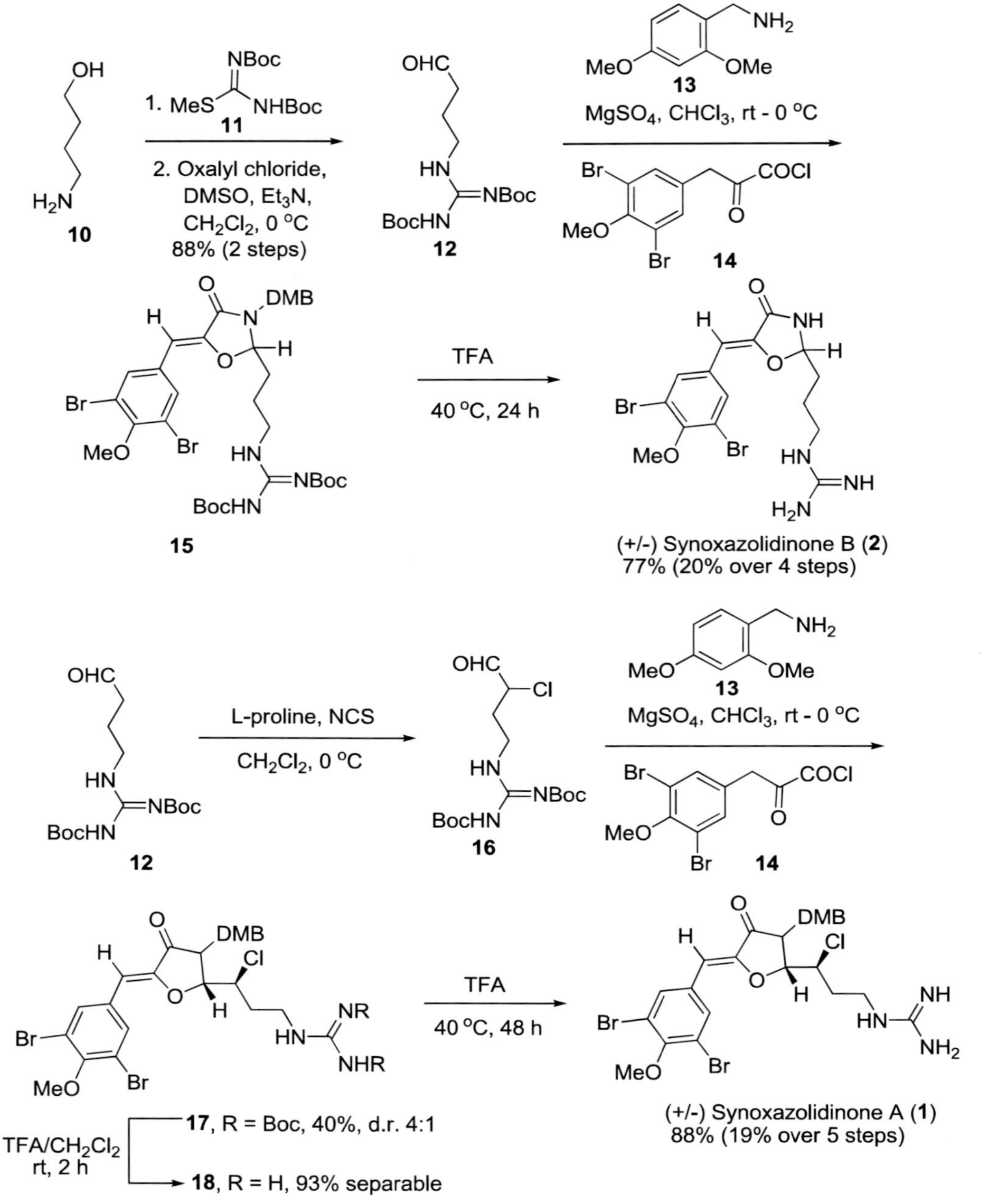

Scheme 2. Synthesis of synoxazolidinones A (**1**) and B (**2**).

The synthesis was started from 4-aminobutanol **10**, which on treatment with *N*-Boc-protected *S*-methylisothiourea **11** followed by oxidation under Swern condition yielded aldehyde **12**. 2,4-Dimethoxybenzyl amine **13** was treated with aldehyde **12** to make imine and then it was treated with acyl chloride **14** to perform acylation /cyclization cascade reaction to make protected synoxazolidinone B scaffold **15**, which on global deprotection yielded synoxazolidinone B (**2**) with reasonably good overall yield (20% in four steps) [14].

While for the synthesis of synoxazolidinone A (**1**), aldehyde **12** was chlorinated by using NCS (*N*-chlorosuccinimide) and stoichiometric amount of proline to obtained product **16** (Scheme 2). 2,4-Dimethoxybenzyl amine **13** was treated with chlorinated aldehyde **16** to make imine and then it was treated with acyl chloride **14** to perform acylation /cyclization cascade reaction. It yielded cyclized product **17** in a diastereomeric ratio of 4:1, which upon deprotection with TFA gave synoxazolidinone A (**1**). Further, analogues of synoxazolidinones A (**1**) and B (**2**) were prepared and screened for their antimicrobial activities as well [14].

2. PULMONARINS A AND B

Further, continuous studies on *S. pulmonaria* by Tadesse and coworkers isolated two more new alkaloids pulmonarins A (**19**) and B (**20**) (Figure 2) [15]. The structures were determined as dibrominated quaternary ammonium salts by spectroscopic methods and confirmed by their synthesis as well. They found acetylcholinesterase (AChE) inhibitors and the most active inhibitor was pulmonarin B (**20**) with the inhibition constant value of 20 μM [15].

Tadesse and coworkers assumed that pulmonarins and synoxazolidinones share the common biosynthetic route in the organism [15]. Pulmonarin A (**19**) had been considered as a dibrominated form of *p*-hydroxybenzoylcholine, which biosynthetically derived from *p*-hydroxybenzyl alcohol [16, 17]. According to the proposed biosynthesis dibromo hydroxybenzyl alcohol **21** get oxidized and methylated before coupling with choline **24** (Scheme 3). The biosynthesis of pulmonarin B (**20**) was assumed *via* the coupling of known metabolites 2-(3,5-dibromo-4-methoxyphenyl)acetic acid **25** [11] and ascophylline **26** [18].

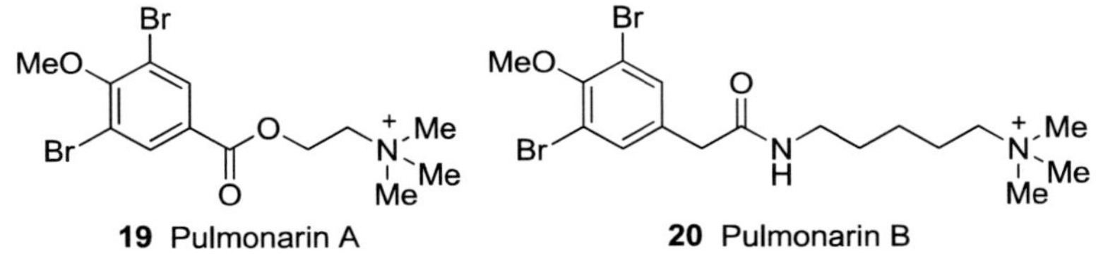

Figure 2. Structures of pulmonarins A and B.

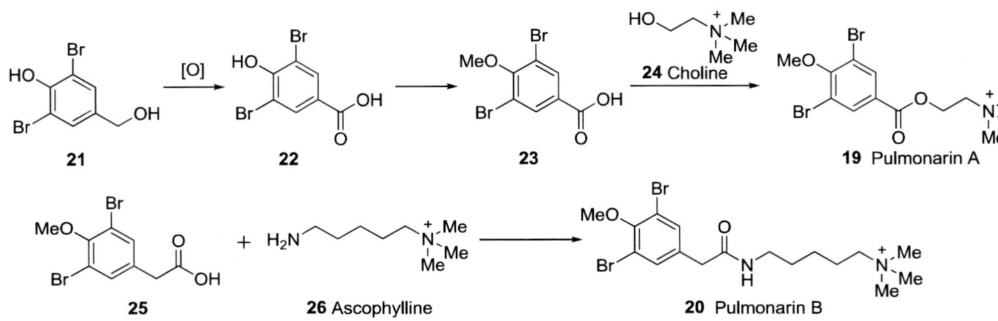

Scheme 3. Proposed biosynthesis of pulmonarins A (**19**) and B (**20**).

3. PSAMMAPLYSINS A-Y

Marine sponges of the order of Verongida have proven a rich source of alkaloids derived from bromotyrosine and some of them were found to possess a range of biological activities [19, 20]. In past three decades, a series of psammaplysins A-Y and its derivatives have been isolated from them (Figure 3) [21-29]. The uniqueness of these natural products is to possess spirooxepinisoxazoline (spiro[4.6]dioxazundecane) moiety. Firstly, psammaplysins A and B were isolated from *Psammaplysilla purpurea* in 1982 [30] and their detailed study was published in 1983 by Kashman and coworkers [31]. They assigned these structures based on the spectroscopic analysis and chemical degradation s studies as it posses spiro[4.5] oxazadecane skeleton. After two years, in 1985, Roll et al. in their full studies, based on extensive NMR experiments and single-crystal X-ray analysis of diacetyl derivative of psammaplysin A, revised the structures of psammaplysins A (**27**) and B (**28**) and identified as they possess spirooxepinisoxazoline (spiro[4.6]dioxazundecane) skeleton (Figure 3) [21]. Psammaplysins A (**27**) and B (**28**) exhibited in vitro antibacterial activity against both Gram-positive and Gram-negative bacteria [30] as well as in vitro cytotoxicity against human colon tumor cell-line HCT116 with the IC_{50} value of 6 μg/mL [22]. Psammaplysin C (**29**) was isolated from *Druinella purpurea* and structurally it is *N*-methylated derivative of psammaplysin B (**28**) [22]. Psammaplysin C (**29**) exhibited in vitro cytotoxic activity against human colon tumor cell-line HCT116 with the IC_{50} value of 3 μg/mL. Psammaplysin D (**30**) and E (**32**) were isolated from *Aplysinella* sp. by Ichiba et al. [23]. Psammaplysin D (**30**) is structurally similar to psammaplysin B (**27**) and has isopentadecanoyl residue at its *N*-terminus. Psammaplysin D (**30**) exhibited anti-HIV activity against the Haitian RF strain of HIV-I and it was detected 51% inhibition at 0.1 μg/mL. Psammaplysin E (**32**) is structurally similar to psammaplysin A (**27**) and has diketocyclopentenylidene enamine functionality at its *N*-terminus. It was found cytotoxic against KB cells (human oral, epidermoid carcinoma) as well as LoVo cells (human colon, adrenocarcinoma) at 5 μg/mL concentration and it also exhibited immunosuppressive activity [23]. Psammaplysin F (**31**) was isolated from *Aplysinella* sp. of sponge from Chuuk, Federated States of Micronesia [24] and structurally it is *N*-methylated derivative of psammaplysin A (**27**). Psammaplysin G (**34**) was isolated from *Hyattella* sp. through mass -directed fractionation and it has *N*-methylurea moiety at the terminal end (Figure 3) [25]. Psammaplysin F (**31**) was found active against parasite *Plasmodium falciparum* and it has IC_{50} values of 1.4 and 0.87 μM, respectively against the two different strains (Dd2 and 3D7) [25].

Further in depth studies on the biological activities of psammaplysin F (**31**) found that it actually inhibits equal partitioning of DNA into each daughter cell of bacteria [32] and its unique character keeping this natural product in the unique category of prokaryotic cell division inhibitor. Psammaplysin G (**34**) exhibited activity against the chloroquine-resistant (Dd2) strain of *P. falciparum* (98% inhibition at 40 μM concentration). Psammaplysin H (**35**) was isolated from the sponge of the genus *Pseudoceratina* [26].

It's structure is similar to psammaplysin F (**31**) and has two additional *N*-methyl group s at the terminal amine. Psammaplysin H (**35**) exhibited potent antimalarial activity against the 3D7 line of *P. falciparum* with an IC_{50} value of 0.41 μM and has very low toxicity against mammalian cell line s making it as a parasite specific. Psammaplysins I (**36**) and J (**37**) were isolated from the sponge of *Suberea* sp. [27].

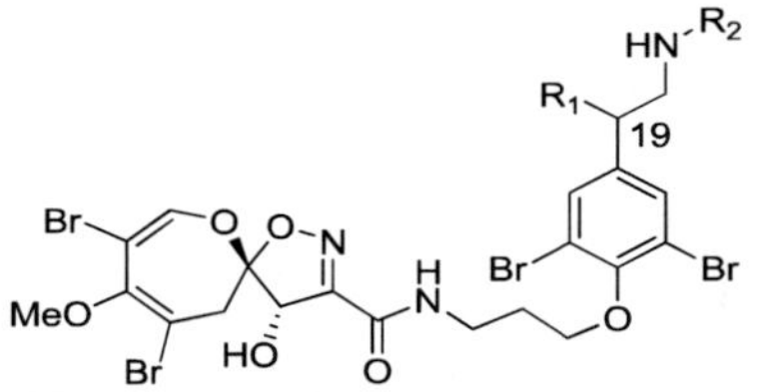

27 $R_1 = R_2 = H$ Psammaplysin A
28 $R_1 = OH$, $R_2 = H$ Psammaplysin B
29 $R_1 = OH$, $R_2 = Me$ Psammaplysin C
30 $R_1 = OH$, $R_2 = CO(CH_2)_{11}CHMe_2$ Psammaplysin D
31 $R_1 = H$, $R_2 = Me$ Psammaplysin F

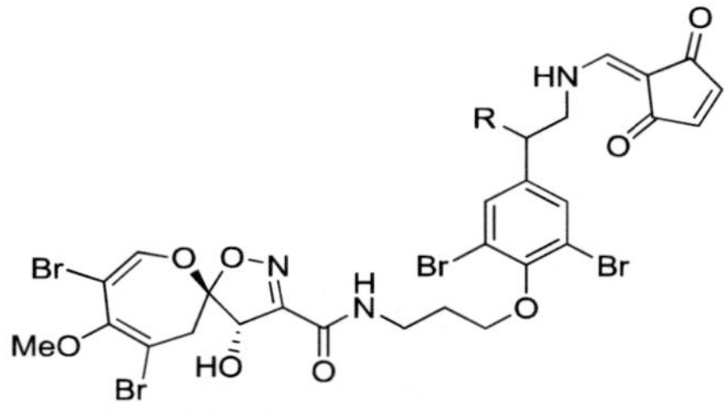

32 R = H Psammaplysin E
33 R = OH 19-Hydroxypsammaplysin E

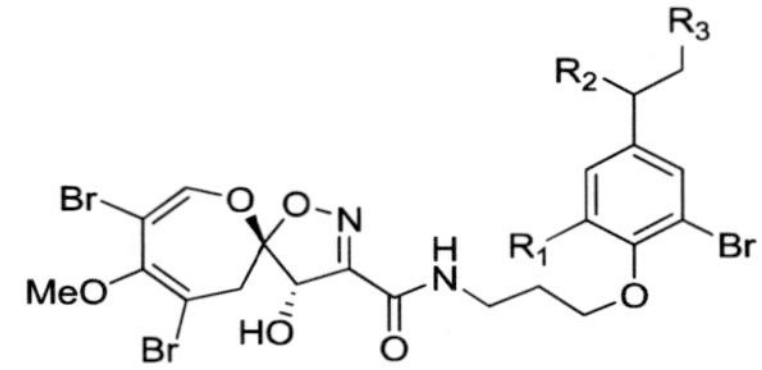

34 $R_1 = Br$, $R_2 = H$, $R_3 = NMeCONH_2$ Psammaplysin G
35 $R_1 = Br$, $R_2 = H$, $R_3 = NMe_3$ Psammaplysin H
36 $R_1 = H$, $R_2 = H$, $R_3 = NH_2$ Psammaplysin I
37 $R_1 = H$, $R_2 = OH$, $R_3 = NH_2$ Psammaplysin J

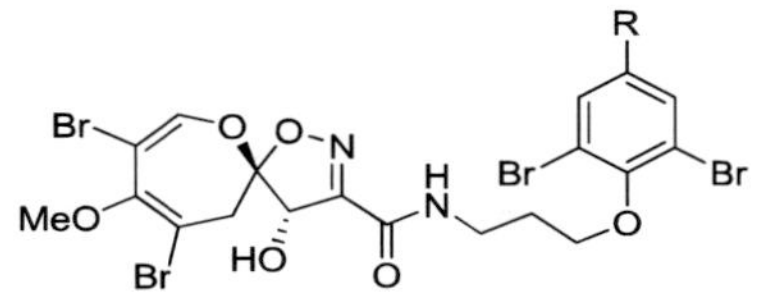

38 R = CHO Psammaplysin K
39 R = 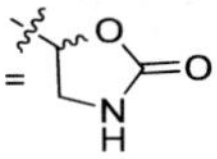 Psammaplysin L

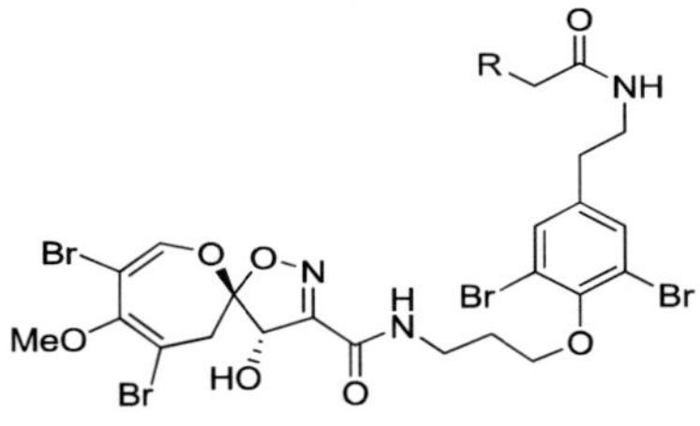

40 R = OH Psammaplysin M
41 $R = (CH_2)_{11}CH(Me)Et$ Psammaplysin N

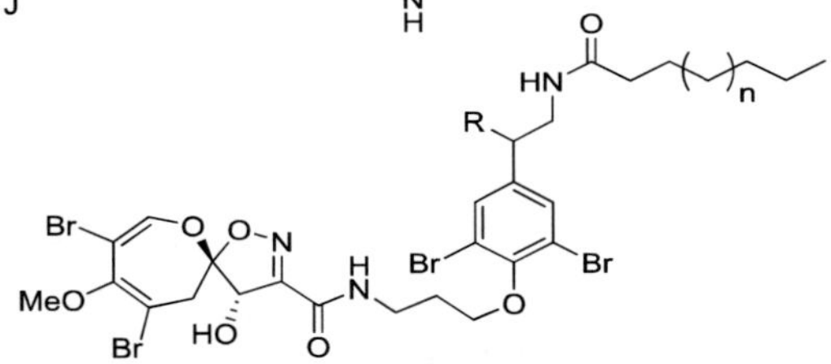

42 **R** = H, n = 10 Psammaplysin O
43 **R** = H, n = 12 Psammaplysin P
44 **R** = H, n = 8 Psammaplysin Q
45 **R** = OH, n = 12 19-Hydroxypsammaplysin P
46 **R** = OH, n = 8 19-Hydroxypsammaplysin Q

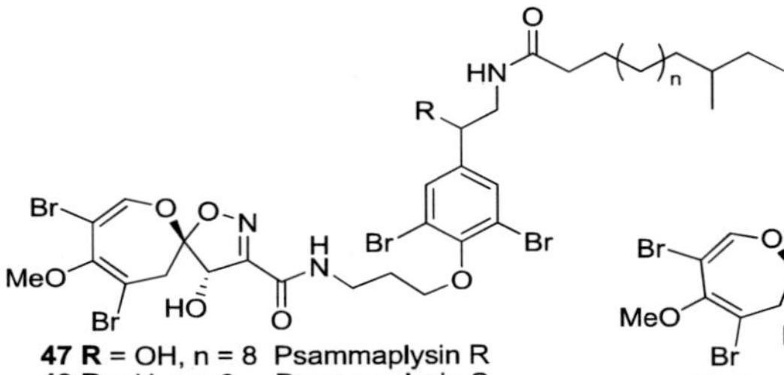

47 **R** = OH, n = 8 Psammaplysin R
48 **R** = H, n = 9 Psammaplysin S
49 **R** = H, n = 11 Psammaplysin T
50 **R** = OH, n = 9 19-Hydroxypsammaplysin S
51 **R** = OH, n = 11 19-Hydroxypsammaplysin T

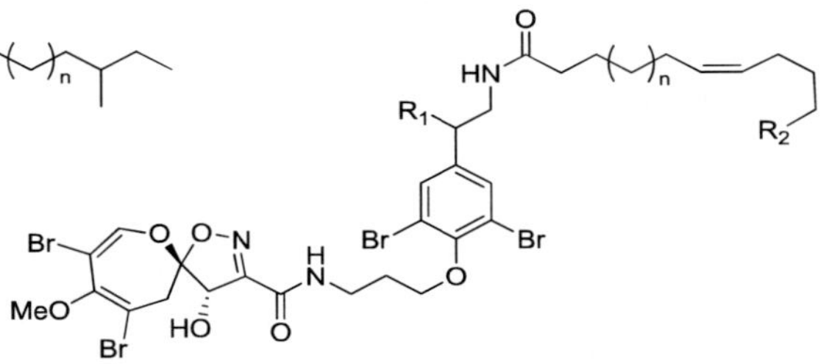

52 $R_1 = H$, $R_2 = CHMe_2$, n = 5 Psammaplysin U
53 $R_1 = OH$, $R_2 = CHMe_2$, n = 5 19-Hydroxypsammaplysin U
54 $R_1 = H$, $R_2 = Et$, n = 5 Psammaplysin V
55 $R_1 = H$, $R_2 = Et$, n = 8 Psammaplysin W
56 $R_1 = OH$, $R_2 = Et$, n = 8 19-Hydroxypsammaplysin W

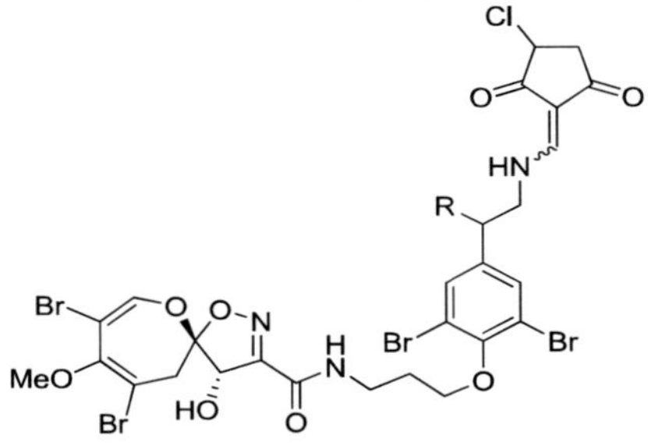

57 R = H Psammaplysin X
58 R = OH 19-Hydroxypsammaplysin X

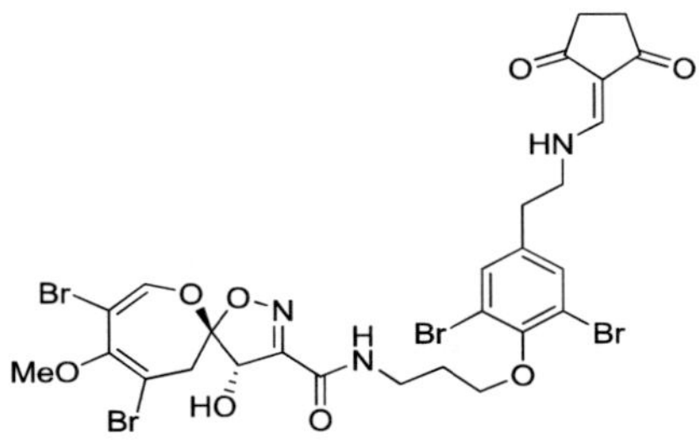

59 Psammaplysin Y

Figure 3. Structures of psammaplysins A-Y.

Their structures resembled to psammaplysins A (**27**) and B (**28**), respectively, except unusual bromotyramine moiety rather than the dibromophenyl moiety. Psammaplysins I (**36**) and J (**37**) were not evaluated for their biological activities. A series of psammaplysins K-W and their derivatives were isolated from the Indonesian marine sponge *Aplysinella strongylata* [28]. Psammaplysin K (**38**) has formyl functionality in place of β-aminoethyl functionality at the dibromophenyl ring. Psammaplysin L (**39**) has 2-oxazolidinone group at the end of dibromophenyl ring. Psammaplysin M (**40**) end up with the glycolamide moiety (Figure 3). Psammaplysin N (**41**) has amide functionality with the 13-methyltetradecanoate fatty acid. Psammaplysins O-Q (**42-44**) have linear fatty acid side chain s connected with the terminal amine as an amide functionality. Psammaplysins P (**43**) and Q (**44**) also have their hydroxyl derivatives as well. Psammaplysins R-T (**47-49**) have *anteiso*-branched saturated fat ty acid side chains connected with the terminal amine as an amide functionality (Figure 3).

Psammaplysins S (**48**) and T (**49**) also have their hydroxyl derivatives as well. Psammaplysins U-W and its hydroxyl derivatives (**52-56**) have monoenoic fatty acid side chain s connected with the terminal amine as an amide functionality. Psammaplysin U (**52**) has *iso*-branched fatty acid side chain, while psammaplysins V (**54**) and W (**55**) have unbranched fatty acid side chains. Psammaplysins X (**57**) and Y (**59**) were isolated from marine sponge of the genus *Suberea* [29]. Psammaplysin X (**57**) has 4-chloro-2-methylenecyclopentane-1,3-dione at the *N*-terminus, while psammaplysin Y (**59**) has unsubstituted 2-methylenecyclopentane-1,3-dione moiety at its *N*-terminus (Figure 3).

Psammaplysin X (**57**) exhibited cytotoxicity against a range of human tumor cell line s with the GI_{50} value of 1.2–6.4 μM. On cytotoxic screening, Lee et al. found that spirooxepinisoxazoline ring system containing compounds exhibited cytotoxic activity, while the compounds don't have spirooxepinisoxazoline ring system have not shown much activity and based on their observations they suggested that psammaplysins might be a good molecular scaffold or platform to develop potent anticancer agents [29]. The biosynthesis of psammaplysins was postulated by Roll et al. [21] and it is depicted in Scheme 4. According to them hydroxylimine derivative **62** derived from tyrosine **60** *via* bromination followed by oxidation, which further converted into epoxide **63** [31]. Ring enlargement by the opening of epoxide ring *via* oxide-oxepin pathway [33-35] leads to carbocation intermediate **64**, which on nucleophilic attack of hydroxyl group of oxime leads spirooxepinisoxazoline skeleton of psammaplysins. Garson and coworkers suggested an alternate path for the biosynthesis of spirooxepinisoxazoline system of psammaplysins [28]. According to them, the skeleton could be generated by a concerted mechanism, first nucleophilic attack of hydroxyl group of oxime on the epoxide and then simultaneous cleavage of C–C bond followed by protonation. Total synthesis of psammaplysin natural products is not reported in literature. Some of the reports discussed only the construction of core spirooxepinisoxazoline skeleton [36, 37].

4. Ceratinamides A and B

Ceratinamides A (**68**) and B (**69**), antifouling bromotyrosine derivatives, were isolated from marine sponge *Pseudoceratina purpurea* through a bioassay -guided isolation (Figure 4) [38]. Ceratinamide A (**68**) is structurally similar to psammaplysin A (**27**) and has one formyl group at the *N*-terminus.

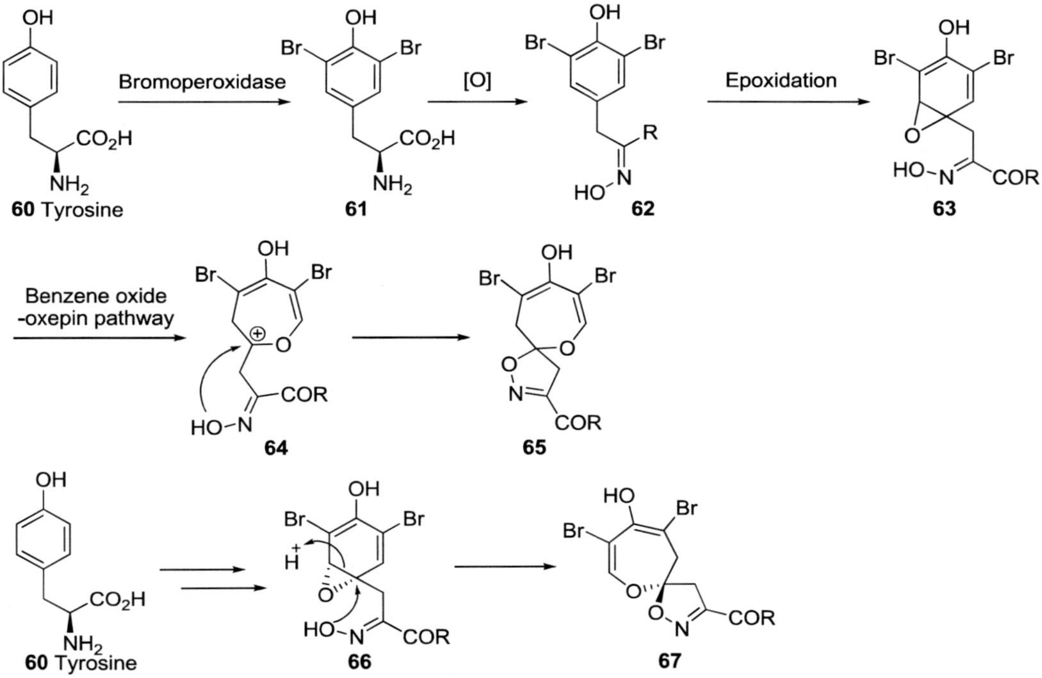

Scheme 4. Proposed biosynthesis of spirooxepinisoxazoline skeleton of psammaplysins.

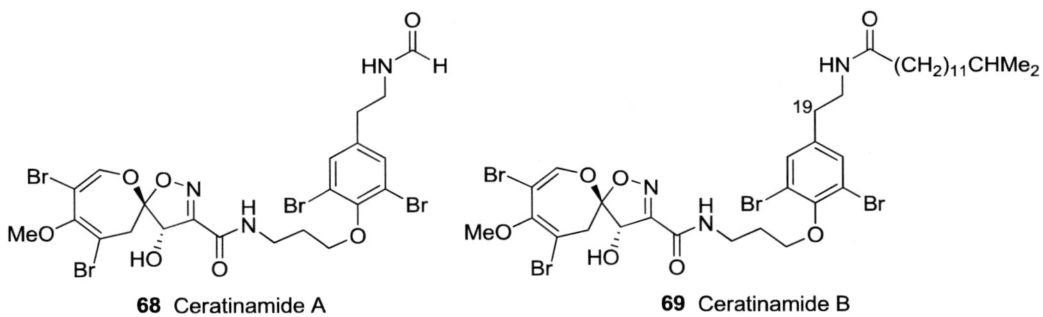

Figure 4. Structures of ceratinamides A and B.

The structure of ceratinamide B (**69**) is similar to psammaplysin D (**30**), except one hydroxyl group at the 19-carbon. Ceratinamides A (**68**) and B (**69**) exhibited antifouling activity against cyprid larvae of the barnacle *Balanus amphitrite* with the ED_{50} values 0.10 and 2.4 μg/mL, respectively, and also exhibited cytotoxic activity against P388 murine leukemia cells with the IC_{50} value of >10 μg/mL [38]. Biogenetically ceratinamides A and B are related to psammaplysins [38].

5. Ianthelliformisamines A-C

Another class of bromotyrosine derived alkaloids is ianthelliformisamines A-C (**70-72**), which were isolated from marine sponge *Suberea ianthelliformis* through a bioassay guided isolation (Figure 5) [39]. This was the first report of natural product isolation from *Suberea ianthelliformis* [39]. All three natural products possess a substituted cinnamyl moiety and a polyamine chain. Ianthelliformisamine A (**70**) has a polyamine chain spermine connected to the substituted cinnamyl moiety, while ianthelliformisamine B (**71**) has a polyamine chain spermidine connected to the substituted cinnamyl moiety. Ianthelliformisamine C (**72**) is a symmetrical molecule and possess substituted cinnamyl moiety on both side of the polyamine chain spermine. Ianthelliformisamines A-C were evaluated for their biological activity as an antibacterial agents against Gram-positive as well as Gram-negative bacteria.

Ianthelliformisamine A (**70**) exhibited activity against *P. aeruginosa* with the MIC value of 35 μM (IC_{50} value 6.8 μM) and exhibited 77% inhibition against *Staph. aureus* at 175 μM. Ianthelliformisamine B (**71**) exhibited 80% inhibition against *P. aeruginosa* at 87.5 μM concentration, while ianthelliformisamine C (**72**) exhibited activity against *P. aeruginosa* with the MIC value of 17.5 μM (IC_{50} value 8.9 μM) and against *Staph. aureus* with the MIC value of 8.75 μM (IC_{50} value 4.1 μM) [39]. Structure activity relationship (SAR) studies of ianthelliformisamines A-C indicated that the difference in the antibacterial activity is may be due to cell permeability of natural products between the Gram-positive and Gram-negative bacteria [39]. Ianthelliformisamines A–C were also tested for in vitro growth inhibitory activity against chloroquine-sensitive *Plasmodium falciparum* line (3D7) and mammalian cell toxicity using the neonatal foreskin fibroblast (NFF) cell line [40]. Ianthelliformisamines A–C exhibited antimalarial activity with the IC_{50} values of 14.5, 12.1 and 4.4 μM, respectively, while their cytotoxic activity was with the IC_{50} values of 100, 100 and 17.5 μM, respectively. Biosynthetically, ianthelliformisamines A–C might be derived by the combination of spermine or spermidine and bromotyrosine derived metabolite.

First synthesis of ianthelliformisamines A-C was reported by Khan et al. [41] and it is depicted in Scheme 5. They have synthesized ianthelliformisamines A-C by the condensation of (*E*)-3-(3,5-dibromo-4-methoxyphenyl)acrylic acid **73** and Boc-protected spermine or spermidine followed by deprotection of Boc-group with TFA (trifluoroacetic acid).

By applying the same reaction protocol, Khan et al. have also synthesized a library of 39 analogues of ianthelliformisamines A-C [41]. The analogues are different in size of polyamine chain, phenyl ring substitution and double bond geometry.

Further, ianthelliformisamines A-C and their 39 analogues were evaluated for their antibacterial activity against Gram-positive (*Staphylococcus aureus*) and Gram-negative (*Escherichia coli*) strains. Ianthelliformisamines A and B, and their synthetic analogues were found active against both strains, while ianthelliformisamine C and its analogues were found inactive against both strains.

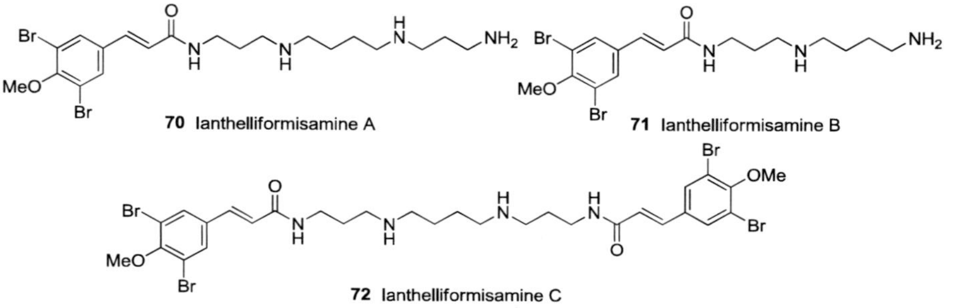

Figure 5. Structures of ianthelliformisamines A-C.

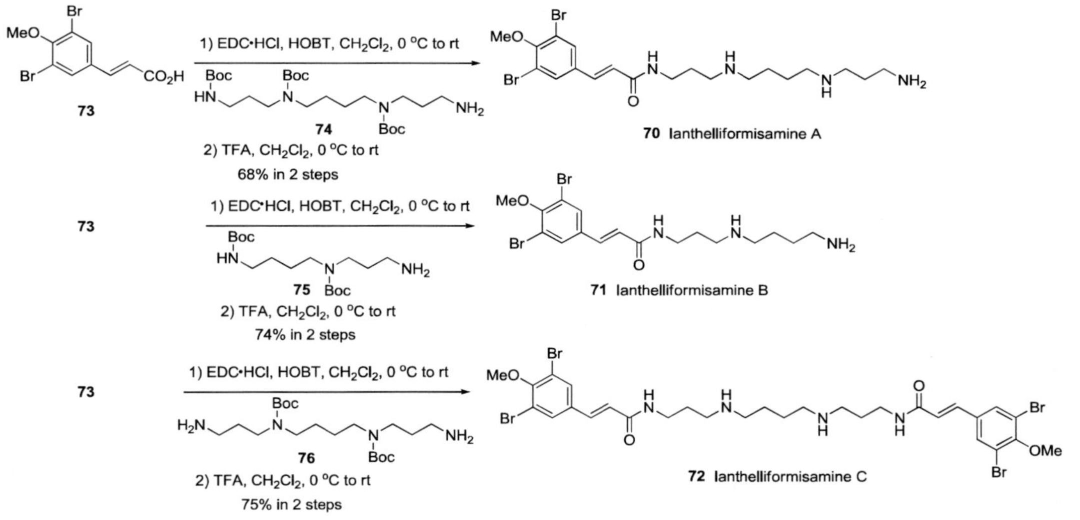

Scheme 5. Syntheses of ianthelliformisamines A-C by Khan et al.

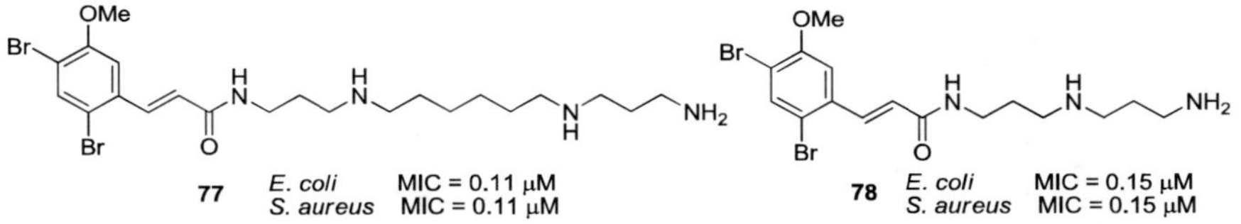

Figure 6. Potent synthetic analogues of library.

Most potent analogues of the library (compounds **77** and **78**) are represented in Figure 6, which were found active against both strains [41].

Another synthesis of ianthelliformisamines A-C was reported by Pieri et al. [42] and it is depicted in Scheme 6. They have synthesized ianthelliformisamines A and C by the condensation of (*E*)-3-(3,5-dibromo-4-methoxyphenyl)acrylic acid **73** and spermine **79**. Ianthelliformisamine B was synthesized by taking Boc-protected spermidine **75** and acid **73** [42]. By utilizing the same synthetic approach, further Pieri et al. have synthesized ianthelliformisamines derivatives (total 12 derivatives). Ianthelliformisamines A-C and its synthetic derivatives were evaluated against resistant Gram-negative bacteria and analyzed their mode of action as well [42].

6. Wilsoniamines A and B

Wilsoniamines A (**80**) and B (**81**) were isolated from Australian marine bryozoan *Amathia wilsoni* Kirkpatrick (Figure 7) [43].

Their structures were determined by extensive NMR experiments and assigned as dibromomethoxybenzyl substituent is attached with bicyclic hexahydropyrrolo[1,2-*c*]imidazol-1-one ring. (−)-Dysibetaine PP (**82**) is the structurally closest natural product, which also has the same core ring skeleton and was isolated from Micronesian marine sponge *Dysidea herbacea* (Figure 7) [44].

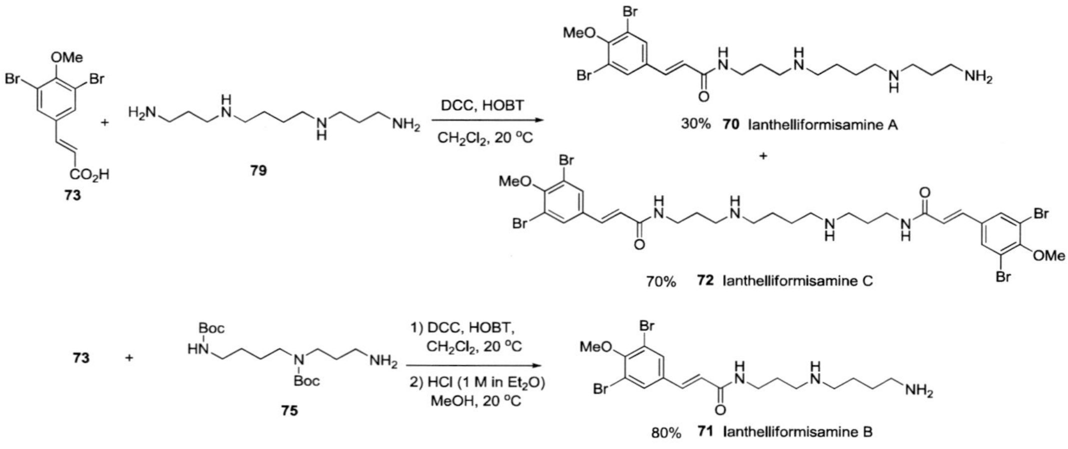

Scheme 6. Syntheses of ianthelliformisamines A-C by Pieri et al.

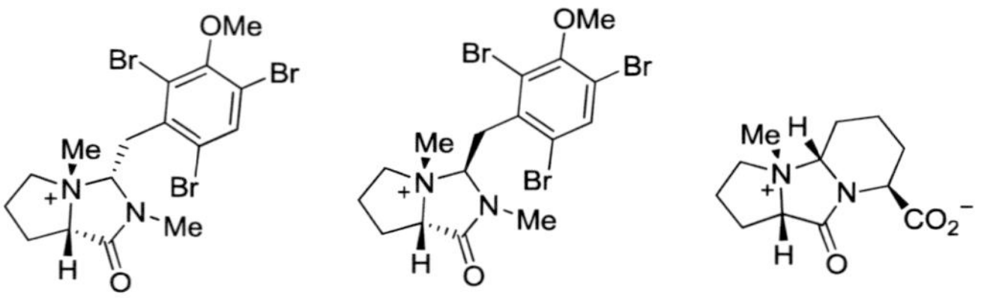

Figure 7. Structures of wilsoniamines A, B and (–)-dysibetaine PP.

Wilsoniamines A (**80**) and B (**81**) were evaluated for their antimicrobial activities against malarial parasite *Plasmodium falciparum* and protozoan parasite *Typanosoma brucei brucei* and the cytotoxicity using normal HEK-293 mammalian cell line and HeLa cancerous cell line. Wilsoniamines A and B exhibited antimicrobial activities at very higher concentration dose and did not show any cytotoxicity.

Based on the observations Carroll et al. suggested that might be the cell permeability of charged molecules is responsible for lower biological activity [43]. Biogenetically wilsoniamines A (**80**) and B (**81**) might be derived from amathamides.

Syntheses of wilsoniamines A (**80**) and B (**81**) were reported by Khan and Ahmad [45] and it is depicted in Scheme 7.

They started synthesis by taking 2-(2,4,6-tribromo-3-methoxyphenyl)acetaldehyde **83** and (*S*)-*N*-methylpyrrolidine-2-carboxamide **84** and achieved in two steps. Condensation of aldehyde **83** and amide **84** in anhydrous methanol yielded cyclize products **85** and **86**. Stereoselective methyl quaternization of compound **86** yielded wilsoniamine A (**80**).

While condensation of aldehyde **83** and amide **84** under TFA/toluene /reflux condition yielded only cyclize product **85**, which on methyl quaternization with Me_3OBF_4 yielded wilsoniamine B (**81**) (Scheme 7) [45].

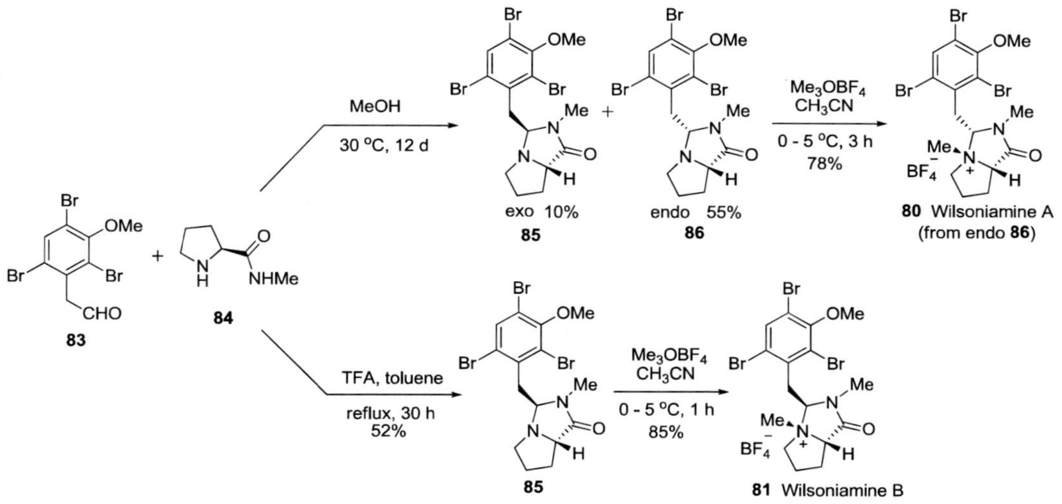

Scheme 7. Syntheses of wilsoniamines A and B.

7. CONVOLUTAMINES A-J AND LUTAMIDES A, C

Convolutamines A-E (**87-96**) were isolated from Floridian marine bryozoan *Amathia convoluta* by Kamano and coworkers (Figure 8) [46]. Their structures were elucidated through extensive spectroscopic analysis and found that all the molecules possess a substituted phenyl ring and a *β*-aminoethyl side chain. Convolutamines A-E (**87-96**) were evaluated for their biological activity as a growth inhibitor against murine P388 lymphocytic leukemia. Out of which convolutamines A (**87**), B (**88**) and D (**90**) were found active with the IC_{50} values of 10.6, 4.8 and 8.6 μg/mL. Kamano and coworkers suggested that the biogenesis of convolutamines A-E (**87-96**) is related with the amathamides [46]. In their continuous study on Floridian marine bryozoan *Amathia convoluta*, Kamano and coworkers isolated two new alkaloids of convolutamine family, convolutamines F (**92**) and G (**93**) (Figure 8) [47].

Convolutamine F (**92**) was found active against human carcinoma KB cells (IC_{50} value 27 μg/mL) and its vincristine-resistant KB/VJ-300 cells (IC_{50} value 9.6 μg/mL), and lymphocytic leukemia U937 cells (IC_{50} value 13 μg/mL). Convolutamine F (**92**) also inhibited cell division of fertilized sea urchin eggs (IC_{50} value 82 μg/mL) [47]. While convolutamine G (**93**) was not evaluated for its biological activity due to less isolated quantity. Convolutamine H (**94**) was isolated from marine bryozoan *Amathia convoluta*, which was collected from Tasmania's east coast [48]. Convolutamine H (**94**) has a fully substituted phenyl ring with a set of three adjoining bromine atoms, a pair of contiguous methoxy groups and a *β*-aminoethyl side-chain (Figure 8).

Convolutamine H (**94**) exhibited nematocidal activity against *Haemonchus contortus* with the LD_{99} value of 0.20 μg/mL and it was found more potent than the commercially available anthelmentic, levamisole. It was suggested that convolutamine H (**94**) might be a precursor of amathamide G (**118**) [48]. Convolutamines I (**95**) and J (**96**) were isolated from marine bryozoan *Amathia tortusa* through a mass directed isoaltion (Figure 8) [49]. Both the molecules possess tribromosubstituted phenyl ring and an extended *β*-aminoethyl side-chain.

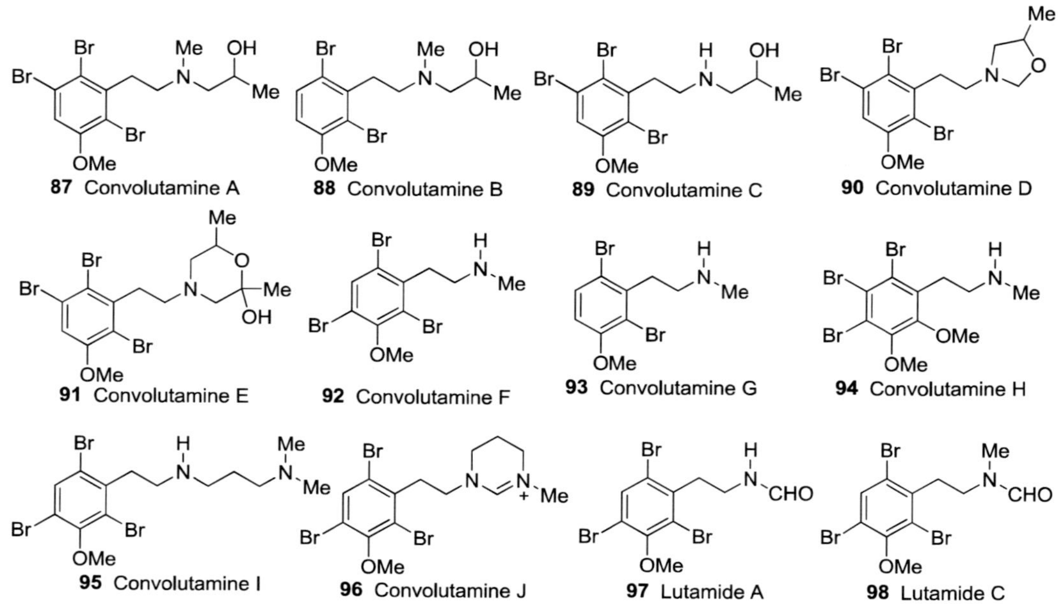

Figure 8. Structures of convolutamines A–J and lutamides A, C.

Convolutamines I (**95**) and J (**96**) exhibited activity against parasite *Trypanosoma brucei brucei* with the IC_{50} values of 1.1 and 13.7 μM, respectively. Convolutamine I (**95**) also exhibited cytotoxic activity with the IC_{50} value of 22.0 μM against human embryonic kidney cell line, HEK293 [49].

Out of total 10 convolutamines, syntheses of convolutamines A (**87**), C (**89**), F (**92**), H (**94**) and I (**95**) have been reported in literature. Convolutamines A (**87**), C (**89**) and F (**92**) were synthesized by Hashima et al. starting from 3-hydroxyphenethylamines [50]. Convolutamine F (**92**) has also been synthesized by Khan and Ahmad from 2,4,6-tribromo-3-methoxybenzaldehyde (Scheme 8) [51]. More intresting alkaloid convolutamine H (**94**), possessing fully substituted phenyl ring and *β*-aminoethyl functionality has been synthesized by Khan and Ahmad (Scheme 8) [51]. They started synthesis of convolutamine H (**94**) by taking substituted norbornene diol **100**, which was converted into monoketo norbornene **102** by a sequence of reactions. Compound **102** was treated with *p*-TsOH to afford Grob-type fragmentation aromatization reaction followed by methylation with diazomethane to afford a fully substituted phenyl ring **103**. Methylester group of compound **103** was converted into aldehyde **104** and then it was homologated to aldehyde **105**. Finally, convolutamine H (**94**) was obtained from aldehyde **105** *via* reductive amination with methylamine (Scheme 8) [51].

Convolutamine I (**95**) has been synthesized by Pham et al. and it is depicted in Scheme 9 [52]. In the first route, 2,4,6-tribromo-3-methoxybenzaldehyde **99** was homologated by performing Darzens condensation followed by reductive amination with N^1,N^1-dimethylpropane-1,3-diamine to get convolutamine I (**95**).

In the second route, 3-methoxy phenylethanamine **106** was brominated and then it was *N*-alkylated to get convolutamine I (**95**) [52]. By employing the same synthetic strategies, Pham et al. prepared a variety of analogues of convolutamine I (**95**) and evaluated for anti-trypanosomal activity against *Trypanosoma brucei brucei* and few of the analogues were found more potent than parent natural product [52].

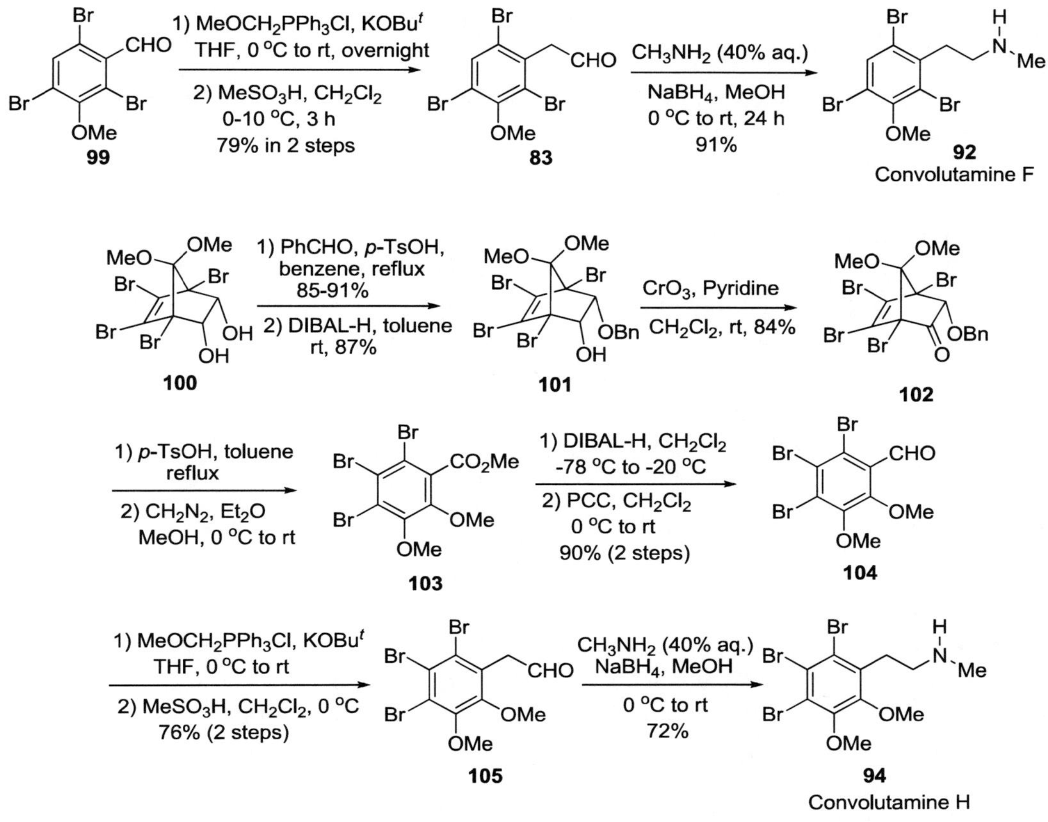

Scheme 8. Syntheses of convolutamines F and H by Khan and Ahmad.

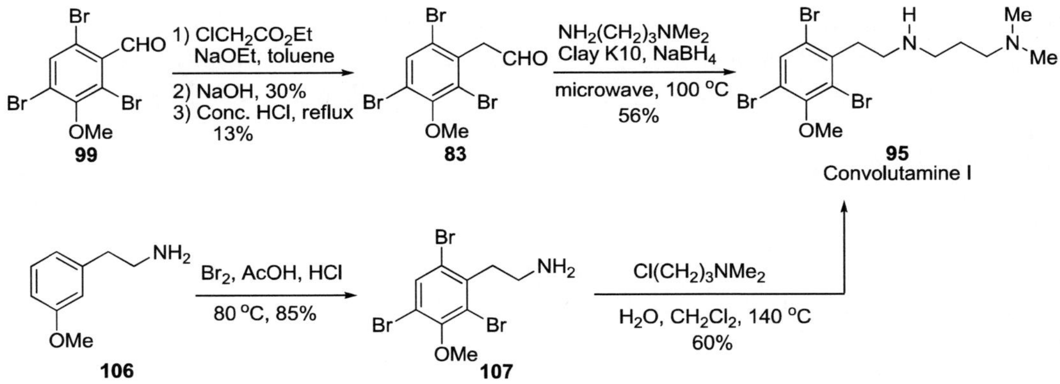

Scheme 9. Synthesis of convolutamine I by Pham et al.

Lutamides A-C were isolated from Floridian marine bryozoan *Amathia convoluta* and possess a tribrominated phenyl ring with formylated β-aminoethyl side chain (Figure 8) [50]. Lutamides A (**97**) and C (**98**) were evaluated for cytotoxic activity in which lutamide C (**98**) was found active against vincristine (VCR)-resistant P388/VCR cells with the IC_{50} value of 4.8 μg/mL. Lutamides A (**97**) and C (**98**) have been synthesized *via* formylation of corresponding amines [50, 51].

8. AMATHAMIDES A-H

Marine bryozoans of *Amathia* genus have proven a rich source of brominated alkaloids, which exhibited a range of biological activities. In past three decades a variety of natural products have been isolated from them, which includes amathamides, amathaspiramides, convolutamines, wilsoniamines and kororamide A. Amathamides A (**108**) and B (**109**) were isolated from the marine bryozoan *Amathia wilsoni* Kirkpatrick by Blackman and Matthews (Figure 9) [53]. Both the molecules possess a dibromomethoxyphenyl moiety, which is connected with L-proline through enamide functionality. Amathamides C-F (**110-117**) were also isolated from *Amathia wilsoni* [54]. Their structures were found similar to amathamides A (**108**) and B (**109**) except pentasubstituted phenyl ring rather than tetrasubstituted. Their structures were assigned as phenyl ring possessing three bromines contiguously or the natural products possess 2,3,4-tribromo-5-methoxyphenyl group (**110**, **112**, **114** and **116**). The structure of amathamide C (**110**) was revised to compound (**111**) by Carroll and coworkers on its re-isolation and the structure of amathamide D (**112**) was revised to compound (**113**) by Khan and coworker by its sequential synthesis and they find that the actual natural products possess symmetrical tribromo substitution on phenyl ring rather than contiguous tribromo phenyl system (Figure 9) [43, 51]. On the ongoing synthesis of amathamides, Khan and Ahmad revised the structures of amathamides E (**114**) and F (**116**) to compounds (**115**) and (**117**), respectively [55]. Amathamide G (**118**) was isolated from Tasmanian bryozoan *Amathia convoluta* and its structure was assigned as 2,3,4-tribromo-5,6-dimethoxy substituted phenyl ring connected with pyrrolidine moiety *via E*-enamide functionality (Figure 9) [56]. Amathamide H (**119**) was also isolated from Australian bryozoan *Amathia wilsoni* and its structure was assigned as the *Z*-isomer of amathamide C (**111**) [43]. Amathamides A-G were not extensively screened for their biological activity. Amathamide H (**119**) and re-isolated amathamide C (**111**) were evaluated for their biological activity against malarial parasite *Plasmodium falciparum* (using chloroquine sensitive (3D7) and chloroquine resistant (Dd2) strains) and protozoan parasite *Typanosoma brucei brucei* as well as for the cytotoxic activity by using HeLa cancerous cell line and normal mammalian cell line HEK-293 [43]. Amathamide H (**119**) and C (**111**) exhibited activity against chloroquine sensitive (3D7) strain with the IC_{50} values of 28.0 and 14.9 μM, respectively, against chloroquine resistant (Dd2) strains with the IC_{50} values of 10.2 and 8.0 μM, respectively and against *Typanosoma brucei brucei* with the IC_{50} values of 32 and 57.1 μM, respectively. While amathamide H (**119**) and C (**111**) did not exhibit any significant cytotoxic activity [43]. It was suggested that amathamides might be derived from proline and phenylalanine. Since, in all amathamides methoxy group is at the *meta*-position rather than *para*-position of phenyl ring with respect to β-ethylamine substituent, suggesting that amathamides are not derived from tyrosine [53]. This fact was further proved with the isolation of a β-phenylethylamine (2-(2,4dibromo-5-methoxyphenyl)-ethanamine) from *Amathia wilsoni* [57]. More interesting skeleton is amathamide G (**118**), possess a fully substituted phenyl ring and it was assumed that amathamide G (**118**) is derived from convolutamine H (**94**) [48]. It is the question of interest that how nature makes a fully substituted phenyl ring of amathamide G (**118**) or convolutamine H (**94**) and more interestingly how nature placed three contiguous bromine atoms on their phenyl rings ?

Out of total eight amathamides, only the synthesis of amathamides A (**108**), B (**109**) and D (**113**) is reported in literature.

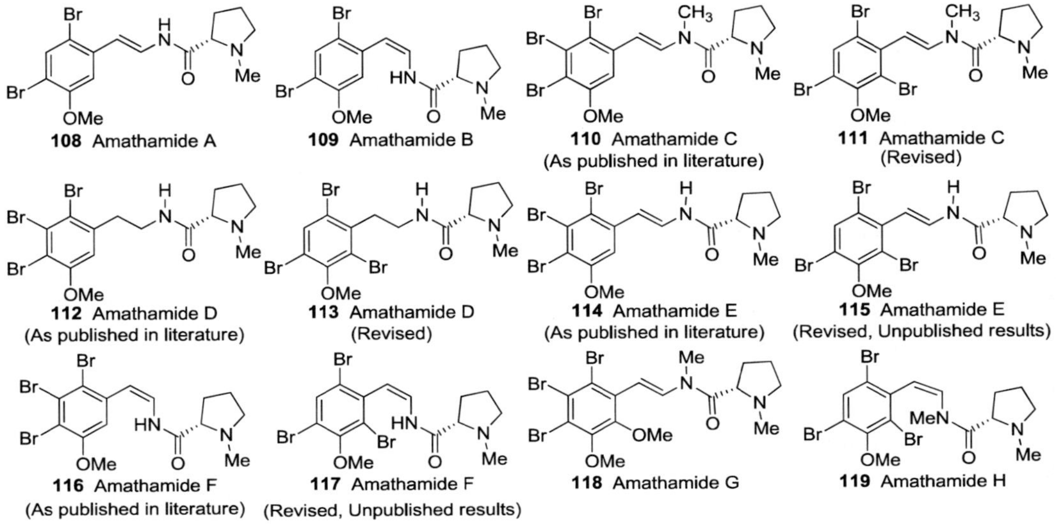

Figure 9. Structures of amathamides A-H.

Amathamides A (**108**) and B (**109**) were synthesized by Osuna et al. from 3-hydroxybenzaldehyde **120** and it is depicted in Scheme 10 [58]. In the synthetic sequence, aldehyde **120** was converted into required dibromomethoxybenzaldehyde **121** *via* bromination followed by methylation. Then aldehyde **121** was converted into amine **123** on nitroolefination followed by Michael addition and nitro group reduction (Scheme 10) [58]. Amine **123** was coupled with *N*-methyl-L-proline to get amide **124**.

Finally, thiophenol group was eliminated to get amathamides A (**108**) and B (**109**). By applying the same synthetic strategy, Osuna and coworkers have also synthesized non-brominated analogues of amathamide A as well as monobrominated analogues of amathamides [59, 60]. They evaluated non-brominated synthetic analogues of amathamide A against antimicrobial activity. They found that natural product amathamide A is weak antibacterial active, while surprisingly its non-brominated analogues are inactive. Based on their observations Osuna et al. suggested that bromine substitution on phenyl ring is necessary for the biological activity [59].

Synthesis of amathamide D (**112**) and its structure revision into compound (**113**) was reported by Khan and Ahmad, and it is depicted in Scheme 11 [51]. In the synthetic sequence, they obtained the required core substituted phenyl skeleton **126** from 1,4,5,6-tetrabromo-7,7-dimethoxybicyclo[2.2.1]hept-5-en-2-one **125** by applying Grob-type fragmentation aromatization strategy. Compound **126** was converted into *α,β*-unsaturated nitro compound **127** and then it was transformed into Boc-protected amine **128**. Boc-group of amine **128** was deprotected and then it was coupled with *N*-methyl-L-proline to get amathamide D (**112**), as structure reported in literature. But NMR (^{1}H and ^{13}C NMR) values of compound **112** was not in conformity with the literature reported values. Then they revised the structure of amathamide D (**112**) by taking 2,4,6-tribromo-3-methoxybenzaldehyde **99**. Aldehyde **99** was transformed into amine **107** by routine functional group transformations then it was coupled with *N*-methyl-L-proline to get compound **113**. Compound **113** had all spectroscopic data (including ^{1}H and ^{13}C NMR) in accordance with the literature reported values.

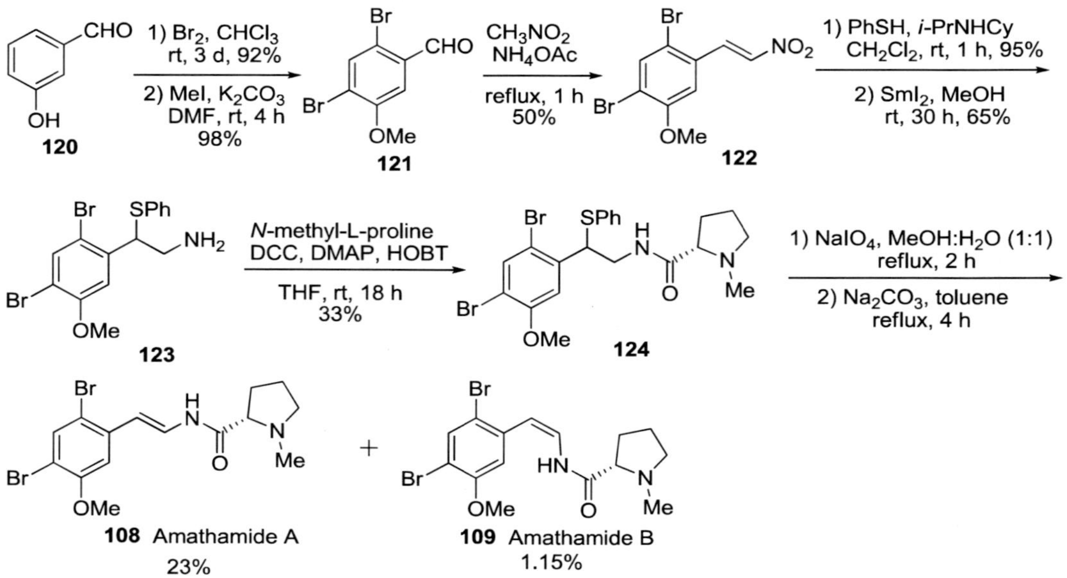

Scheme 10. Syntheses of amathamides A and B.

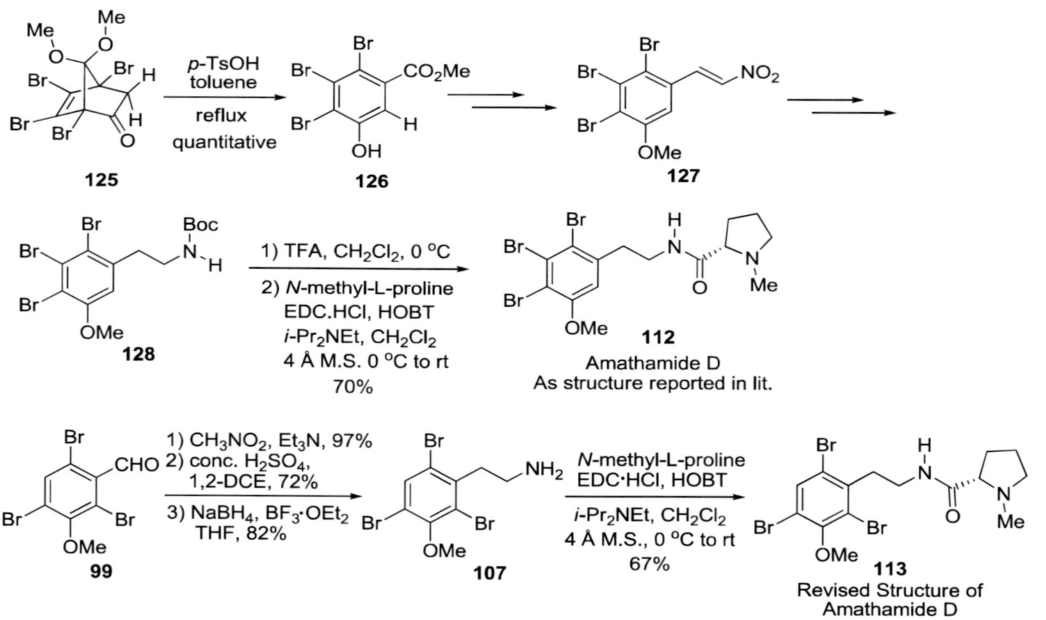

Scheme 11. Synthesis of amathamide D and its structure revision.

Based on the facts and their observations, they concluded that natural amathamide D possess a symmetrical tribromo substitution on phenyl ring, rather than contiguous tribromo substitution on phenyl ring (Scheme 11) [51].

9. AMATHASPIRAMIDES A-F

Amathaspiramides A-F (**129-134**) were isolated from marine bryozoan *Amathia wilsoni* by Morris and Prinsep (Figure 10) [61]. Their structures as well as absolute configuration were determined by extensive NMR experiments and X-ray crystallography and were assigned as a dibromomethoxyphenyl ring connected with the spirobicyclic core consisting of pyrrolidine and pyrrolidinone rings. Amathaspiramides A (**129**) and C (**131**) are similar to each other except *N*-methyl group in the pyrrolidine ring. Amathaspiramides B (**130**) and D (**132**) are similar to each other except *N*-methyl group in the pyrrolidine ring. Amathaspiramide E (**133**) has double bond at the pyrrolidine ring, while amathaspiramide F (**134**) is epimeric of amathaspiramide C (**131**) at hydroxyl group [61].

Out of six amathaspiramides, only four amathaspiramides A-C (**129-131**) and E (**133**) were evaluated for their biological activity as a cytotoxic, antiviral, antimicrobial and for P388 murine leukemic. Amathaspiramides A (**129**) and E (**133**) exhibited moderate cytotoxic activity against the BSC-1 cells (1+, 40 μg/well), while amathaspiramides B (**130**) and C (**131**) were found inactive at the tested concentrations. Amathaspiramide E (**133**) was found highly active (4+, 40 μg/well) against the *Polio* virus Type 1 (Pfizer vaccine strain).

Amathaspiramides A (**129**) and E (**133**) were found mild active (1 mm inhibition zone, 60 μg/disk) against Gram-positive bacterium *Bacillus subtilis* and fungus *Trichophyton mentagrophytes*, while amathaspiramides B (**130**) and C (**131**) did not exhibit any activity at the tested concentrations. Amathaspiramides did not exhibit significant activity in the P388 assay [61]. It was assumed that amathaspiramides A-F (**129-134**) are biosynthetically related to amathamides [62].

Because of their unique structures and skeleton, various groups have targeted to synthesize amathaspiramides. First total synthesis of (−)-amathaspiramide F (**134**) was reported by Hughes and Trauner, and it is depicted in Scheme 12 [63]. The synthesis was started by taking *N,N*-acetal **136**, which was synthesized from L-proline **135**.

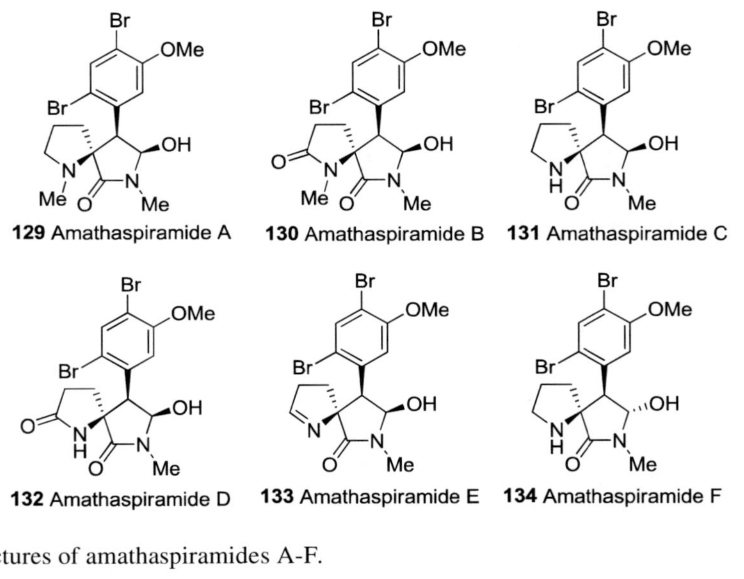

Figure 10. Structures of amathaspiramides A-F.

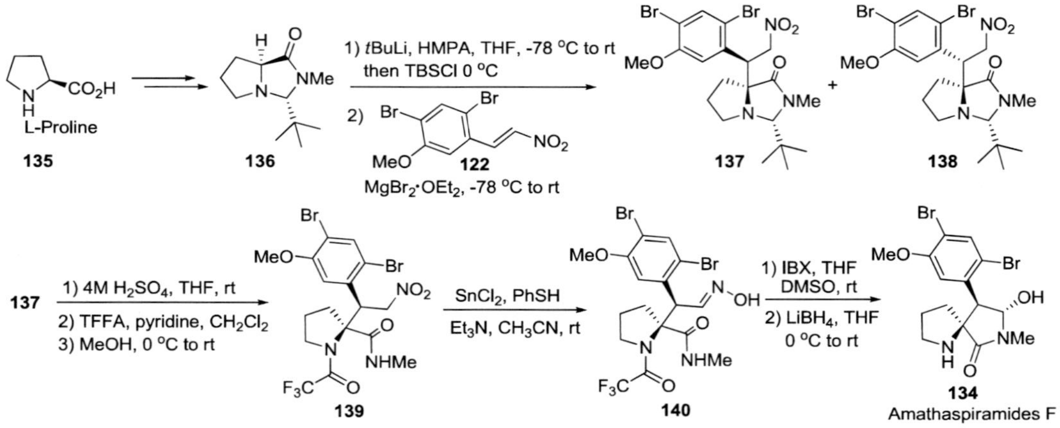

Scheme 12. Syntheses of (−)-amathaspiramide F by Hughes and Trauner.

Acetal **136** on enolization followed by conjugate addition with nitro olefin **122** gave alkylated products **137** and **138** in a diastereomeric mixture of 10:1. Compound **137** was proceeded for amathaspiramide F synthesis, which on hydrolysis followed by *N*-trifluoroacetyl protection gave amide **139**. Nitro group of compound **139** was converted into oxime **140**. Hydrolysis of oxime followed by cyclization and then deprotection of *N*-trifluoroacetyl group gave amathaspiramide F (**134**) Scheme 12 [63].

(−)-Amathaspiramide F has also been synthesized by Sakaguchi and coworkers [64, 65], and they started the synthesis by taking α-hydroxy-α-ethynylsilane as a starting material. The key step of the synthesis was enolate Claisen rearrangement. While (±)-amathaspiramide F has been synthesized by Soheili and Tambar [66]. They started the synthesis by taking substituted phenyl allyl carbonate and *N*-substituted proline ester.

The key step of the synthesis was palladium -catalyzed allylic amination/[2,3]-Stevens rearrangement. Amathaspiramides A-F (**129-134**) were synthesized by Chiyoda et al. [67]. They started the synthesis by taking 3-methoxyphenacyl bromide and the key step was the reduction of lactam moiety to the cyclic imine by Schwartz's reagent, [$Cp_2Zr(H)Cl$].

10. Kororamide A and Convolutindole A

Kororamide A (**141**) was isolated from Australian bryozoan *Amathia tortuosa* by Carroll et al. (Figure 11) [68]. The structure was determined by extensive NMR spectroscopic techniques and it was assigned as 2,6,7-tri-bromo substituted indole ring connected to a proline moiety through *Z*-enamide functionality.

It was also found that kororamide A exists as a mixture of *cis–trans* isomers (ratio 4:5) about the amide bond. Kororamide A (**141**) was screened for its biological activity as a cytotoxic and antimalarial.

It was found inactive against breast and pancreatic cancer ous cells and normal human embryonic cells (HEK) at the tested concentration (40 μM). While it was active against chloroquine resistant strain of the malarial parasite *Plasmodium falciparum* at 20 μM (50% inhibition) and against chloroquine-sensitive strain at 20 μM (70% inhibition) [68]. Biosynthesis of kororamide A (**141**) is not reported in literature.

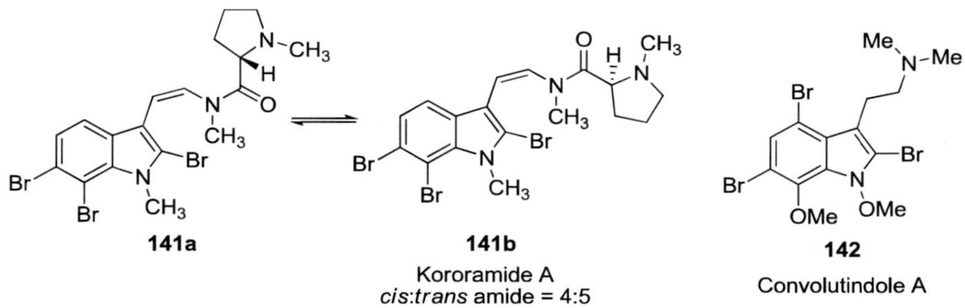

Figure 11. Structures of kororamide A and convolutindole A.

But it can be assumed that kororamide A is derived from the combinations of brominated tryptophan derived metabolite and proline derivative. Synthesis of kororamide A (**141**) is not reported in literature.

Convolutindole A (**142**) was isolated from marine bryozoan *Amathia convoluta* along with convolutamine H (**94**) (Figure 11) [48]. The structure of convolutindole A (**142**) was assigned as tribrominated indole possessing a methoxy group on indole-nitrogen, which is rarely seen in marine natural products. Convolutindole A (**142**) exhibited nematocidal activity against *Haemonchus contortus* and *Bacillus subtilis* with the LD_{99} values of 0.39 and 50 μg/mL, respectively [48]. Its biosynthesis as well as its synthesis is not reported in literature.

11. Chartellines A-C and Chartellamides A, B

Chartelline A (**143**) was isolated form marine bryozoan *Chartella papyracea* by Christophersen and coworkers (Figure 12) [69]. The structure of chartelline A (**143**) was determined by extensive NMR experiments and X-ray crystallography and was assigned as a tetrabrominated alkaloid, possessing indolenine ring (indole derivative) with three contiguous bromines, imidazole ring with one bromine substituent and a *β*-lactam ring, which are connected to each other through a 10-membered ring. Chartelline A did not show any antimicrobial activity against Gram-positive as well as Gram-negative bacteria and molds, and also was found inactive in NCI's leukemia screen (3PS31) at the concentration of 5.60 mg/kg. While it exhibited activity in the KB and PS tests with the ED_{50} values of 29 and 32 mcg/mL, respectively [70]. Further, continuous studies on marine bryozoan *Chartella papyracea* by Christophersen and coworkers isolated two more new alkaloids chartellines B (**144**) and C (**145**) (Figure 12) [70]. Structures of chartellines B (**144**) and C (**145**) were found similar to chartelline A (**143**) except number of bromine substitutions on the indolenine ring. Further, Christophersen and coworkers isolated chartellamides A (**146**) and B (**147**) from bryozoan *Chartella papyracea* (Figure 12) [71]. Their structures were assigned as brominated indole-imidazole ring possessing *β*-lactam functionality. Chartellines A-C as well as chartellamides A, B were not extensively evaluated for their biological activity.

A formal synthesis of chartelline A (**143**) was reported by Weinreb and coworkers and it is depicted in Scheme 13 [72]. The synthesis was started by taking 5-nitroisatin **148**, which was transformed into *β*-lactam product **149** and then converted into amine **150**. Now, three contiguous bromines were installed by the treatment of benzyltrimethylammonium tribromide

($BnMe_3NBr_3$) followed by modified Sandmeyer reaction. The obtained compound **151** was coupled with alkyne **152** to get acetonide adduct **153**.

Acetonide group was hydrolyzed and cleaved to get aldehyde **154**, which on treatment with *p*-toluenesulfonic acid gave seven-membered-ring compound **155** rather than macrocyclic enamide required for chartelline A (**143**) [72].

Further, in continuation to the synthesis of chartellines, an approach was developed by Baran et al. to construct macrocyclic skeleton and *β*-lactam functionality required for chartellines, securamines and securines [73].

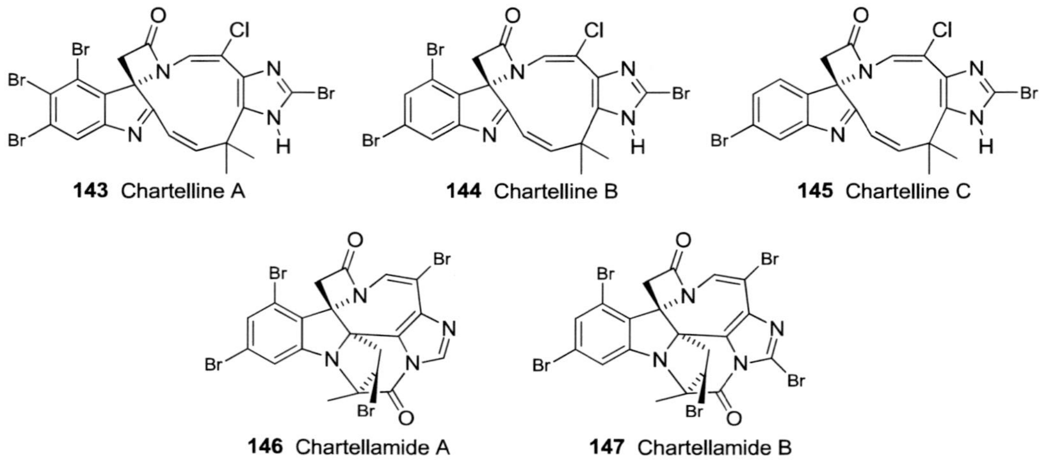

Figure 12. Structures of chartellines A-C and chartellamides A, B.

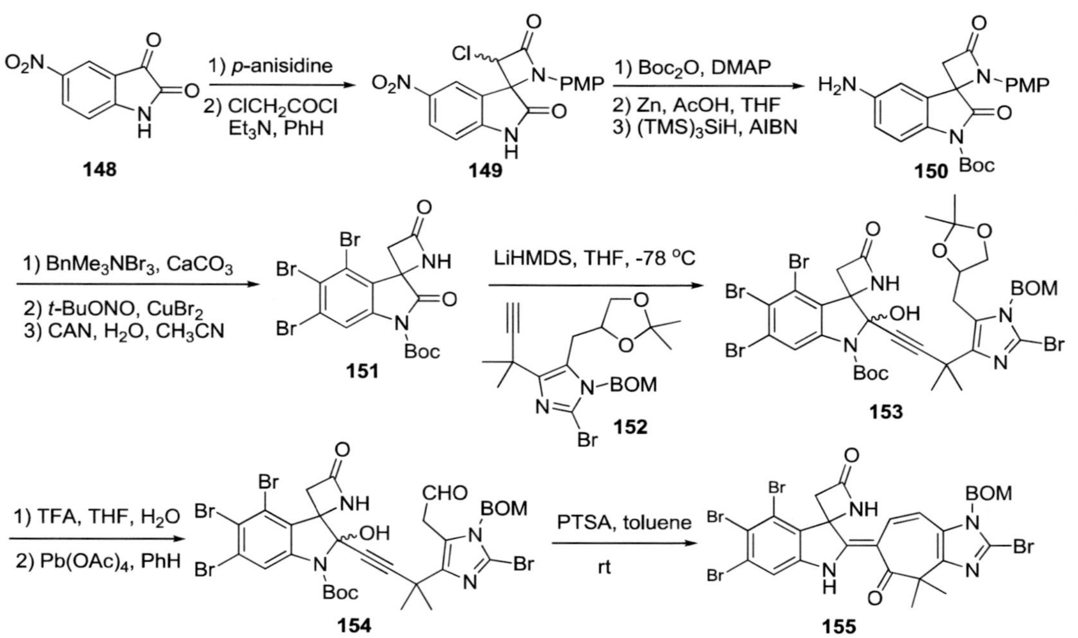

Scheme 13. Synthesis towards chartelline A (**143**) by Weinreb and coworkers.

They explored this strategy for the synthesis of chartelline C (**145**) and it was started by taking indole **156** and alkyne **157**, which after Heck-Sonogashira coupling yielded compound **158** (Scheme 14) [74].

Compound **158** was transformed into aldehyde **159** by performing a sequence of reactions and then it was converted into compound **161**. Compound **161** on bromination followed by key transformations gave core skeleton **163**, which after TMSE deprotection and decarboxylation gave natural product chartelline C (**145**) (Scheme 13) [74]. Synthetic studies as well as synthesis of precursors of chartellines have also been reported in literature [75-80].

12. Securamines A-G and Securines A, B

Securamines A-G (**164-170**) were isolated from marine bryozoan *Securiflustra securifrons* by Christophersen and coworkers (Figure 13) [81, 82] and their structures were determined by extensive spectroscopic techniques. Securamines A (**164**) and B (**165**) on dissolving in DMSO-d_6 resulted securines A (**171**) and B (**172**), respectively [81]. A tricyclic pyrroloindole core and a macrocyclic ring is common in all securamines A-G (**164-170**). Securamines A (**164**) and B (**165**) as well as securamines C (**166**) and D (**167**) are differ to each other only by a bromine substitution at the indole ring [81]. Securamine E (**168**) is a dibrominated form of securamine C (**166**), while securamine F (**169**) is a diastereomer of securamine C (**166**) which differs in the configuration at chlorine attached carbon. Securamine G (**170**) has only one double bond and it can be considered as hydrogenated form of securamine C (**166**) (Figure 13) [82]. Securamines A-G (**164-170**) were not evaluated for their biological activity. Synthesis of securamines A-G also did not reported in literature, only carbon skeleton of securine A (**171**) was constructed by Korakas et al. [83].

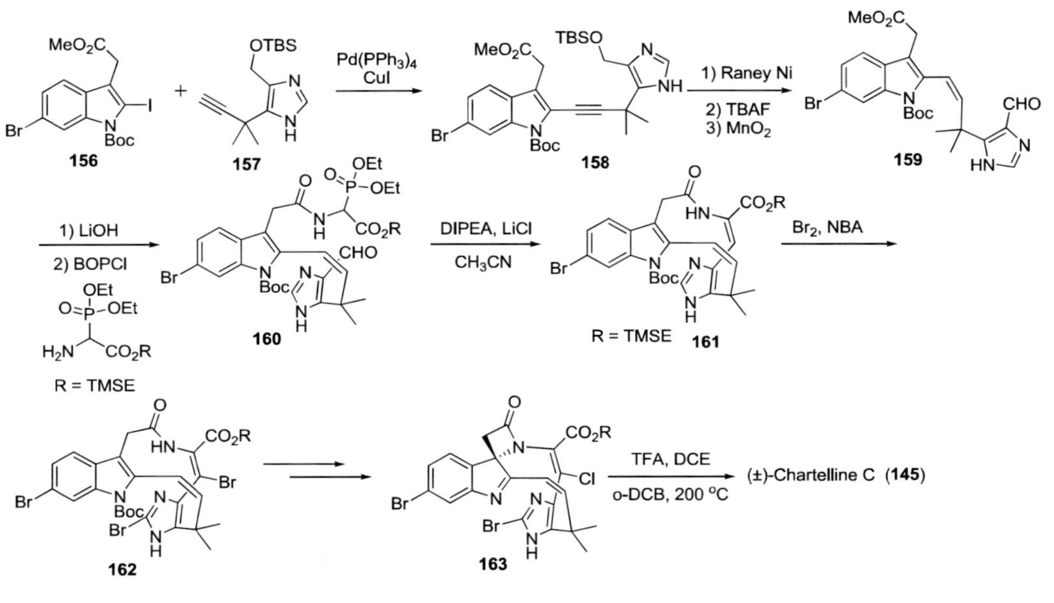

Scheme 14. Synthesis of chartelline C (**145**) by Baran and Shenvi.

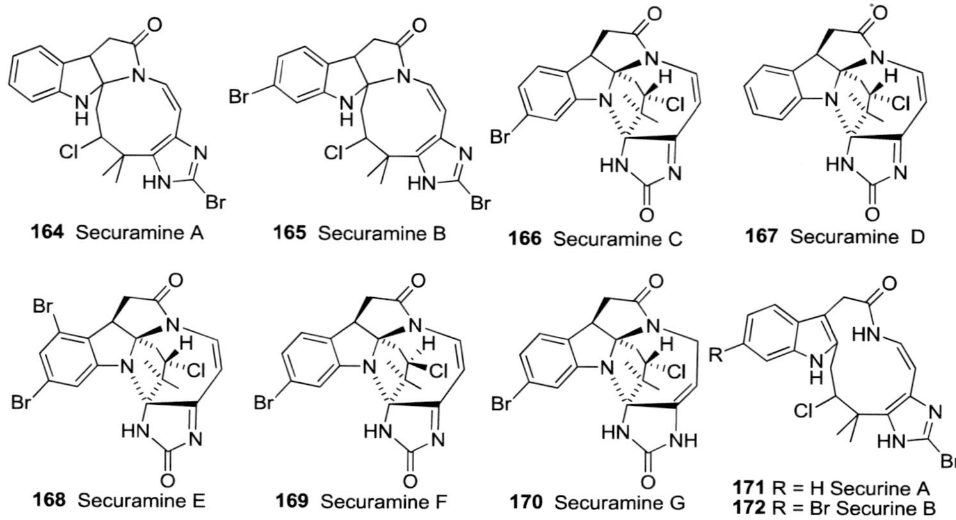

Figure 13. Structures of securamines A-G and securines A, B.

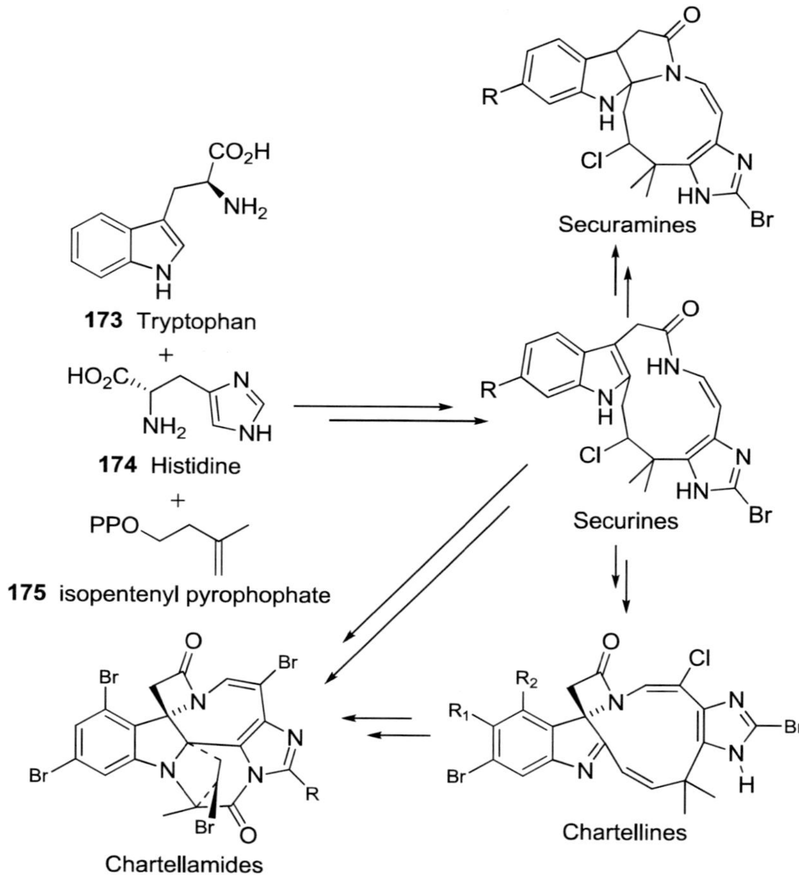

Scheme 15. Postulated biosynthesis of securines, securamines, chartellines and chartellamides.

The biosynthesis of securines, securamines, chartellines and chartellamides is not experimental studied so far and no evidence is reported for the installation of three contiguous bromines on indole ring of chartelline A (**143**). The biosynthesis of securines, securamines, chartellines and chartellamides is postulated in Scheme 15 [74, 76]. It was assumed that initially securines were derived from the combinations of tryptophan **173**, histidine **174** and isoprene unit **175** and then securamines, chartellines and chartellamides are derived from securines.

References

[1] Pomponi, S. A. The oceans and human health: the discovery and development of marine-derived drugs. *Oceanography* 2001, 14, 78-87.

[2] Dembitsky, V. M. Bromo- and iodo-containing alkaloids from marine microorganisms and sponges. *Russ. J. Bioorg. Chem.* 2002, 28, 170-182.

[3] Gribble, G. W. Natural organohalogens: a new frontier for medicinal agents? *J. Chem. Educ.* 2004, 81, 1441-1449.

[4] Gribble, G. W. The diversity of naturally occurring organobromine compounds. *Chem. Soc. Rev.* 1999, 28, 335-346.

[5] Gribble, G. W., *Naturally Occurring Organohalogen Compounds–A Comprehensive Update*, Progress in the Chemistry of Organic Natural Products, Springer Wien New York, 2010; Vol. 91.

[6] Gribble, G. W. Recently discovered naturally occurring heterocyclic organohalogen compounds. *Heterocycles* 2012, 84, 157-207.

[7] Wang, B. G., Gloer, J. B., Ji, N. Y., Zhao, J. C. Halogenated organic molecules of rhodomelaceae origin: chemistry and biology. *Chem. Rev.* 2013, 113, 3632-3685.

[8] Tadesse, M., Strøm, M. B., Svenson, J., Jaspars, M., Milne, B. F., Tørfoss, V., Andersen, J. H., Hansen, E., Stensvåg, K., Haug, T. Synoxazolidinones A and B: novel bioactive alkaloids from the ascidian synoicum pulmonaria. *Org. Lett.* 2010, 12, 4752-4755.

[9] Tadesse, M., Svenson, J., Jaspars, M., Strøm, M. B., Abdelrahman, M. H., Andersen, J. H., Hansen, E., Kristiansen, P. E., Stensvåg, K., Haug, T. Synoxazolidinone C; a bicyclic member of the synoxazolidinone family with antibacterial and anticancer activities. *Tetrahedron Lett.* 2011, 52, 1804-1806.

[10] Hopmann, K. H., Šebestík, J., Novotná, J., Stensen, W., Urbanová, M., Svenson, J., Svendsen, J. S., Bouř, P., Ruud, K. Determining the absolute configuration of two marine compounds using vibrational chiroptical spectroscopy. *J. Org. Chem.* 2012, 77, 858-869.

[11] Chantraine, J.-M., Combaut, G., Teste, J. Phenols bromes d'une algue rouge, *Halopytis incurvus*: Acides carboxyliques. *Phytochemistry* 1973, 12, 1793-1796.

[12] Greve, H., Meis, S., Kassack, M. U., Kehraus, S., Krick, A., Wright, A. D., Konig, G. M. New iantherans from the marine sponge *Ianthella quadrangulata*: novel agonists of the $P2Y_{11}$ receptor. *J. Med. Chem.* 2007, 50, 5600-5607.

[13] Horyn, O., Luhovyy, B., Lazarow, A., Daikin, Y., Nissim, I., Yudkoff, M., Itzhak, N. Biosynthesis of agmatine in isolated mitochondria and perfused rat liver: studies with ^{15}N-labelled arginine. *Biochem. J.* 2005, 388, 419-425.

[14] Shymanska, N. V., An, I. H., Pierce, J. G. A rapid synthesis of 4-oxazolidinones: total synthesis of synoxazolidinones A and B. *Angew. Chem. Int. Ed.* 2014, 53, 5401-5404.

[15] Tadesse, M., Svenson, J., Sepčić, K., Trembleau, L., Engqvist, M., Andersen, J. H., Jaspars, M., Stensvåg, K., Haug, T. Isolation and synthesis of pulmonarins A and B, acetylcholinesterase inhibitor s from the colonial ascidian *Synoicum pulmonaria*, *J. Nat. Prod.* 2014, 77, 364-369.

[16] Liu, L. F., Liu, T., Li, G. X., Wang, Q., Ng, T. Isolation and determination of p-hydroxybenzoylcholine in traditional Chinese medicine Semen sinapis Albae. *Anal. Bioanal. Chem.* 2003, 376, 854-858.

[17] Clausen, S., Olsen, O., Sorensen, H. 4-Hydroxybenzoylcholine: A natural product present in *Sinapis alba. Phytochemistry* 1982, 21, 917-922.

[18] Blunden, G., Gordon, S. M., Smith, B. E., Fletcher, R. L. Quaternary ammonium compounds in species of the Fucaceae (Phaeophyceae) from Britain. *Br. Phycol. J.* 1985, 20, 105-108.

[19] Kornprobst, J.-M. *Encyclopedia of Marine Natural Products*; Wiley-Blackwell: Weinheim, 2010; Vol. 2, Chapter 19, pp. 796-805.

[20] Blunt, J. W., Copp, B. R., Keyzers, R. A., Munro, M. H. G., Prinsep, M. R. Marine natural products. *Nat. Prod. Rep.* 2012, 29, 144-222.

[21] Roll, D. M., Chang, C. W. J., Scheuer, P. J., Gray, G. A., Shoolery, J. N., Matsumoto, G. K., Van Duyne, S. D., Clardy, J. Structure of the psammaplysins. *J. Am. Chem. Soc.* 1985, 107, 2916-2920.

[22] Copp, B. R., Ireland, C. M., Barrow, L. R. Psammaplysin C: a new cytotoxic dibromotyrosine-derived metabolite from the marine sponge *Druinella* (= *Psammaplysilla*) *purpurea*. *J. Nat. Prod.* 1992, 55, 822-823.

[23] Ichiba, T., Scheuer, P. J., Kelly-Borges, M. Three bromotyrosine derivatives, one terminating in an unprecedented diketocyclopentenylidene enamine. *J. Org. Chem.* 1993, 58, 4149-4150.

[24] Liu, S., Fu, X., Schmitz, F. J., Kelly-Borges, M. Psammaplysin F, a new bromotyrosine derivative from a sponge, *Aplysinella* sp. *J. Nat. Prod.* 1997, 60, 614-615.

[25] Yang, X., Davis, R. A., Buchanan, M. S., Duffy, S., Avery, V. M., Camp, D., Quinn, R. J. Antimalarial bromotyrosine derivatives from the Australian marine sponge *Hyattella* sp. *J. Nat. Prod.* 2010, 73, 985-987.

[26] Xu, M., Andrews, K. T., Birrell, G. W., Tran, T. L., Camp, D., Davis, R. A., Quinn, R. J. Psammaplysin H, a new antimalarial bromotyrosine alkaloid from a marine sponge of the genus *Pseudoceratina*. *Bioorg. Med. Chem. Lett.* 2011, 21, 846-848.

[27] Wright, A. D., Schupp, P. J., Schrör, J.-P., Engemann, A., Rohde, S., Kelman, D., de Voogd, N., Carroll, A., Motti, C. A. Twilight zone sponges from Guam yield theonellin isocyanate and psammaplysins I and J. *J. Nat. Prod.* 2012, 75, 502-506.

[28] Mudianta, I. W., Skinner-Adams, T., Andrews, K. T., Davis, R. A., Hadi, T. A., Hayes, P. Y., Garson, M. J. Psammaplysin derivatives from the Balinese marine sponge *Aplysinella strongylata*. *J. Nat. Prod.* 2012, 75, 2132-2143.

[29] Lee, Y.-J., Han, S., Lee, H.-S., Kang, J. S., Yun, J., Sim, C. J., Shin, H. J., Lee, J. S. Cytotoxic psammaplysin analogues from a *Suberea* sp. marine sponge and the role of the spirooxepinisoxazoline in their activity. *J. Nat. Prod.* 2013, 76, 1731-1736.

[30] Kashman, Y., Groweiss, A., Carmely, S., Kinamoni, Z., Czarkie, D., Rotem, M. *Pure and Appl. Chem.* 1982, 54, 1995-2010.

[31] Rotem, M., Carmely, S., Kashman, Y., Loya, Y. *Tetrahedron* 1983, 39, 667-676.

[32] Ramsey, D. M., Islam, M. A., Turnbull, L., Davis, R. A., Whitchurch, C. B., McAlpine, S. R. Psammaplysin F: a unique inhibitor of bacterial chromosomal partitioning, *Bioorg. Med. Chem. Lett.* 2013, 23, 4862-4866.

[33] Vogel, E., Günther, H. Benzene oxide-oxepin valence tautomerism. *Angew. Chem. Int. Ed. Engl.* 1967, 6, 385-401.

[34] Neuss, N., Nagarajan, R., Molloy, B. B., Huckstep, L. L. Aranotin and related metabolites. II. Isolation, characterization, and structures of two new metabolites. *Tetrahedron Lett.* 1968, 9, 4467-4471.

[35] Brannon, D. R., Mabe, J. A., Molloy, B. B., Day, W. A. Biosynthesis of dithiadiketopiperazine antibiotics: Comparison of possible aromatic amino acid precursors. *Biochem. Biophys. Res. Commun.* 1971, 43, 588-594.

[36] Das, P., Omollo, A. O., McClendon, E., Sitole, L., Valente, E. J., Walker, L. R., Raucher, D., Hamme, A. T. II. Biological evaluation and synthesis of spiro-isoxazolines *via* intramolecular cyclization. *Abstracts of Papers, 245th ACS National Meeting and Exposition,* New Orleans, LA, United States, April 7-11, 2013.

[37] Hamme, A. T. II, Das, P., Omollo, A. O., Sitole, L., McClendon, E., Valente, E. J., Raucher, D., Walker, L. R. Design and synthesis of spiro-isoxazolines: Novel inhibitors of MCF-7, MDA-MB -231, PC-3, and DU-145. *Abstracts of Papers, 247th ACS National Meeting and Exposition,* Dallas, TX, United States, March 16-20, 2014.

[38] Tsukamoto, S., Kato, H., Hirota, H., Fusetani, N. Ceratinamides A and B: New antifouling dibromotyrosine derivatives from the marine sponge *Pseudoceratina purpurea. Tetrahedron* 1996, 52, 8181-8186.

[39] Xu, M., Davis, R. A., Feng, Y., Sykes, M. L., Shelper, T., Avery, V. M., Camp, D., Quinn, R. J. Ianthelliformisamines A−C, antibacterial bromotyrosine-derived metabolites from the marine sponge *Suberea ianthelliformis. J. Nat. Prod.* 2012, 75, 1001-1005.

[40] Choomuenwai, V., Schwartz, B. D., Beattie, K. D., Andrews, K. T., Khokhar, S., Davis, R. A. The discovery, synthesis and antimalarial evaluation of natural product-based polyamine alkaloids. *Tetrahedron Lett.* 2013, 54, 5188-5191.

[41] Khan, F. A., Ahmad, S., Kodipelli, N., Shivange, G., Anindya, R. Syntheses of a library of molecules on the marine natural product ianthelliformisamines platform and their biological evaluation. *Org. Biomol. Chem.* 2014, 12, 3847-3865.

[42] Pieri, C., Borselli, D., Di Giorgio, C., De Méo, M., Bolla, J.-M., Vidal, N., Combes, S., Brunel, J. M. New ianthelliformisamine derivatives as antibiotic enhancers against resistant Gram-Negative bacteria. *J. Med. Chem.* 2014, 57, 4263-4272.

[43] Carroll, A. R., Duffy, S., Sykes, M., Avery, V. M. Wilsoniamines A and B: novel alkaloids from the temperate Australian bryozoan, *Amathia wilsoni. Org. Biomol. Chem.* 2011, 9, 604-609.

[44] Sakai, R., Suzuki, K., Shimamoto, K., Kamiya, H. Novel betaines from a micronesian sponge *Dysidea herbacea. J. Org. Chem.* 2004, 69, 1180-1185.

[45] Khan, F. A., Ahmad, S. Synthesis of wilsoniamines A and B. *Tetrahedron Lett.* 2013, 54, 2996-2998.

[46] Zhang, H.-P., Kamano, Y., Kizu, H., Itokawa, H., Pettit, G. R., Herald, C. L. Convolutamines A-E, novel β-phenylethylamine alkaloids from marine bryozoan *Amathia convoluta*. *Chem. Lett.* 1994, 2271-2274.

[47] Kamano, Y., Kotake, A., Hashima, H., Hayakawa, I., Hiraide, H., Zhang, H.-P., Kizu, H., Komiyama, K., Hayashi, M., Pettit, G. R. Three new alkaloids, convolutamines F and G, and convolutamydine E, from the Floridian marine bryozoan *Amathia convoluta*. *Collect. Czech. Chem. Commun.* 1999, 64, 1147-1153.

[48] Narkowicz, C. K., Blackman, A. J., Lacey, E., Gill, J. H., Heiland, K. Convolutindole A and convolutamine H, new nematocidal brominated alkaloids from the marine bryozoan *Amathia convoluta*. *J. Nat. Prod.* 2002, 65, 938-941.

[49] Davis, R. A., Sykes, M., Avery, V. M., Camp, D., Quinn, R. J. Convolutamines I and J, antitrypanosomal alkaloids from the bryozoan *Amathia tortusa*. *Bioorg. Med. Chem.* 2011, 19, 6615-6619.

[50] Hashima, H., Hayashi, M., Kamano, Y., Sato, N. Synthesis and biological activities of the marine bryozoan alkaloids convolutamines A, C and F, and lutamides A and C. *Bioorg. Med. Chem.* 2000, 8, 1757-1766.

[51] Khan, F. A., Ahmad, S. Synthesis of reported and revised structures of amathamide D and synthesis of convolutamine F, H and lutamide A, C. *J. Org. Chem.* 2012, 77, 2389-2397.

[52] Pham, N. B., Deydier, S., Labaied, M., Monnerat, S., Stuart, K., Quinn, R. J. N^1, N^1-Dimethyl-N^3-(3-(trifluoromethyl)phenethyl)propane -1,3-diamine, a new lead for the treatment of human African trypanosomiasis. *Eur. J. Med. Chem.* 2014, 74, 541-551.

[53] Blackman, A. J., Matthews, D. J. Amathamide alkaloids from the marine bryozoan *Amathia wilsoni* Kirkpatrick. *Heterocycles* 1985, 23, 2829-2833.

[54] Blackman, A. J., Green, R. D. Further amathamide alkaloids from the bryozoan *Amathia wilsoni*. *Aust. J. Chem.* 1987, 40, 1655-1662.

[55] Ahmad, S., Choudhury, S., Khan, F. A. Syntheses and structure revisions of marine brominated alkaloids amathamides E and F: a palladium -catalyzed enamide synthesis (Manuscript in preparation).

[56] Blackman, A. J., Eldershaw, T. P. D., Garland, S. M. Alkaloids from two further *Amathia* bryozoan species. *Aust. J. Chem.* 1993, 46, 401-405.

[57] Blackman, A. J., Fu, S.-L. A β-phenylethylamine-derived possible biosynthetic precursor to the amathamides, alkaloids from the bryozoan *Amathia wilsoni*. *J. Nat. Prod.* 1989, 52, 436-438.

[58] Osuna, M. R., Aguirre, G., Somanathan, R., Molins, E. Asymmetric synthesis of amathamides A and B: novel alkaloids isolated from *Amathia wilsoni*. *Tetrahedron: Asymmetry* 2002, 13, 2261-2266.

[59] Osuna, M. R., Chávez, D., Hernández, L., Molins E., Somanathan, R., Aguirre, G. Synthesis of analogs of amathamide A and their preliminary antimicrobial activity. *Molecules* 2005, 10, 295-301.

[60] Osuna, M. R., Garayzar, O. F. N., Somanathan, R., Aguirre, G. Syntheses of three mono-brominated enamide analogs of natural alkaloids isolated from the Tasmanian marine bryozoan *Amathia wilsoni*. *J. Mex. Chem. Soc.* 2011, 55, 57-61.

[61] Morris, B. D., Prinsep, M. R. Amathaspiramides A-F, novel brominated alkaloids from the marine bryozoan *Amathia wilsoni*. *J. Nat. Prod.* 1999, 62, 688-693.

[62] Kerr, R. G., Kohl, A. C., Boehnlein, J. M. Bioactive compounds from bryozoans. *Recent Advances in Marine Biotechnology (2001),* 6(Bio-Organic Compounds: Chemistry and Biomedical Applications), 149-164.

[63] Hughes C. C., Trauner, D. The total synthesis of (−)-amathaspiramide F. *Angew. Chem. Int. Ed.* 2002, 41, 4556-4559.

[64] Sakaguchi, K., Ayabe, M., Watanabe, Y., Okada, T., Kawamura, K., Shiada, T., Ohfune, Y. Total synthesis of (−)-amathaspiramide F. *Org. Lett.* 2008, 10, 5449-5452.

[65] Sakaguchi, K., Ayabe, M., Watanabe, Y., Okada, T., Kawamura, K.; Shiada, T.; Ohfune, Y. Total synthesis of (−)-amathaspiramide F. *Tetrahedron* 2009, 65, 10355-10364.

[66] Soheili, A., Tambar, U. K. Synthesis of (±)-amathaspiramide F and discovery of an unusual stereocontrolling element for the [2,3]-Stevens rearrangement. *Org. Lett.* 2013, 15, 5138-5141.

[67] Chiyoda, K., Shimokawa, J., Fukuyama, T. Total syntheses of all the amathaspiramides. *Angew. Chem. Int. Ed.* 2012, 51, 2505-2508.

[68] Carroll, A. R., Wild, S. J., Duffy, S., Avery, V. M. Kororamide A, a new tribrominated indole alkaloid from the Australian bryozoan *Amathia tortuosa*. *Tetrahedron Lett.* 2012, 53, 2873-2875.

[69] Chevolot, L., Chevolot, A.-M., Gajhede, M., Larsen, C., Anthoni, U., Christophersen, C. Chartelline A: a pentahalogenated alkaloid from the marine bryozoan *Chartella papyracea*. *J. Am. Chem. Soc.* 1985, 107, 4542-4543.

[70] Anthoni, U., Chevolot, L., Larsen, C., Nielsen, P. H., Christophersen, C. Marine alkaloid s. 12. chartellines, halogenated *β*-lactam alkaloids from the marine bryozoan *Chartella papyracea*. *J. Org. Chem.* 1987, 52, 4709-4712.

[71] Anthoni, U., Block, K., Chevolot, L., Larsen, C., Nielsen, P. H., Christophersen, C. Chartellamide A and B, halogenated *β*-lactam indole-imidazole alkaloids from the marine bryozoan *Chartella papyracea*. *J. Org. Chem.* 1987, 52, 5638-5639.

[72] Sun, C., Lin, X., Weinreb, S. M. Explorations on the total synthesis of the unusual marine alkaloid chartelline A. *J. Org. Chem.* 2006, 71, 3159-3166.

[73] Baran, P. S., Shenvi, R. A., Mitsos, C. A. A remarkable ring contraction en route to the chartelline alkaloids. *Angew. Chem. Int. Ed.* 2005, 44, 3714-3717.

[74] Baran, Phil S., Shenvi, R. A. Total synthesis of (±)-chartelline C. *J. Am. Chem. Soc.* 2006, 128, 14028-14029.

[75] Black, P. J., Hecker, E. A., Magnus, P. Studies towards the synthesis of the marine alkaloid chartelline C. *Tetrahedron Lett.* 2007, 48, 6364-6367.

[76] Kajii, S., Nishikawa, T., Isobe, M. Synthetic studies and biosynthetic speculation on marine alkaloid chartelline. *Chem. Commun.* 2008, 27, 3121-3123.

[77] Iwasaki, K., Kanno, R., Morimoto, T., Yamashita, T., Yokoshima, S., Fukuyama, T. Synthetic studies on chartelline C: stereoselective construction of the core skeleton *Angew. Chem. Int. Ed.* 2012, 51, 9160-9163.

[78] Nishikawa, T., Kajii, S., Isobe, M. Novel synthesis of bromoindolenine with spiro-β-lactam in chartelline. *Synlett* 2004, 11, 2025-2027.

[79] Nishikawa, T., Kajii, S., Isobe, M. Synthesis of model compound containing an indole spiro-β-lactam moiety with vinylchloride in chartellines. *Chem. Lett.* 2004, 33, 440-441.

[80] Kajii, S., Nishikawa, T., Isobe, M. Synthesis of N-hydroxyenamide, a potential precursor of chartelline. *Tetrahedron Lett.* 2008, 49, 594-597.

[81] Rahbaek, L., Anthoni, U., Christophersen, C., Nielsen, P. H., Petersen, B. O. Marine alkaloid s. 18. securamines and securines, halogenated indole-imidazole alkaloids from the marine bryozoan *Securiflustra securifrons. J. Org. Chem.* 1996, 61, 887-889.

[82] Rahbaek, L., Christophersen, C. Marine alkaloid s. 19. three new alkaloids, securamines E-G, from the marine bryozoan *Securiflustra securifrons. J. Nat. Prod.* 1997, 60, 175-177.

[83] Korakas, P., Chaffee, S., Shotwell, J. B., Duque, P., Wood, J. L. Efficient construction of the securine A carbon skeleton. *Proc. Natl. Acad. Sci. US* 2004, 101, 12054-12057.

In: Alkaloids
Editor: Eduardo Sobarzo-Sánchez

ISBN: 978-1-63482-074-5

Chapter 5

ERYTHRINA ALKALOIDS: RECENT ADVANCES IN THEIR SYNTHESIS, ISOLATION AND PHARMACOLOGY

Olimpo García-Beltrán
and Miguel Moreno-Palacios
Facultad de Ciencias Naturales y Matemáticas,
Universidad de Ibagué, Ibagué, Colombia

ABSTRACT

The genus *Erythrina* (subfamily Papilionoidae) comprises between 110 and 118 species with a mainly pantropical distribution. However, some species are also found in subtropical regions such as South Africa, Southeast United States and the Himalayas. Out of 118 species, 43 occur in the Old World, while 75 are distributed only in the New World. They inhabit a broad range of ecosystems, spanning from lowland dry and humid forests, including floodplains, to Andean highland humid forest above 3000 m. *Erythrina* species are widely grown as ornamental plants, the edible fruits of some are sought after for their broad nutritional spectrum. The trees are planted as soil and watershed protectors, the leaves serve as high quality forage, are used as pet food and widely used in folk medicine in various parts of the world. This genus is known to produce various secondary metabolites found between C-prenylated flavanones, isoflavones, isoflavanones, pterocarpans, indole alkaloids and erythrinane alkaloids. However, the erythrinane skeleton plays an important role because of its curare-like, hypnotic and pharmacological effects associated with it including sedative, hypotensive, neuromuscular blocking and CNS activity. This chapter summarizes the literature relating to the alkaloids isolated from the genus Erythrina, synthetic approaches since 2010, like its pharmacology.

Keywords: Leguminosae-Fabaceae, *Erythrina,* erythrinane alkaloids, synthesis, biosynthesis, pharmacology

1. Introduction

The genus *Erythrina* Mart. (Leguminoseae-Fabaceae) is common in tropical and subtropical regions, and also in some temperate regions of the world [1-3]. The trees are planted as soil and watershed protectors, the leaves serve as high quality forage, are used as pet food and widely used in folk medicine in various parts of the world [4-6]. The erythrinane skeleton plays an important role because of its curare-like [7], hypnotic and pharmacological effects associated with it including sedative, hypotensive, neuromuscular blocking and CNS activity [8-15]. Other uses are in patients with epilepsy or schizophrenia, to treat female infertility, pain, stomach ailments, gonorrhea [16], inflammation, heart disorders and insomnia [17], as anxiolytics [18], toxics [19], and antioxidants [20]. This genus is known to produce a broad range of secondary metabolites including *C*-prenylated flavanones, isoflavones, isoflavanones, pterocarpans [21], and erythrinan alkaloids [22]. The erythrinan alkaloids constitute a medium-sized group of natural products sharing a 1,2,3,4,5,6,8,9-octahydroindolo[7a,1a]isoquinoline (Figure 1) skeleton that are limited to the plant genera *Erythrina* (Fabaceae), *Cocculus*, *Hyperbaena* and *Pachygone* (Menispermaceae). However, this chapter will only treat the alkaloids isolated from *Erythrina*.

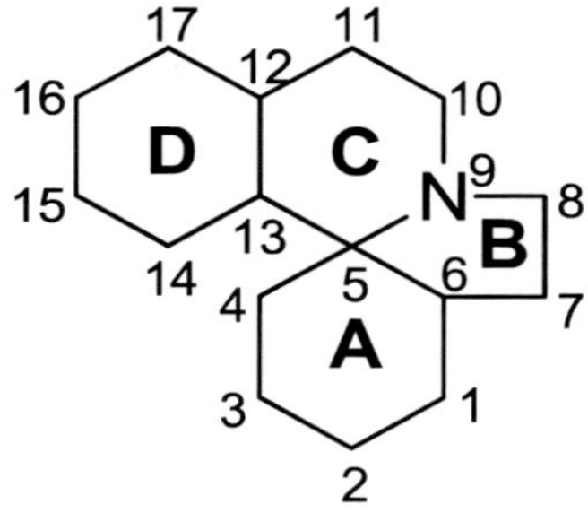

Figure 1. General structure of the erythrinan alkaloids.

2. Botanical Aspects of the Genus *Erythrina*

The family Leguminosae (Fabaceae) is one of the most diverse Angiosperm groups in the world. With 730 genera and 19,500 species [23], this family is only surpassed in number by the Asteraceae and Orchidaceae [24]. Within this large family, the genus *Erythrina* (subfamily Papilionoidae) comprises between 110 and 118 species [25-28] representing a monophyletic group, distinguishable from others within the tribe Phaseoleae [27,29-30]. Its members present hard red to orange, odorless flowers, diurnal anthesis and copious nectar characterized by its content of sucrose and hexoses, which reflects an ornithophilous syndrome, being pollinated by passerine birds (Passeriformes) and hummingbirds (Apodiformes: Trochilidae) [29,31].

According to Gentry (1993) [26], out of 118 *Erythrina* species, 43 occur in the Old World, while 75 are distributed only in the New World. Species of *Erythrina* inhabit a broad range of ecosystems, from lowland dry and humid forests, including floodplains, to Andean highland humid forest above 3000 meters [32].

"Genus Erythrina is constituted by trees and shrubs frequently armed with conical or recurved thorns. Young branches terete, occasionally lenticeled. Leaves alternate, petioled, compound, trifoliate, pubescent when young and glabrous when mature; conpicuos petiole pulvinus. Leaflets rhomboid, ovate, elliptic or suborbicular, apex acute, acuminate or obtuse, obtuse, truncate or acute base, margin entire, pubescent or glabrous, simple trichomes occasionally dendritic-echinoids, papery to chartaceous or coriaceous consistency, 6-8 pairs of secondary straight, parallel veins. Asymmetric lateral leaflets, generally smaller than terminal leaflet, stipels glandular at base of petiolules. Stipules free, deciduous. Inflorescences consist of terminal, pseudo-terminal or axillary racemes, also simple flowers or fasciculate in groups (2-3,6). Calyx gamosepalous is cylindrical or campanulate, notched or entire, sometimes with prominent lobes, glabrous or pubescent. Floral bracts are persistent or deciduous; deciduous bracteoles. Standard (vexillum) is always higher than the keel, generally elliptical, ovate or obovate, erect or reflexed. In stamens 10, one vexilar is generally shorter; filaments connate at the base into a tube, the free portion of filaments ca. 2/3 of the total length of the stamens; anthers dorsifixed, longitudinal dehiscence. Ovary stipitate is usually pubescent; puberulous style at base; stigma usually capitate, small. Fruit pedicellate, usually dehiscent legume or bivalve pods, stipitate, linear-oblong, moniliforme, or inflated, lightly or heavily compressed between seeds, mucronate apex; pubescent or glabrous. Seeds subreniform are red to orange or bicolor (red with black), brown or brown-marbled; epigeal germination ".

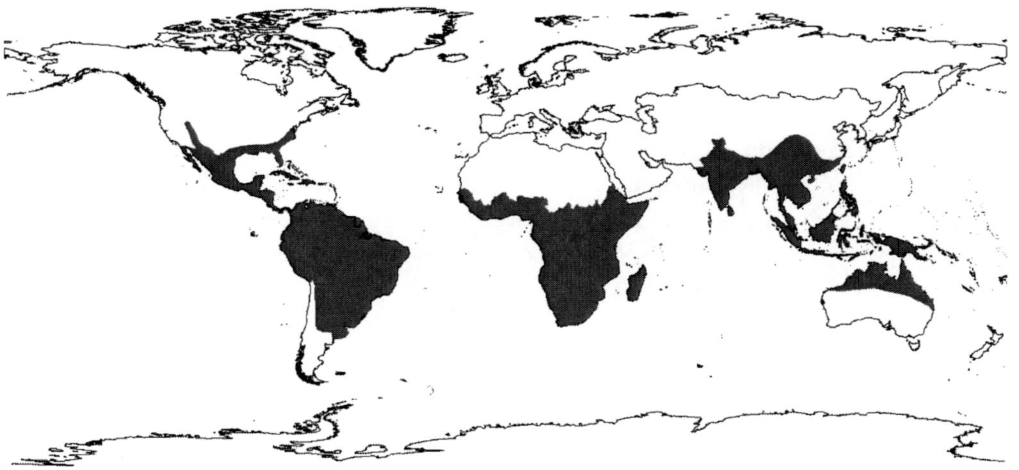

Figure 2. *Erythrina* worldwide distribution.

3. Structural Classification of *Erythrina* Alkaloids

The *Erythrina* alkaloids contain four linked rings, labeled A, B, C, and D (Figure 1), and can be divided or classified into two main groups: alkaloids possessessing a 6,5,6,6 indoloisoquinoline ring system called erythrinans, and those with a 6,5,7,6 ring system, the homoerythrinan alkaloids (Figure 3).

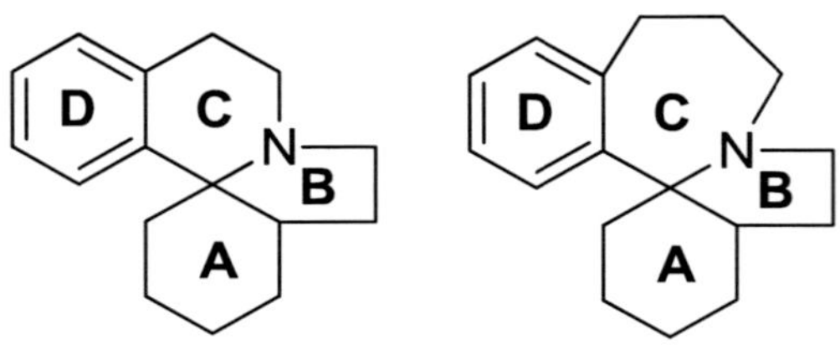

Figure 3. Ring systems of the *Erythrina* alkaloids: erythrinanes and homoerythrinans.

The erythrinanes depending of the nature of ring D, both for erythrinanes of the families Fabaceae and Menispermaceae, as homoerythrinane found in the family Phellinaceae and Colchicaceae can be subdivided into aromatic and non-aromatic alkaloids. The lattergroup includes the D ring oxa compounds, and are known as lactonic alkaloids such as (+)-8-oxo-β-erythoidine (**1**). In *Erythrina* these alkaloids are of the pyran type, unlike those found in Menispermeaceae, where the lactone ring is of the furan type. In both series alkaloids with a pyridine ring have been isolated, such as (+)-erymelanthine (**2**) [33,34].

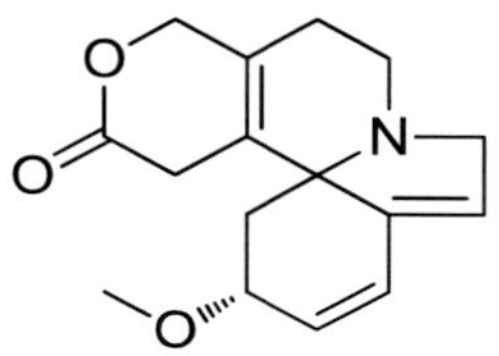

1

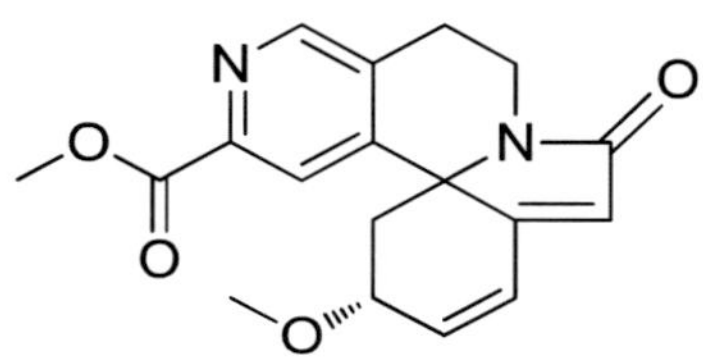

2

Likewise, the position and number of double bonds in rings A and B have led to another subdivision of the alkaloids: dienoid, alkenoid, and recently saturated analogues have been reported. The dienoid alkaloids are characterized by a conjugated system spanning C1-C2 and C6-C7, as shown in erythascine (**3**)**,** unlike the alkenoid alkaloids that only present one double bond at positions C1-C2, C1-C6 or C6-C7, for example dihydroerisotrine (**4**)**,** the aliphatic oxygenation patterns present in the C2 and C3, among which (+)-10,11-dioxoepierythratidine can be mentioned (**5**) [35].

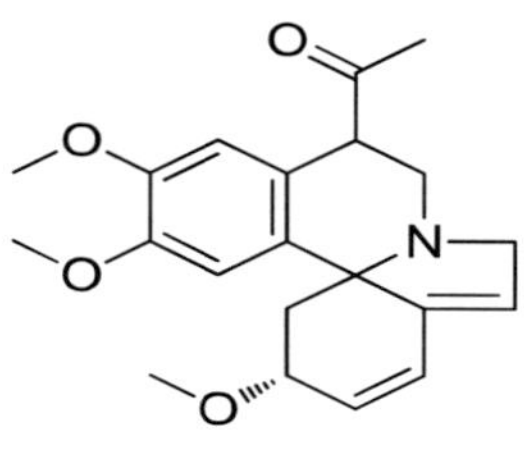

3

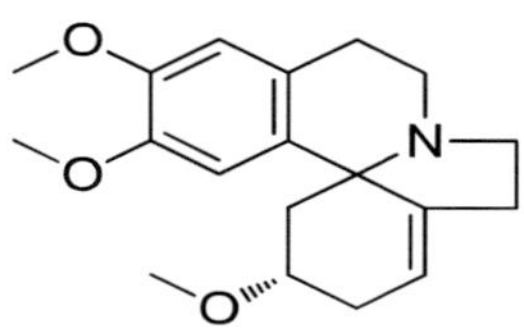

4

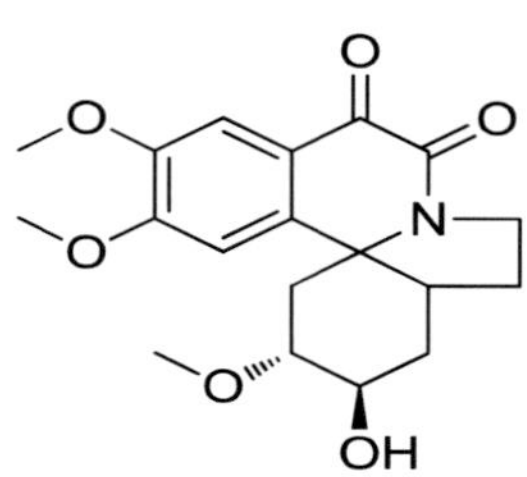

5

These compounds in turn have defined oxygenation patterns at C3, C15 and C16 mainly, and besides there is a group of alkaloids deoxygenated at C-16. The saturated compounds in both series of alkaloids have a *cis* configuration in the A/B ring fusion. Generally these alkaloids are dextrorotatory and their absolute configuration at C5 is (*S*) with respect to the basic skeleton [34].

There are many reports which illustrate the variety of chemical modifications rings A, B and C may undergo. Examples of this is are erythratidinone (**6**), 8-oxoerythraline epoxide (**7**) and erythrinarbine (**8**), modified in ring A; crystamidine (**9**) and erysotrine *N*-oxide (**10**) as examples of ring B modifications; and finally (+)-10,11-dioxoerythratine (**11**), and (+)-10,11-dioxoerythratidinone (**12**), exponents of ring C modification [36-39].

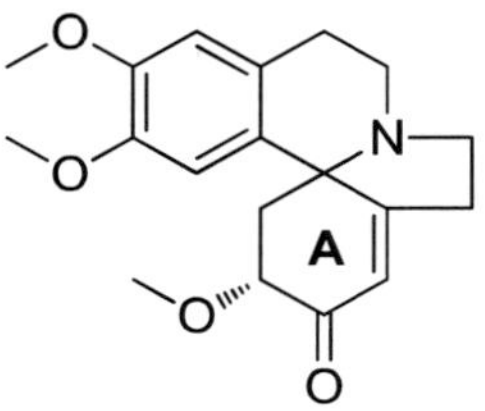

6

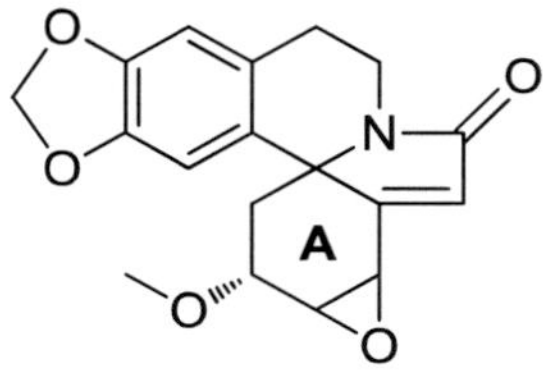

7

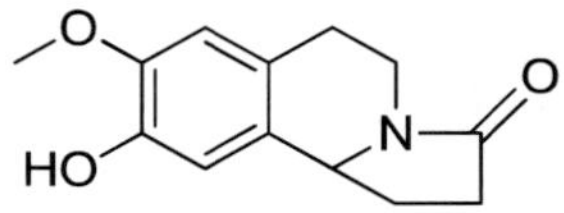

8

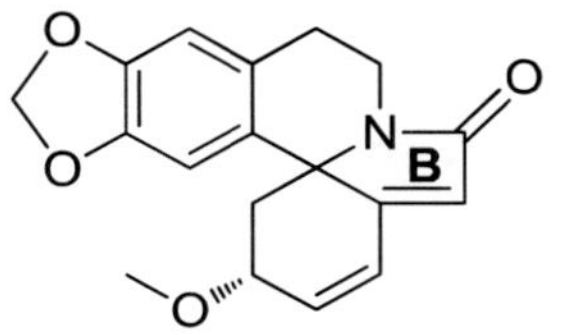

9

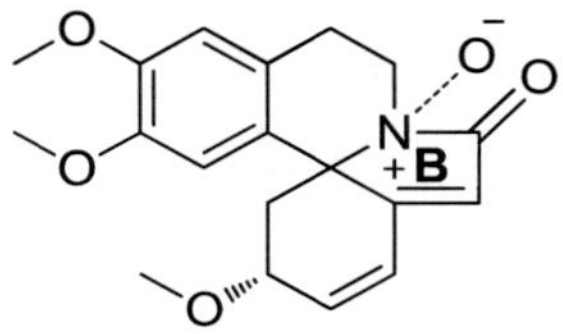

10

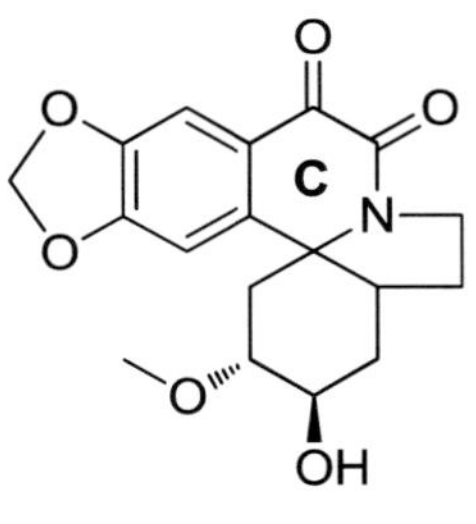

11

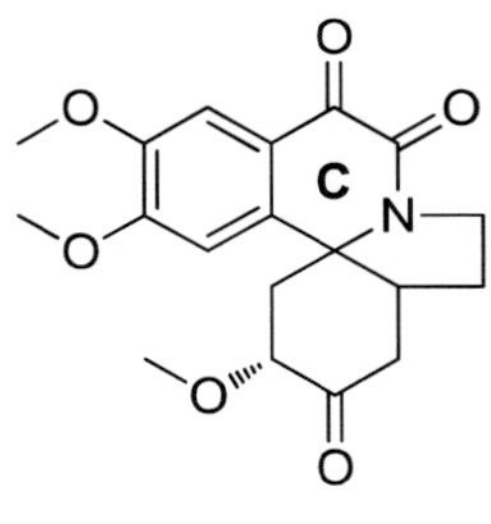

12

4. Biosynthesis of Erythrinans

Studies performed to elucidate the biosynthesis of erythrinoid type alkaloids are very few. The first studies were done by Barton and coworkers [40,41] who (Scheme 1), through a series of experiments *in vivo* and *in vitro,* demonstrated that an isoquinoline derivative, (*S*)-norprotosinomenine (**13**), is the main biosynthetic precursor of these alkaloids. They then postulated that his compound is cyclized through a *p-p* coupling to a neoproaporphine derivative (**14**), which undergoes a rearrangement to form a dibenzazonine derivative (**15**) whose hydrogenation product (**16**) is oxidized to the corresponding diphenoquinone (**17**). Finally a Michael type intramolecular addition type occurs to form the erysodienone (**18**) whose structure corresponds to the erythrinan alkaloid type [40-43].

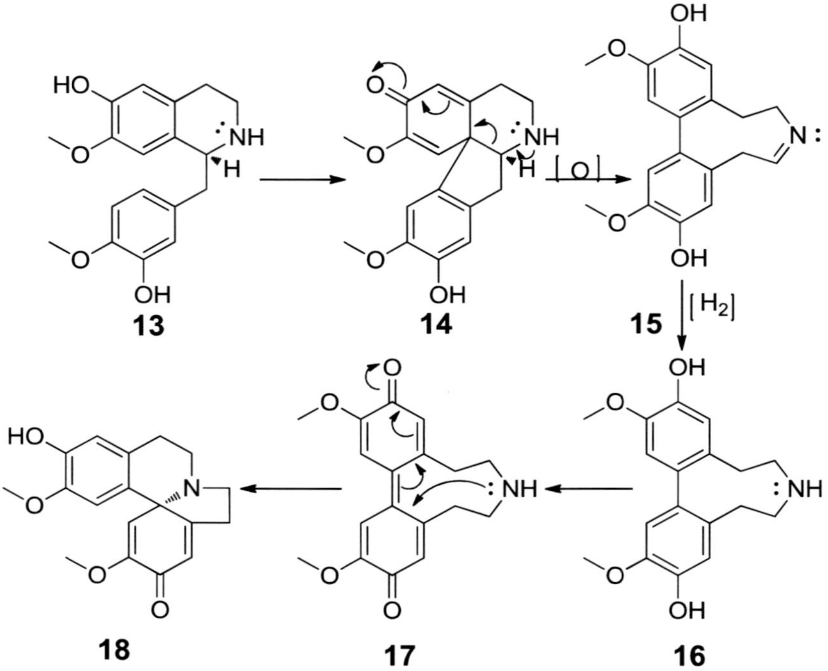

Scheme 1. Biogenesis of the *Erythrina* alkaloid erysodienone proposed by Barton [43].

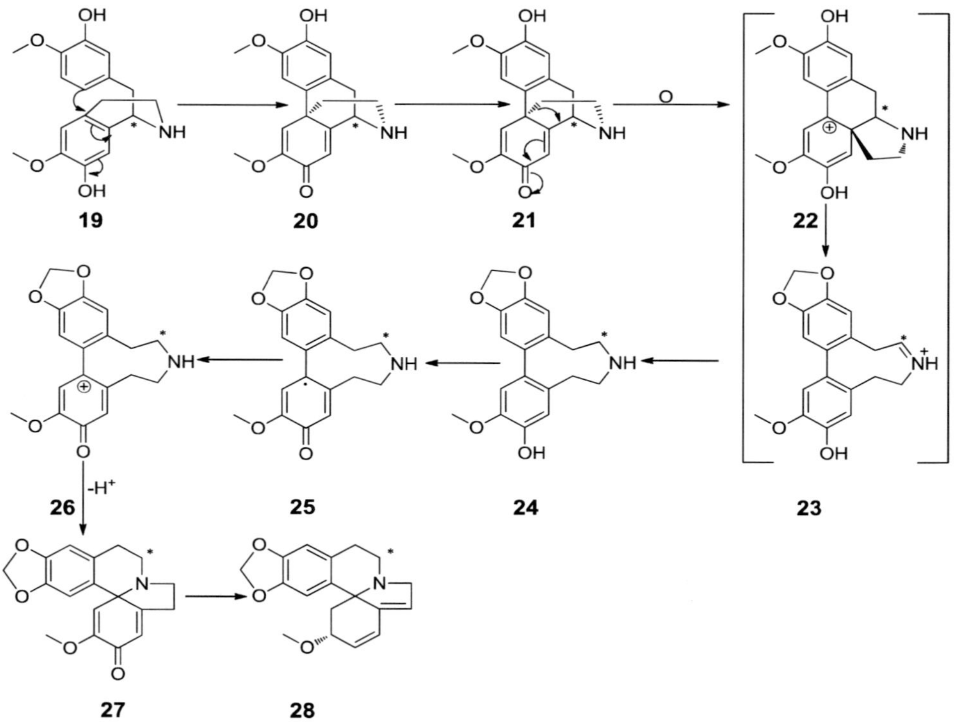

Scheme 2. Revised biogenesis of the *Erythrina* alkaloid erythraline [44].

The intermediate compounds **13** and **16** are natural products. This hypothesis was reinvestigated [44-47] showing that the relative rate of incorporation of **1** into erythraline is small. It was later demonstrated that alkaloids are biosynthesized in the wall tissue of the fruit [47]. Additional research led to the development of an improved delivery system for radio - and stable isotope precursors of *Erythrina* alkaloids in *Erythrina crista-galli* L. This work allowed the discovery that compound **13**, previously assumed to be a precursor, was not incorporated into the *Erythrina* alkaloids. However, it showed that [1-^{13}C]-labeled (*S*)-norreticuline (**19**) was metabolized to give erythraline (**28**) with selective incorporation of ^{13}C into C-10. This led to the proposal of another mechanism for *Erythrina* alkaloid biosynthesis where phenolic coupling of (*S*)-norreticuline (**19**) is implicated. An old study [48] indicated that morphinandienone and norisosalutaridine (**20**) are incorporated, generating the *Erythrina* tetracyclic ring system **27** and its subsequent transformation to erythraline (**28**) (Scheme 2), lending support to Barton's proposal [43] (Scheme 1).

5. Occurrence and Isolation

Reports identifying new erythrinane alkaloids in reviews and during the last 15 years rare and have been framed in the synthesis of these compounds. Therefore, in this revision we centre on reports of new alkaloids from the year 2000 onwards.

The methanol extract of *E. herbacea* flowers was subjected to the conventional extraction and isolation procedures for alkaloids. A new *Erythrina* alkaloid, 10-hydroxy-11-oxoerysotrine (**29**), was isolated together with five known compounds. The structure of the new compound was elucidated on the basis of its spectral data, including 2-D NMR and mass spectra (MS). The new compound is a rare C-10 ox ygenated *Erythrina* alkaloid [49].

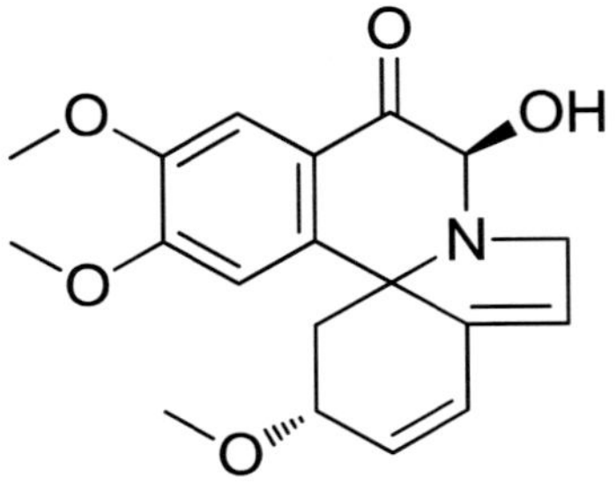

29

From the seeds of *E. velutina* were identified four new erythrinan alkaloids called sodium (2*R*)-2,11-dimethoxy-2,6,8,9-tetrahydro-1*H*-indolo[7a,1-*a*]isoquinolin-12-yl sulfate (**30**), sodium (2*R*)-11-hydroxy-2-methoxy-2,6,8,9-tetrahydro-1*H*-indolo[7a,1-*a*]isoquinolin-12-yl sulfate (**31**), (2*R*)-2,12-dimethoxy-11-(2*S*,3*R*,4*S*,6*R*)-3,4,5-trihydroxy-6-hydroxymethyl) tetrahydro-2*H*-pyran-2-yl)oxy)-8,9-dihydro-1*H*-indolo[7a,1-*a*]isoquinolin-6(2*H*)-one (**32**) and sodium (2*R*)-11-hydroxy-2-methoxy-7-oxido-1,2,6,7,8,9-hexahydroindolo[7a,1-*a*] isoquinolin-12-yl sulfate (**33**). The obtained fractions were further purified using a silica -gel column and silica-gel HPLC to afford four new compounds and the structures of these new were elucidated by spectroscopic methods including 2D-NMR. Three of the four were found to be novel sulfated erythrinan alkaloids [50].

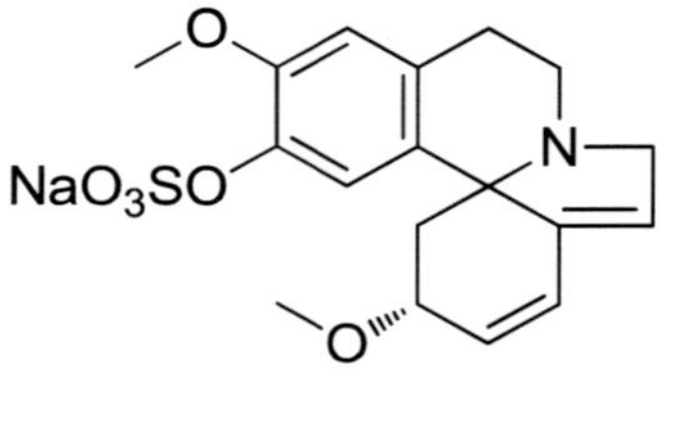

30

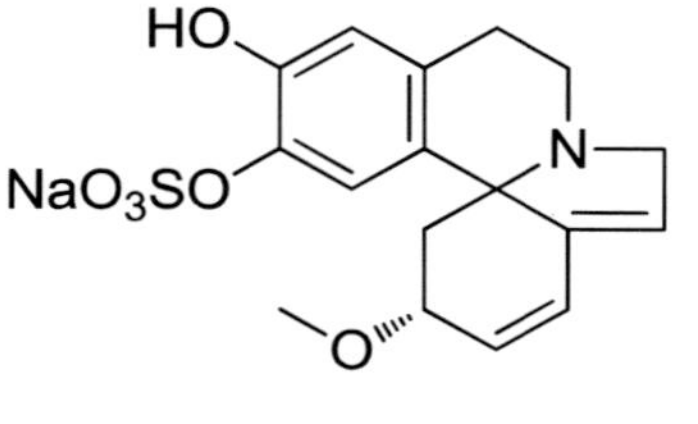

31

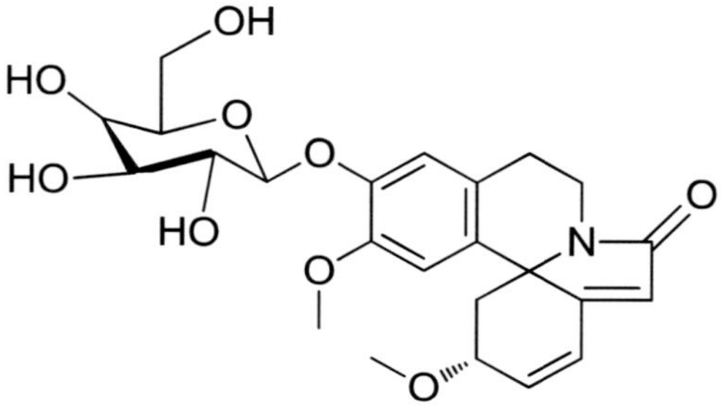

32

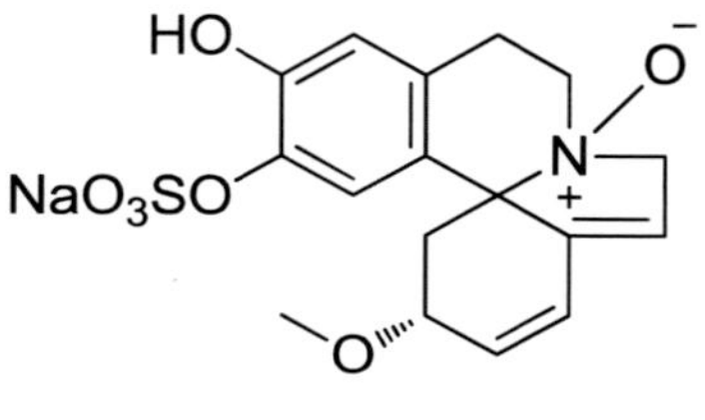

33

A novel alkaloid was isolated from the stem of *E. arborescens* Roxb. Its structure and stereochemistry were elucidated on the basis of the 1D and 2D homonuclear and heteronuclear NMR spectroscopic data, and it was characterized as 10-hydroxy-11-methoxy-benzo[g]hexahydroindolizin-3-one and named erythrinarbine [36].

The structures of two novel glycodienoid alkaloids isolated from *E. latissima* seeds have been established as (+)-16β-D-glucoerysopine (**34**) and (+)-15β-D-glucoerysopine (**35**). NOE experiments established that 1 and 2 are positional isomers. Their structures were elucidated from their 1D and 2D homonuclear and heteronuclear NMR spectroscopic data. Another seven known alkaloids were purified and identified [51].

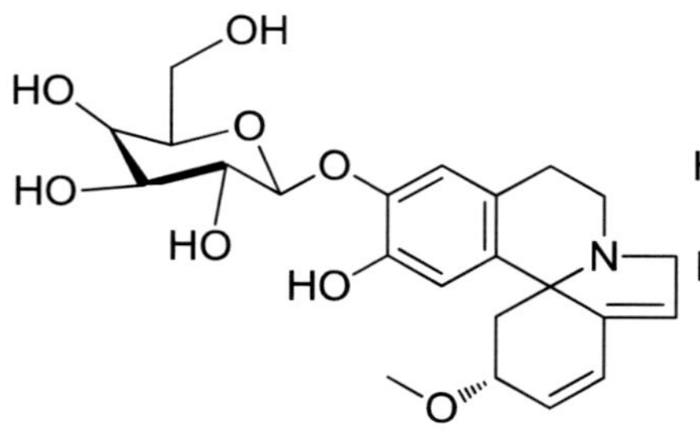

34

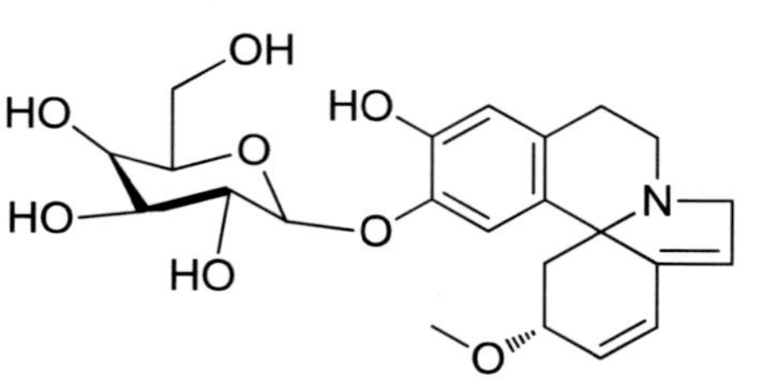

35

From the flowers and pods of *E. lysistemon* were isolated four alkaloids reported for the first time in nature, and five others found for the first time in this species, plus others that were re-isolated. The new compounds are (+)-11β-hydroxyerysotramidine (**36**), (+)-11β-methoxyerysotramidine (**37**), (+)-11β-hydroxyerysotrine N-oxide (**38**) and (+)-11β-hydroxyerysotrine (**39**) [52].

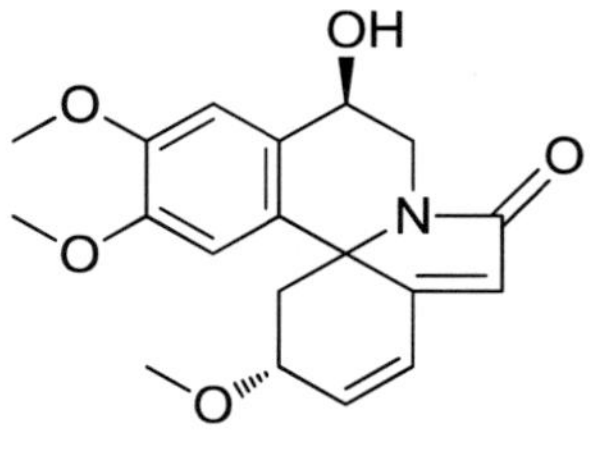

36

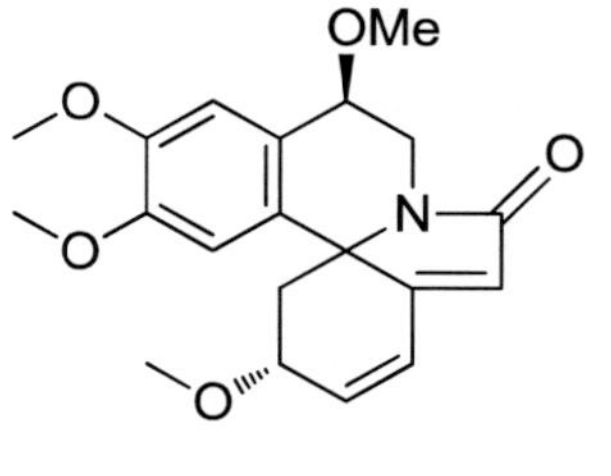

37

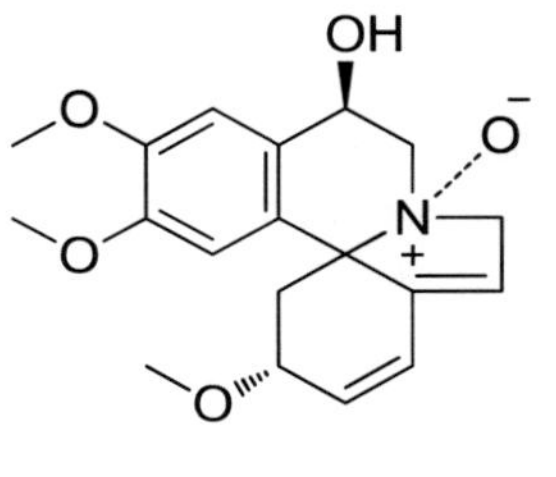

38

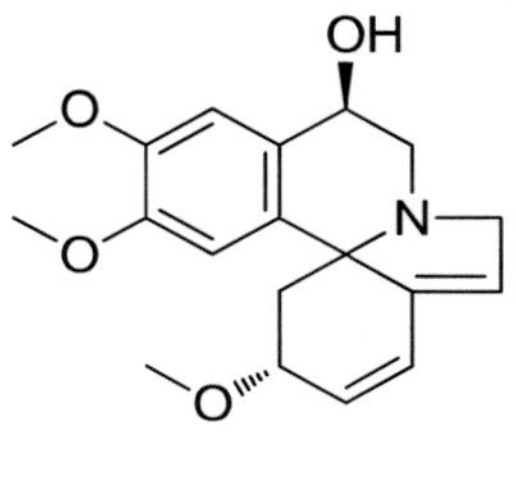

39

From the bark of *E. subumbrans* were isolated three new rythrina alkaloids, (+)-10,11-dioxoerythratine (**11**), (+)-10,11-dioxoerythratidinone (**12**) and (+)-10,11-dioxoepierythratidine (**40**). Other compounds of interest isolated were a new pterocarpan, 1-methoxyerythrabyssin II, and seven known pterocarpans [53].

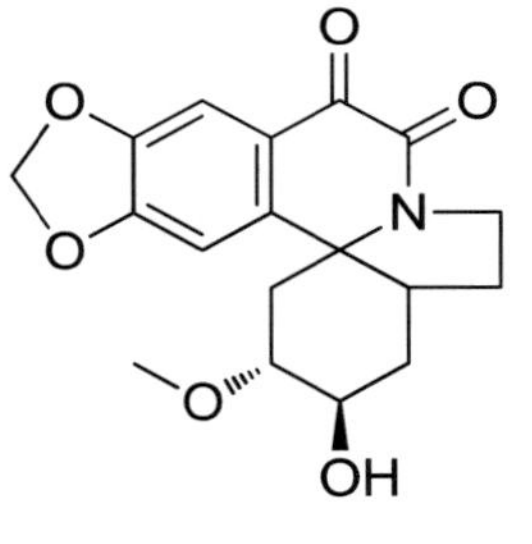

11

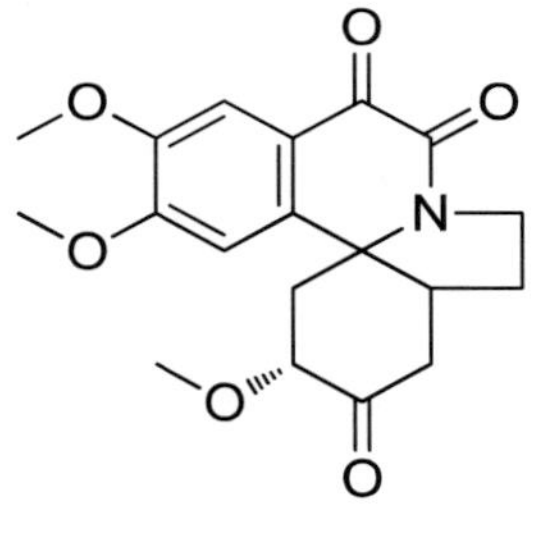

12

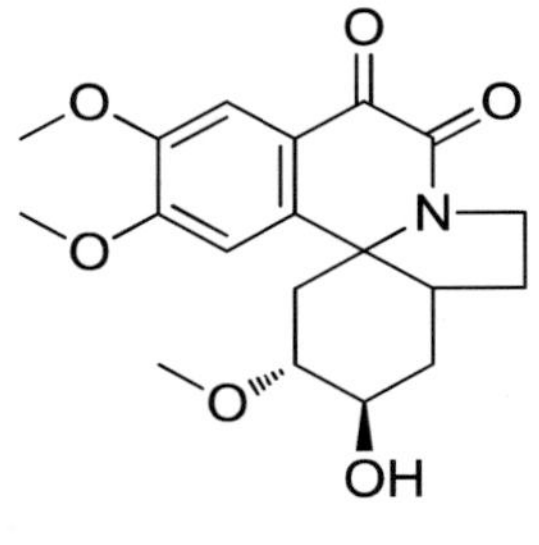

40

A new phytochemical study of the alkaloidal $CHCl_3$ and n-BuOH extracts of the bark of *E. cristagalli* L. yielded two new erythrinan alkaloids, cristanines A (**41**) and B (**42**), together with nine known erythrinan alkaloids and an indole alkaloid. The molecular formulas of cristanine A (**41**) and B (**42**) were determined by high resolution electron ionization mass spectrometry (HR-EI-MS) and the IR spectrum implied the presence of an aromatic moiety and a conjugated olefin, as well as ^{1}H- and ^{13}C-NMR and ^{1}H-detected heteronuclear single quantum coherence (HSQC) [54].

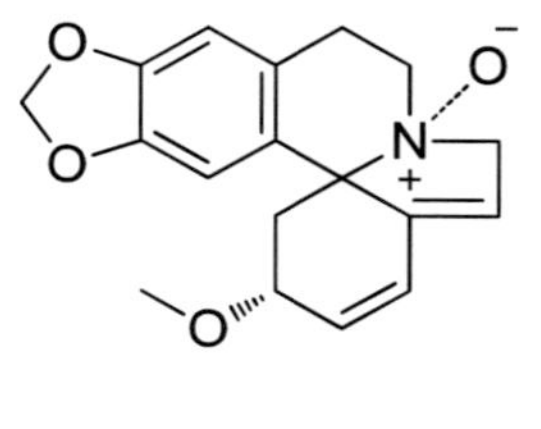

41

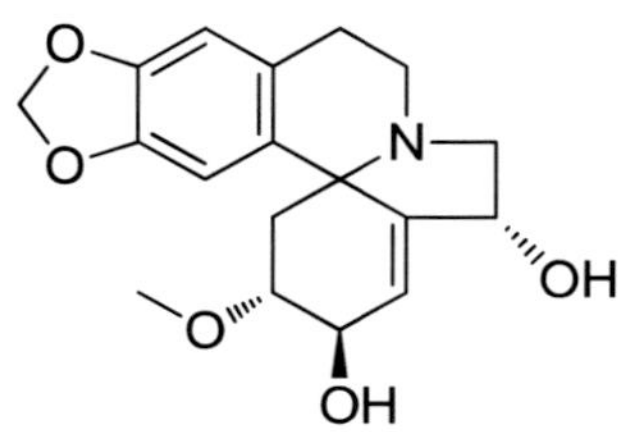

42

6. SYNTHESIS

The synthetic approaches to the aromatic D-ring *Erythrina* alkaloids is organized according to the ring formed in the last synthetic step (Scheme 3). Final formation of the C ring is the most common strategy, and our discussion is separated into two subsections, based on whether the formation of the C ring is performed under acid conditions or not. For the synthesis of the B and C rings a multicomponent reaction is typically used. Considering that three of the rings, A, B and C, can be joined in a final step, two synthetic strategies have been developed: the formation of one of the mentioned rings (route A, B, or C) or the simultaneous or sequential formation of more than one ring, for example A/B, A/C, B/C or A/B/C in one step (called routes A/B, B/C, etc., Schemes 3 and 4). According to this two different routes can be generated for the simultaneous construction of the B and C rings (= route B/C (a) or B/C (b) [55]

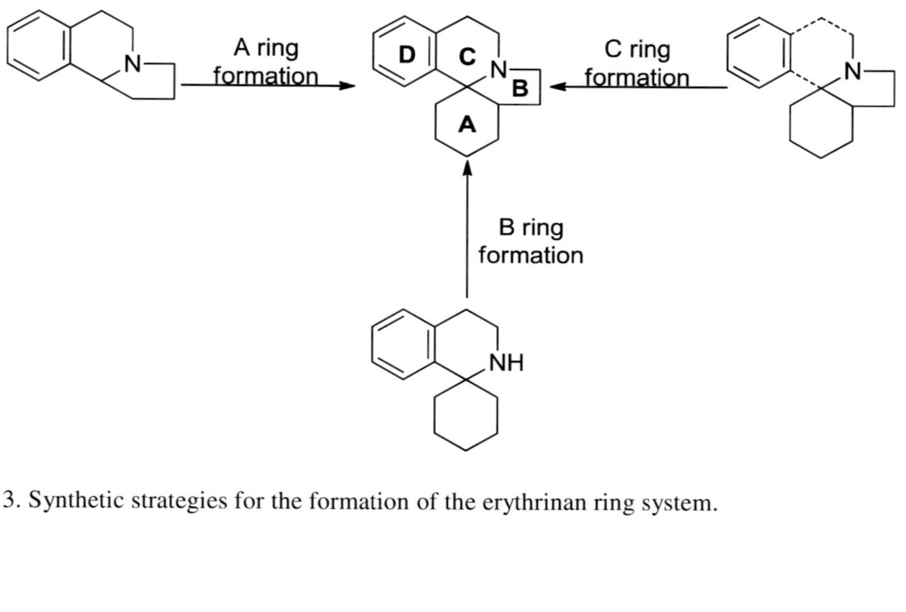

Scheme 3. Synthetic strategies for the formation of the erythrinan ring system.

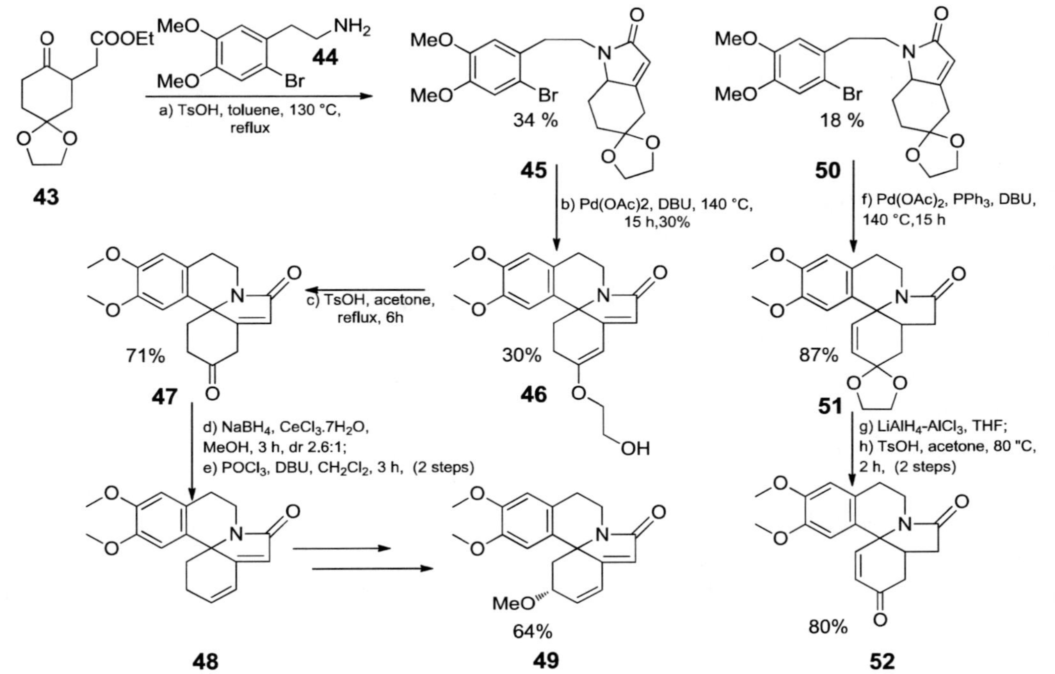

Scheme 4. Synthetic strategies for erysotramidine and iso-13-demethoxyerythratidinone.

For the synthetic application of this method to *Erythrina* alkaloids, Anh Tuan and Kim [56] reinvestigated the palladium catalyzed cyclization of the requisite precursor which could be formed from the condensation of ketoester **43** and bromo-arylamine **44**. In the reaction, two main products were formed. When the intermediate **45** was treated with $Pd(OAc)_2$ in DBU, an enol intermediate (**46**) was formed in 30% yield through γ-lactam enolate formation followed by cyclization. Some amount of the corresponding ketal compound, the precursor of the enol intermediate, could be obtained if the reaction process was quenched earlier. Treatment of **46** with TsOH in acetone afforded compound **47** in 71% yield. Reduction of the carbonyl compound **47** under Luche condition, affording a mixture of diasteromers in a 2.6: 1 ratio, was followed by elimination to afford the known intermediate (**48**) for erysotramidine (**49**) in 64% yield.

On the other hand, when intermediate **50** was subjected to the conventional Heck reaction, it gave the 6-membered quaternary structure **51** in 87% yield rather than a 7-membered ring compound. Reduction of the amide group of **51** by LAH/$AlCl_3$ to amine was followed by deprotection of the ketal by acid to iso-13-demethoxyerythratidinone (**52**) in 80% yield in two steps (Scheme 4).

8-Oxoerymelanthine is one of the two *Erythrina* alkaloids containing a pyridine ring isolated from nature. The characteristic 6-5-6-6-membered ring system was constructed using a method previously developed for the synthesis of aromatic erythrinan alkaloids. Arylethylamine **54**, synthesized from 2,3,4-trimethoxybenzaldehyde **53** was condensed with methyl malonyl chloride under Schotten-Baumann conditions and then the Bischler-Napieralski reaction of the intermediary using polyphosphate ester produced the C-ring closure product (**55**), which was then treated with oxalyl chloride to give the desired dioxopyrroline (**56**). The A-ring possessing the desired stereochemistry was constructed using the Diels-Alder reaction. A solution of **56** and 1-methoxy-3-trimethylsilyloxy-butadiene reacted to afford tetracyclic ring products. The crude products, without purification, were reduced with $LiBH_4$ and were then treated with HCl to afford compound **57**. This was followed by the chemo- and regioselective ozonolytic cleavage of the o-dimethoxybenzene system to the corresponding muconate (**58**) and was satisfactorily controlled by addition of BF_3-etherate. Treatment of **58** in acetic acid in a sealed tube provided a ring-closure product, which was then treated with diazomethane. Then the intermediate was treated with $MgCl_2$ and KI at 110 °C, followed by elimination of mesylate, to produce dienone **59**. This compound was reduced regio- and stereoselectively using $NaBH_4$ in the presence of $CeCl_3$ to afford the alcohol intermedite. The latter was converted to the corresponding methyl ether (**60**) by treatment with diazomethane in the presence of neutral silica gel. Aminolysis was achieved using *p*-methoxybenzylamine in methanol, whereby both the pyrone and the methyl ester were transformed to the corresponding pyridone and *p*-methoxybenzyl amide (**61**). Then the compound was heated with TFA at 100 °C for 4 h in a sealed tube to give the deprotected intermediate which was then converted to the corresponding methyl ester (**62**) by refluxing in methanol in the presence of TMSCl. To complete the total synthesis of 8-oxoerymelanthine (**63**) [57], several more stages were required (Scheme 5).

The erythrinan skeleton has been obtained starting from simple furans. For this purpose furyl aldehyde (**64**) was used as substrate. In the first part of the one-pot was condensed with amine and furyl aldehyde to furnish the enamide intermediary which cyclized spontaneously to afford the fused bicycle and being obtained of compound **65**. This was then treated *in situ* with Lewis acids, wherewith the reaction immediately took on the desired profile. Using BF_3•OEt_2 the reactants were stirred for 16 h to obtain the tetracyclic compound (3*R*,13b*S*)-3-hydroxy-11,12-dimethoxy-3,4,4a,5,8,9-hexahydro-1*H*-indolo[7a,1-*a*]isoquinolin-6(2*H*)-one (**66**) in 57% yield. In addition it was found that $AlCl_3$ (4 equiv) was also a successful mediator of the desired reaction, but affording (*S*)-11,12-dimethoxy-2,3,8,9-tetrahydro-1*H*-indolo[7a,1-*a*]isoquinolin-6(5*H*)-one (**67**), as shown in scheme 6 [58].

The synthesis of compound **77** has as starting reagents phenol **68** and the necessary amine, which were joined through an amide linkage to produce **69** in 84% yield using a reaction promoted by an aluminum salt. Compound **69** was treated in methanol with diacetoxyiodobenzene (DIB) to induce the first oxidative dearomatization, affording dienone **70** in 62% yield. **70** was treated with TMS-OTf followed by BF_3•Et_2O which led directly to indolinone **72**. The transformation of **70** into **72** represents a new two-step route to indolinone derivatives. In a second phase, aromatic activation mediated by bis(trifluoroacetoxy)-iodobenzene (PIFA) led to the desired dienone (**73**). The main tetracyclic core (**74**) was obtained under acidic conditions in 71% yield. The transformation resembled a Pictet−Spengler cyclization, which is a popular method of producing the aza-spiran moiety of the *Erythrina* alkaloids [59]. The dienoid system appearing in the structure of erythrinans was first obtained bytreatment of compound **74** with KHMDS, leading to the flattened structure **75** through an E1cB mechanism in 82% yield. The synthesis of erysotramidine (**76**) culminated with a mild and quantitative methylation in the presence of iodomethane in 86% yield [60].

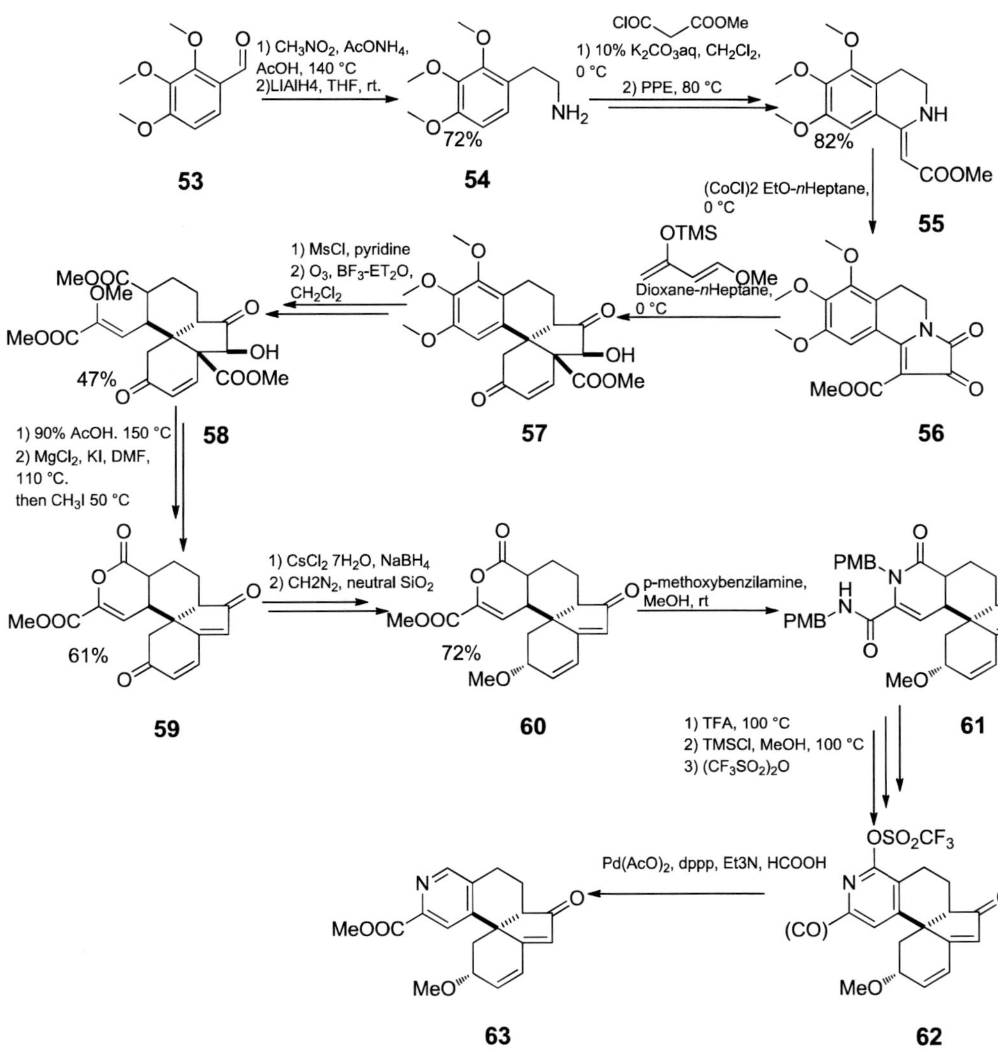

Scheme 5. Synthetic strategy for obtaining 8-oxoerymelanthine (63).

Treatment of 4-aminophenol (**77**) with 2-(3,4-dimethoxyphenyl)acetyl chloride (**78**) gave amide **79**, converted to amine **80** by a reduction with lithium aluminium hydride. By treatment of **80** with ethyl 3-chloro-3-oxopropanoate, ethyl 3-((4-hydroxyphenethyl)(4-hydroxyphenyl)amino)-3-oxopropanoate (**81**) was obtained in 50% yield.

An optimized ''one pot'' procedure was established, and the *Erythrina* -type compound **82** was obtained in 82% isolated yield after an oxidative carbon –carbon coupling (PIFA) followed by a Michael addition (K_2CO_3) in trifluoroethanol.

Compound **82** was further manipulated *via* hydrogenation followed by decarboxylation, to afford a known intermediate for the synthesis of natural *Erythrina* alkaloids (**86**) [61].

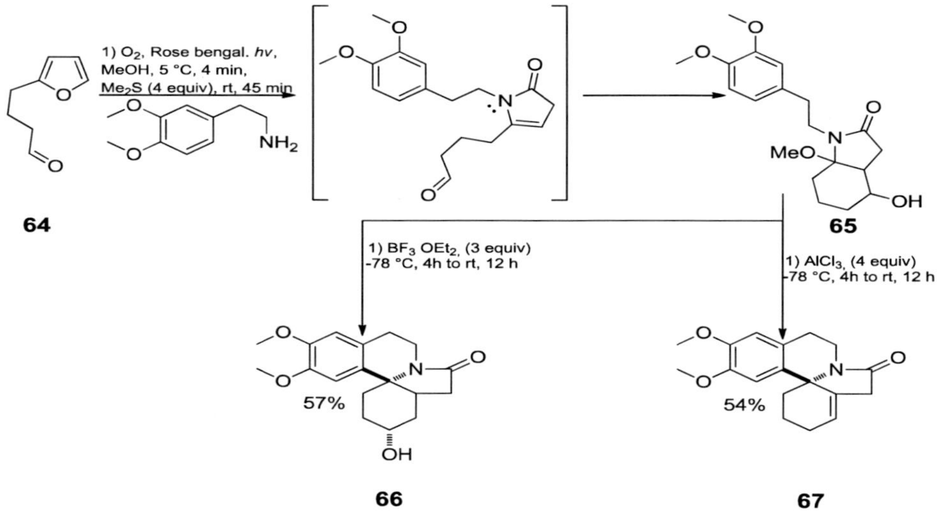

Scheme 6. Synthetic strategies for (3*R*,13b*S*)-3-hydroxy-11,12-dimethoxy-3,4,4a,5,8,9-hexahydro-1*H*-indolo[7a,1-*a*]isoquinolin-6(2*H*)-one and (*S*)-11,12-dimethoxy-2,3,8,9-tetrahydro-1*H*-indolo[7a,1-*a*]isoquinolin-6(5*H*)-one.

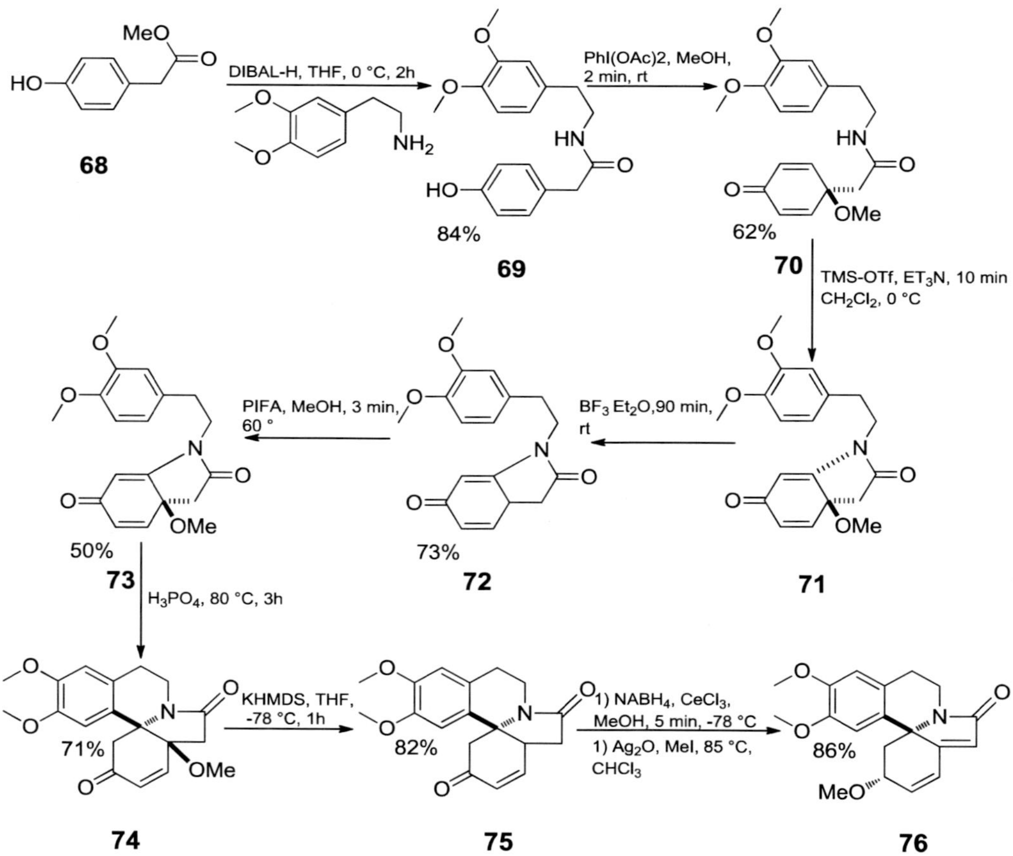

Scheme 7. Synthetic strategy for obtaining erysotramidine.

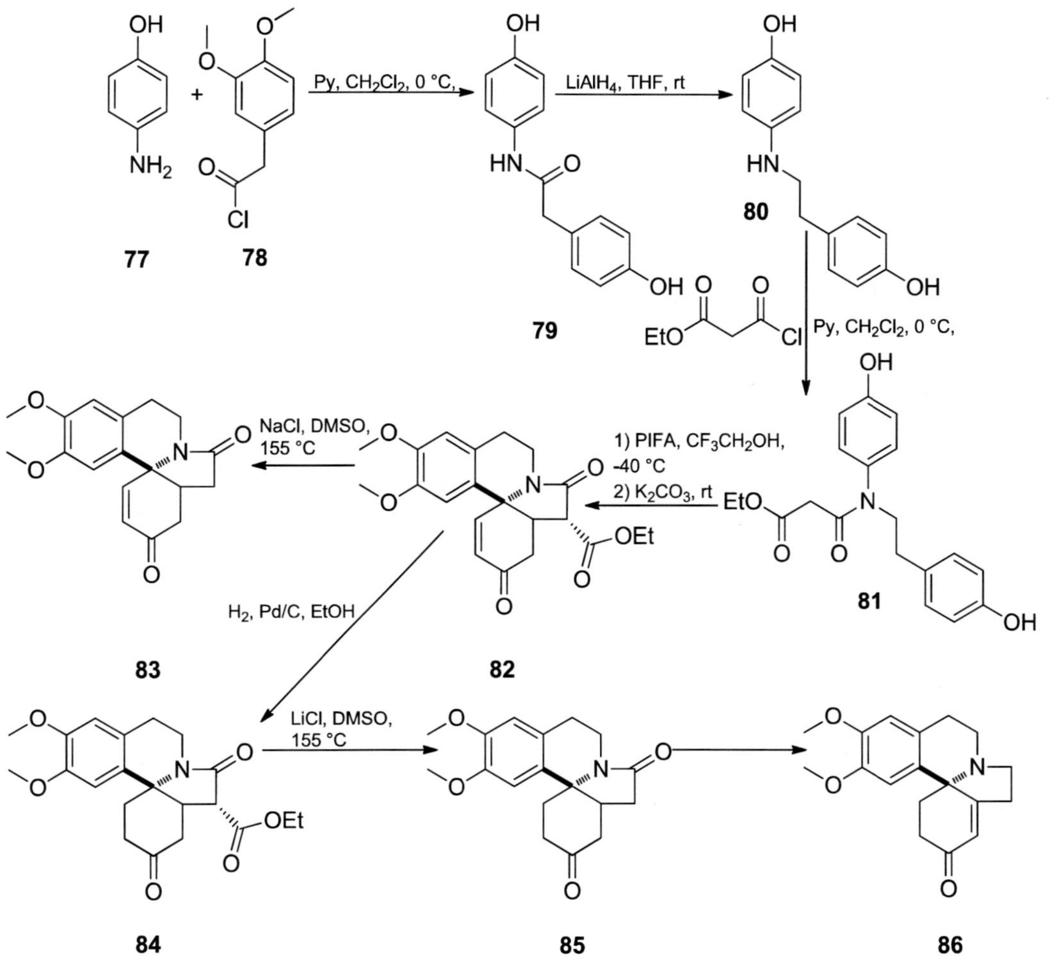

Scheme 8. Synthetic strategy for obtaining *Erythrina* alkaloid 86.

For the synthesis of compound **93** ((±)-3-demethoxyerythratidinone), iodoisoquinoline **87** and *N*-methoxy-*N*-methyl-3-(2-methyl-1,3-dioxolan-2-yl)propanamide (**88**) are used as starting reagents. These precursors, in three steps, give the benzindolizinone **91**, *via* ketone **89** which, treated with ethynylmagnesium bromide, afforded the propargylic tertiary alcohol **90** and followed was cycloisomerized to **91**. Various reduction protocols under carefully controlled conditions, could be effected with deprotection of the ketone group in a one-pot process to give **92**, who subsequently was then converted into (±)-3-demethoxyerythratidinone (**93**) [62] by a base -mediated aldol condensation as previously reported by Simpkins and co-workers [63] (Scheme 9).

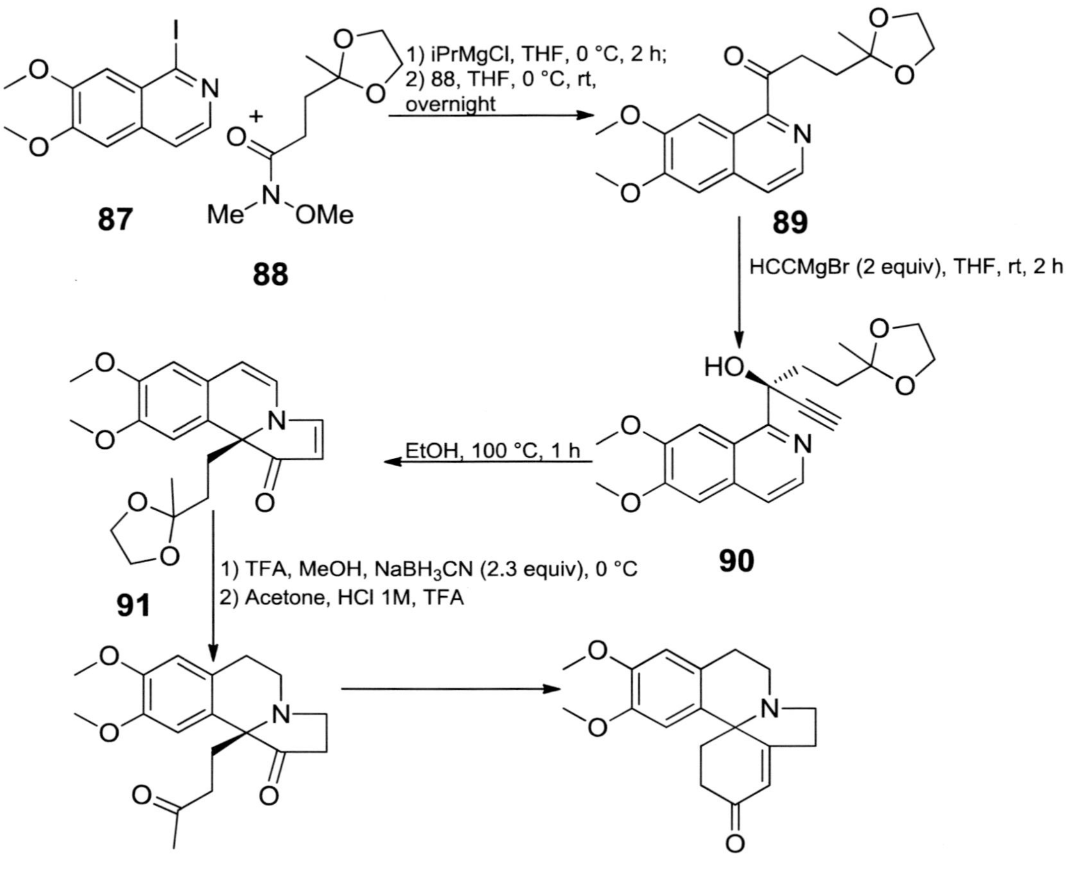

Scheme 9. Synthetic strategy for obtaining the *Erythrina* alkaloid 93.

The enantioselective synthesis of pyrrolinoisoquinolines as precursors of *Erythrina* alkaloids has been described. Derivative from **94** was reduced with $LiBH_4$, bis-diol lactam **95** being obtained in 95% yield. The secondary hydroxyl group of **95** was readily protected in 83% yield via 3 steps; tert-butyldiphenylsilyl (TBDPS) protection, methoxymethyl (MOM) protection, and deprotection of the TBDPS group giving rise to compound **96**. Subsequently, oxidation with tetrapropylammonium perruthenate (TPAP) resulted in the desired aldehyde gave compound **97** in 85% yield. Then, the introduction of the C-2 subunit was achieved *via* Wittig reaction with $Ph_3PCH(OCH_3)CH_3Cl$=nBuLi, which provided **98**. Thereafter, the crude lactam (**98**) was reduced with $LiAlH_4$ followed by removal of the MOM group with AcCl-MeOH and double TPAP oxidation to provide the final product (**99**) in 31% yield (Scheme 10) [64]. This compound was easily converted to (±)-3-demethoxyerythratidinone (**100**) by an intramolecular aldol cyclization, as reported in the literature [65,66].

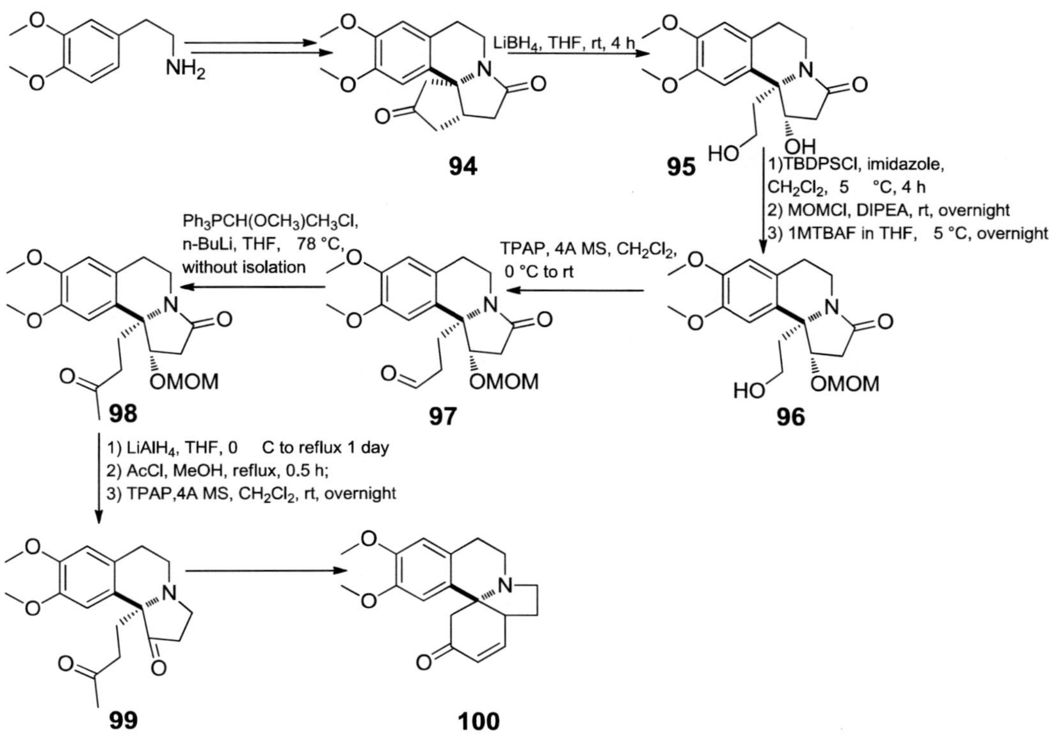

Scheme 10. Synthetic strategy for obtaining (±)-3-demethoxyerythratidinone.

7. PHARMACOLOGY

In recent years the number of pharmacological applications has been increasing. However, the most important activities are related to the CNS. Amongst these investigations the molecular determinants for competitive inhibition of α4β2 nicotinic acetylcholine receptors by dihydro-β-erythroidine (DHβE, **101**) were revealed. We studied the effects of DHβE and a range of aromatic *Erythrina* alkaloids on [^{3}H]cytisine binding and receptor function in conjunction with homology models of the α4β2 nAChR, mutagenesis and functional assays [6,67]. Erysodine (**102**) is also a competitive reversible antagonist of nicotine and the induced release of dopamine [67].

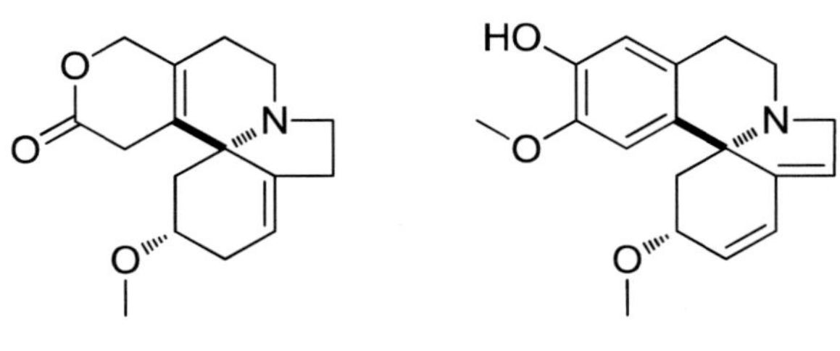

101 **102**

The crystal structure of AChBP from the water snail *Lymnaea stagnalis* (Ls) was determined recently, in complex with DHβE. The structure reveals that binding of DHβE to this AChBP imposes closure of the C-loop as is the case with agonists, but also a shift perpendicular to previously observed C-loop movements. These observations suggest that DHβE may antagonize the receptor *via* a different mechanism compared to prototypical antagonists and toxins [68].

The alkaloid erythraline (**103**), isolated from several *Erythrina* species which are used as Brazilian medicine plants for the treatment of inflammation, suppressed nitric oxide production and induction of inducible nitric oxide synthase expression in RAW264.7 cells stimulated by lipopolysaccharide. Erythraline also inhibits phosphorylation of mitogen - activated protein kinases. These results suggested that erythraline might inhibit the kinase activity of TAK1 [69].

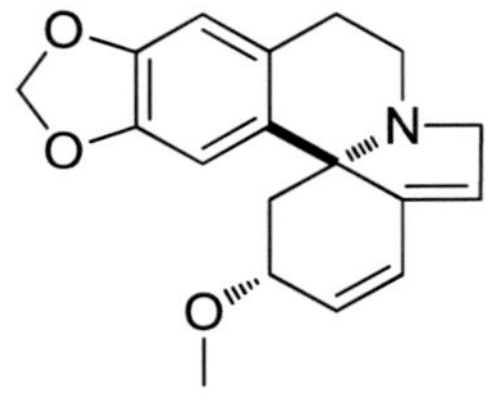

103

The estrogenicity of four erythroidine alkaloids isolated from *E. poeppigiana*, α-erythroidine (**104**), β-erythroidine (**101**), 8-oxo-α-erythroidine (**105**) and 8-oxo-β-erythroidine (**1**). was assayed in various estrogen receptor-(ER)-dependent test systems, including receptor binding affinity, cell culture based ER-dependent reporter gene assays, α-Erythroidine and β-erythroidine showed binding affinity values for ERα, whereas only β-erythroidine bound to ERβ. Both erythroidines exhibited a significant dose-dependent estrogenic stimulation of ER-dependent reporter gene activity in osteosarcoma cells detectable already at 10 nM. [70].

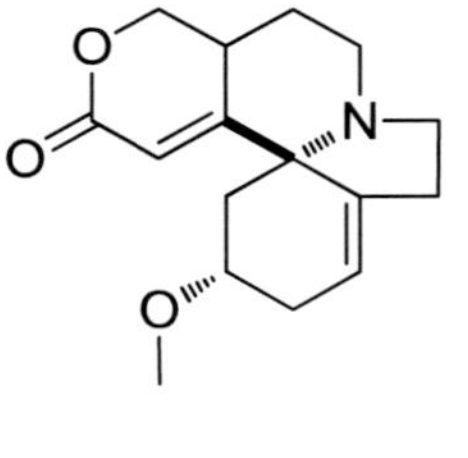

104

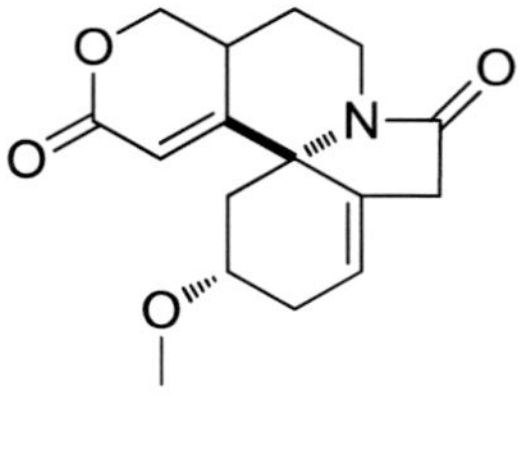

105

From *E. suberosa* were isolated two alkaloids, erysodine (**102**) and erysothrine (**106**). These compounds were investigated for anxiolytic properties. Both alkaloids produced anxiolytic-like effects in mice exposed to a two widely used anxiety tests – the elevated plus-maze and the light -dark transition model. In the elevated plus-maze, only erysodine increased the percentage of open arm entries and tended to increase the percentage of time spent in the open arms, the two conventional measures of anxiety in the elevated plus-maze [71].

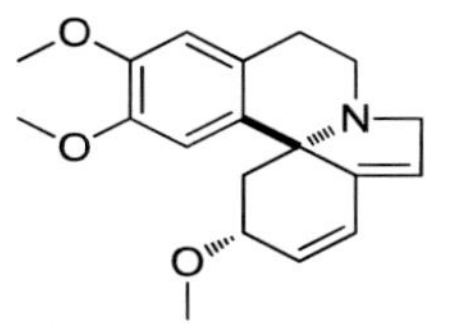

106

From crude extracts of *E. mulungu* were isolated three alkaloids that showed anxiolytic effects in different models [72]. The authors investigated whether these alkaloids could affect nicotinic acetylcholine receptors and if they are selective for different central nervous system subtypes. Screening experiments were performed using a single concentration of the alkaloid co-applied with acetylcholine in whole cell patch-clamp recordings in three different cell models expressing α3 nAChR, α7 nAChR and α4β2 nAChR. For all three receptors, the percent inhibition of acetylcholine-activated currents by (+)-11α-hydroxyerysotrine (**107**) was the lowest, whereas (+)-erythravine (**108**) and (+)-11α-hydroxyerythravine (**109**) inhibited the currents to a greater extent.. These alkaloids may exert their behavioral effects through inhibition of CNS nAChR, particularly the α4β2 subtype.

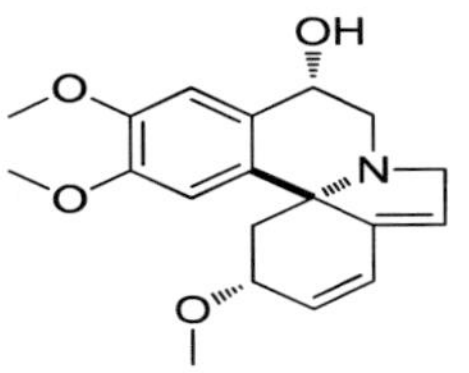

107

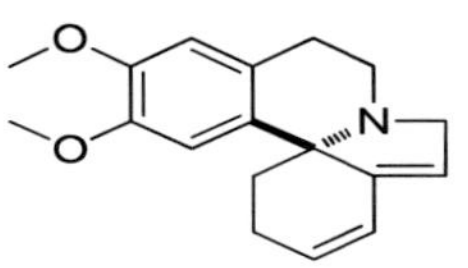

108

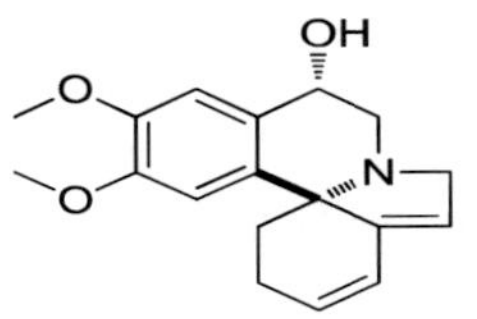

109

REFERENCES

[1] Neill, D. A. Experimental studies on species relationships in *Erythrina* (Leguminosae: Papilionoideae). *Ann. Missouri Bot. Gard.*, 1988, 75: 886-969.

[2] Tanaka, H.; Tanaka, T.; Etoh, H.; Goto, S.; Terada, Y. *Two new erythrinan alkaloids* from *Erythrina* x *bidwillii*. *Heterocycles*, 1999, 51, 2759-2764.

[3] Dyke, S. F.; Quessy, S.N. *The Alkaloids,* Vol. 18; Rodrigo, R.G.A. Academic Press: New York, 1981.

[4] Barrera, N. Etnobotánica y domesticación del Chachafruto *E. edulis* en la zona andina de Colombia. Tesis Doctoral. Universidad Politécnica de Madrid. Escuela Técnica Superior de Ingenieros Agrónomos. 2003.

[5] Sulbarán, G.; Fanky L. Usos de especies del genero Erythrinas. *Memorias IV Congreso Forestal Venezolano*. 2004, 1-5.

[6] Iturriaga-Vásquez, P.; Carbone, A.; García-Beltrán, O.; Livingstone, P. D.; Biggin, P. C; Cassels, B. K.; Wonnacott, S.; Zapata-Torres, G.; Bermudez, I. Molecular determinants for competitive inhibition of a4ß2 nicotinic acetylcholine receptors. *Mol. Pharm.,* 2010, 78, 366-375.

[7] Cornelius, W. W.; Akeng'a, T.; Obiero, G. O.; Lutta, K. P. activities Antifeedant of the Erythrinaline alkaloids from *Erythrina latissima* against *Spodoptera littoralis* (Lepidoptera: Noctuidae). *Rec. Nat. Prod.,* 2009, 32, 96-103.

[8] Padwa, A.; Hennig, R.; Kappe, C.O.; Reger, T. S. A triple cascade sequence as a strategy for the construction of the Erythrinane skeleton. *J. Org. Chem.*, 1998, 63, 1144-1155.

[9] Rigby, J.H.; Deur, C.; Heeg, M. J. Synthetic studies on the *Erythrina* alkaloids. Preparation of (±)-2-epi-erythrinitol. *Tetrahedron Lett.,* 1999, 40, 6887-6890.

[10] Suda, Y. T.; Hosoi, S.; Ishida, K.; Sangai, M. Chiral Synthesis of *Erythrina* Alkaloids. Synthesis of enantio-type Erythrinan alkaloids utilizing asymmetric acylation and kinetic resolution of diastereomers. *Chem. Pharm. Bull.,* 1994, 42, 204-213.

[11] Lete, E.; Egiarte, A.; Sotomayor, N.; Vicente, T.; Villa, M.-J. Synthesis of pyrroloisoquinolones via organolithium additions to N-Phenethylsuccinimides. *Synlett.,* 1993, 41-42.

[12] Manteca, I.; Sotomayor, N.; Villa, M.-J.; Lete, E. Tandem carbophilic addition-N-acyliminium ion cyclization for the synthesis of functionalized pyrrolo[2,1-a]isoquinolones: Key intermediates for the preparation of *Erythrina* -type alkaloids. *Tetrahedron Lett.*, 1996, 37, 7841-7844.

[13] Lee, Y.S.; Kang, D.W.; Lee, S.J.; Park, H. asymmetric synthesis of both enantiomers of pyrrolidinoisoquinoline derivatives from L-Malic Acid and L-Tartaric Acid. *J. Org. Chem.*, 1995, 60, 7149-7152.

[14] Katritzky, A.R.; Mehta, S.; He, H.-Y. Syntheses of Pyrrolo- and Indoloisoquinolinones by Intramolecular Cyclizations of 1-(2-Arylethyl)-5-benzotriazolylpyrrolidin-2-ones and 3-Benzotriazolyl-2-(2-arylethyl)-1-isoindolinones. *J. Org. Chem.* 2001, 66, 148-152.

[15] Garcia, E.; Arrasate, S.; Ardeo, A.; Lete, E.; Sotomayor, N. Enantioselective synthesis of pyrrolo[2,1-a]isoquinolones via stereocontrolled N-acyliminium ion cyclisations. *Tetrahedron Lett.,* 2001, 42, 1511-1513.

[16] Calle, J.; Pinzon R.; Ospina L.; Medina N.; Carrion A.; Bautista E. Alcaloides Isoquinolínicos de la corteza y flores de *Erythrina fusca*. *Rev. Col. Cienc. Quím. Farm.,* 1997, 26, 39-42.

[17] Flausino, O.; Pereira A. M.; da Silva, V.; Nunes, R. Anxiolytic Effects of Erythrinian Alkaloids from *Erythrina mulungu*. *J. Nat. Prod.,* 2007, 70, 48-53.

[18] Flausino, O.; de Avila Santos, L.; Verli, H.; Pereira, A. M.; da Silva, V.; Nunes, R. Effects of Erythrinian alkaloids isolated from *Erythrina mulungu* (Papilionaceae) in mice submitted to animal models of anxiety. *Biol. Pharm. Bull.*, 2007, 30, 375-378.

[19] García-Mateos, R.; Pérez-Pacheco, R.; Rodríguez-Hernández, C.; Soto-Hernández, M. Toxicidad de alcaloides de *Erythrina americana* en larvas de mosquito *Culex quinquefasciatus*. *Rev. Fitotec. Mex.*, 2004, 27, 297-303.

[20] Juma, B.; Majinda, R. Erythrinaline alkaloids from the flowers and pods of *Erythrina lysistemon* and their DPPH radical scavenging properties. *Phytochemistry,* 2004, 65, 1397-1404.

[21] Chacha, M.; Bojase-Moleta, G.; Majinda. R.R.T. Antimicrobial and radical scavenging flavonoids from the stem wood of *Erythrina latissima*. *Phytochemistry.,* 2005, 66, 99-104.

[22] Amer, M. E.; Shamma, M.; Freyer, A.J. The tetracyclic *Erythrina* alkaloids. *J. of Nat. Prod.*, 1991, 54, 329-363.

[23] Simpson, M. G. *Plant Systematics*, 2nd ed. Elsevier-Academic Press: Amsterdam, 2010.

[24] Sing, G. *Plant Systematics: An integrated aproach.* 3rd. ed. Science Publishers Inc.: New Hampshire, 2010.

[25] Raven, P. H. *Erythrina* (Fabaceae: Faboideae): Introduction to Symposium III. *Erythrina* Symposium III. *Ann. Missouri Bot. Gard.,* 1979, 66: 417-421.

[26] Gentry, A. H. *A field guide to the families and Genera of Woody Plants of Northwest South America (Colombia, Ecuador, Perú).* University of Chicago Press: Chicago, 1993.

[27] Buneau, A. Phylogenetic and biogeographical patterns in *Erythrina* (Leguminosae: Phaseoleae) as inferred from morphological and chloroplast DNA characters. *System. Bot.*, 1996, 21, 587-605.

[28] Granados-Tochoy, J.C.; Ruiz, L.K.; Forero, E. Sinopsis de las especies colombianas del género *Erythrina* (Leguminosae: Papilionoideae: Phaseoleae). in: Estudios en Leguminosas Colombianas; Forero, E.; Romero, C.; Academia Colombiana de Ciencias Exactas, Físicas y Naturales, Instituto Humboldt, Instituto de Ciencias Naturales y Red Latinoamericana de Botánica. Ed. Guadalupe Ltda: Bogotá, 2005.

[29] Bruneau, A. Evolution and Homology of Bird Pollination Syndromes in *Erythrina* (Leguminosae). *Am. J. Bot.,* 1997, 84, 54-71.

[30] Neill, D. The genus *Erythrina:* taxonomy, distribution and ecological differentiation. In: *Erythrina* in the New and Old Worlds; Westley, S. B.; Powell, M. H. *Nitrogen Fixing Tree Research Reports*, 1993, Special issue, 15-25.

[31] Baker, I.; Baker, H. G. Constituents of the Nectars of Two *Erythrina* Species and Their Hybrid. *Ann. Missouri Bot. Gard.,* 1979, 66, 446-450.

[32] Granados-Tochoy, J. C. El género *Erythrina* (Faboideae) en Colombia y su importancia económica; In: Libro de Resúmenes, Rangel-Ch., J. O.; Aguirre-C, J.; Andrade, G. M.; VIII Congreso Latinoamericano de Botánica y II Congreso Colombiano de Botánica. Instituto de Ciencias Naturales, Universidad Nacional de Colombia, Bogotá. 2002.

[33] Callejon, D.R.; Riul, T.B.; Feitosa, L.G.P.; Guaratini, T.; Silva, D. B.; Adhikari, A.; Shrestha, L. R.; Marques, L. M. M.; Baruffi, M. D.; Lopes, J. L. C.; Lopes, N. P. Leishmanicidal evaluation of Tetrahydroprotoberberine and spirocyclic *Erythrina* - alkaloids. *Molecules*, 2014, 19, 5692-5703.

[34] Amer, M. E.; Shamma, M.; Freyer, A. The tetracyclic *Erythrina* alkaloids. *J. Nat. Prod.,* 1991, 54, 329-363.

[35] Guaratini, T.; Silva, D.B.; Bizaro, A.C.; Sartori, L. R.; Humpf, H.-U. Lopes, N.P.; Costa-Lotufo, L.V.; Callegari-Lopes, J. L. In vitro metabolism studies of erythraline, the major spiroalkaloid from *Erythrina verna*. *J. Altern. Complement. Med.,* 2014, 14, 61-65.

[36] Yu, D. L.; GUO, J.; Xu, L.Z.; Yang, S. L. Erythrinarbine, a novel nor-A ring *Erythrina* alkaloid from *Erythrina arborescens*. *Chinese Chem. Lett.,* 1999, 10, 139-142.

[37] García-Beltrán, O.; Soto-Delgado, J.; Iturriaga-Vásquez, P.; Areche, C.; Cassels, B.K. Structural reassignment of Epierythratidine based on NMR studies and computational methods. *J. Chilean Chem. Soc.* 2012, 57, 1323-1327.

[38] Juma, B.; Majinda, R. Erythrinaline alkaloids from the flowers and pods of *Erythrina lysistemon* and their DPPH radical scavenging properties. *Phytochemistry,* 2004, 65, 1397-1404.

[39] Tanaka, H.; Tanaka, T.; Etoh, H.; Goto, S.; Terada, Y. *Two new erythrinan alkaloids* from *Erythrina* x *bidwillii. Heterocycles*, 1999, 51, 2759-2764.

[40] Barton, D.H.R; Cohen, T. Some biogenetic aspects of phenol oxidation. In: *Festschrift Artur Stoll*. Birkhäuser: Basel, 1957, 117-143.

[41] Barton, D.H.R.; Boar, R.B.; Widdowson, D.A. Phenol Oxidation and Biosynthesis Part XXI. The Biosynthesis of the *Erythrina* Alkaloids, *J. Chem. Soc. C.,* 1970, 1213-1218.

[42] Barton, D.H.R.; Potter, C.J.; Widdowson, D.A. Phenol oxidation and biosynthesis Part XXIII. On the benzyltetrahydroisoquinoline origins of the *Erytrina* alkaloids. *J. Chem. Soc. Perkin Trans.,* 1974, 1, 346-348.

[43] Barton, D.H.R.; James, R.; Kirby, G.W.; Turner, D.W.; Widdowson, D.A. Phenol oxidation and biosynthesis. Part XVIII. The structure and biosynthesis of *Erythrina* alkaloids. *J. Chem. Soc. C.*, 1968, 1529-1537.

[44] Maier, U.H.; Rodl, W.; Deus-Neumann, B.; Zenk, M.H. Biosynthesis of *Erythrina* alkaloids in *Erythrina crista-galli. Phytochemistry,* 1999, 52, 373-382.

[45] Zenk, M.H. Chasing the enzymes of secondary metabolism: plant cell culture s as a pot of gold. *Phytochemistry,* 1991, 30, 3861-3863.

[46] Maier, U.H.; Zenk, M.H. (S)-Norrreticuline is the precursor for the biosynthesis of *Erythrina* alkaloids. *Chem. Commun.*, 1997, 2313-2314.

[47] Zenk, M.H.; Juenger, M. Evolution and current status of the phytochemistry of nitrogenous compounds. *Phytochemistry,* 2007, 68, 2757-2772.

[48] Franck, B.; Teetz, V. Model reactions for the biosynthesis of *Erythrina* alkaloids. *Angew. Chem.,* 1971, 10, 411-412.

[49] Tanaka, H.; Hattori, H.; Tanaka, T.; Sakai, E.; Tanaka, N.; Kulkarni, A.; Etoh, H. A new *Erythrina* alkaloid from *Erythrina herbácea. J. Nat. Med.,* 2008, 62, 228-231.

[50] Ozawa, M.; Kishida, A.; Ohsaki. A. Erythrinan alkaloids from Seeds of *Erythrina velutina. Chem. Pharm. Bull.,* 2011, 59, 564-567.

[51] Cornelius C.; Wanjala, W.; Majinda, R.R.T. Two novel glucodienoid alkaloids from *Erythrina latissima* seeds. *J. Nat. Prod.,* 2000, 63, 871-873.

[52] Juma, B.F.; Majinda, R.R.T. Erythrinaline alkaloids from the flowers and pods of *Erythrina lysistemon* and their DPPH radical scavenging properties. *Phytochemistry,* 2004, 65, 1397-1404.

[53] Rukachaisirikul, T.; Innok, P.; Suksamrarn, A. *Erythrina* alkaloids and a pterocarpan from the bark of *Erythrina subumbrans. J. Nat. Prod.,* 2008, 71, 156-158.

[54] Ozawa, M.; Kawamata, S.; Etoh, T.; Hayashi, M.; Komiyama, K.; Kishida, A.; Kuroda, C.; Ohsaki, A. Structures of new Erythrinan alkaloids and nitric oxide production inhibitors from *Erythrina crista-galli. Chem. Pharm. Bull.* 2010, 58, 1119-1122.

[55] Kawasaki, T.; Onoda, N.; Watanabe, H.; Kitara, T. Total synthesis of (±) cocculolidine. *Tetrahedron Lett.,* 2001, 42, 8003.

[56] Tuan, L. A.; Kim, G. A short path to *Erythrina* alkaloid derivatives. *Bull. Korean Chem. Soc.,* 2010, 31, 1800-1802.

[57] Yoshida, Y.; Mohri, K.;, Isobe, K.; Itoh, T.; Yamamoto, K. Biomimetic Total Synthesis of (±)-8-Oxoerymelanthine. *J. Org. Chem.*, 2009, 74, 6010–6015.

[58] Kalaitzakis, D.; Montagnon, T.; Antonatou, E.; Vassilikogiannakis, G. One-pot Synthesis of the Tetracyclic Framework of the Aromatic Erythrina Alkaloids from Simple Furans. *Org. Lett.*, 2013, 15, 3714–3717.

[59] Maryanoff, B. E.; Zhang, H.-C.; Cohen, J. H.; Turchi, I. J.; Maryanoff, C. A. Cyclizations of N-Acyliminium Ions. *Chem. Rev.* 2004, 104, 1431-1628.

[60] L'Homme, Ch.; Ménard, M.; Canesi, S. Synthesis of the Erythrina Alkaloid Erysotramidine. *J. Org. Chem.*, 2014, 79, 8481−8485.

[61] Liang, J.; Chen, J.; Liu, J.; Li, L.; Zhang, H. Oxidative dearomatization in the synthesis of erythrina, oxindole and hexahydropyrrolo[2,3-b]indole skeletons, *Chem. Commun.*, 2010, 46, 3666–3668.

[62] Heller, S.T.; Kiho, T.; Narayan, A. R. H.; Sarpong, R. Protic-Solvent-Mediated Cycloisomerization of Quinoline and Isoquinoline Propargylic Alcohols: Syntheses of (±)-3-Demethoxyerythratidinone and (±)-Cocculidine. *Angew. Chem. Int. Ed.*, 2013, 52, 11129 –11133.

[63] Zhang, F.; Simpkins, N. S.; Blake, A. J. *Org. Biomol. Chem.* 2009, 7, 1963.

[64] Moon, J.T.; Jung, J.A.; Ha, S.H; Song, S.H; Park, S.J; Kim, J.; Choo, D.J.; Lee. Y.S.; Lee. J.Y. Synthetic Studies on Erythrina Alkaloids: Formal Total Synthesis of (+)-3-Demethoxyerythratidinone. Synthetic Communications, 2011, 41, 1282–1292.

[65] Wasserman, H. H.; Amici, R. M. The chemistry of vicinal tricarbonyls: A total synthesis of (±)-3-demethoxyerythratidinone. *J. Org. Chem.*, 1989, 54, 5843–5844.

[66] Zhang, F.; Simpkins, N. S.; Wilson, C. An enantiospecific synthesis of (±)-demethoxyerythratidinone from (S)-malic acid: Key observations concerning the diastereocontrol in malic acid–derived N-acyliminium ion cyclisations. *Tetrahedron Lett.*, 2007, 48, 5942–5947.

[67] Decker, M. W.; Anderson, D. J.; Brioni, J.D.; Donnelly-Roberts, D.L.; Kang, C.H.; O'Neill, A. B.; Piattoni-Kaplan, M.; Swanson, S.; Sullivan, J. P. Erysodine, a competitive antagonist at neuronal nicotinic acetylcholine receptors. Eur J Pharmacol 1995, 280, 79–89.

[68] Shahsavar, A.; Kastrup, J.S.; Nielsen, E. Ø.; Kristensen, J.L; Gajhede, M; Balle, T. Crystal Structure of Lymnaea stagnalis AChBP Complexed with the Potent nAChR Antagonist DHβE Suggests a Unique Mode of Antagonism. *PLOS ONE*, 2012, 7, e40757.

[69] Etoh, T.; Kim, Y.P.; Ohsaki, A.; Komiyama, K.; Hayashia, M. Inhibitory Effect of Erythraline on Toll-Like Receptor Signaling Pathway in RAW264.7 Cells. *Biol. Pharm. Bull.*, 2013, 36, 1363–1369.

[70] Djiogue, S.; Halabalaki, M.; Njamen, D.; Kretzschmar, G.; Lambrinidis, G.; Hoepping, J.; Raffaelli, F.M.; Mikros, E.; Skaltsounis, A-L.; Vollmer, G. Erythroidine Alkaloids: A Novel Class of Phytoestrogens. *Planta Med.*, 2014, 80, 861–869.

[71] Rodrigues Serrano, M.A.; de Luca Batista, A.N.; da Silva Bolzani, V.; de Ávila Santos, L.; de Campos Nogueira, P.J.; Nunes-de-Souza, R.L.; e Mohammad Arfan, A.L. Anxiolitic-Like Effects of Erythrinian Alkaloids from *Erythrina suberosa*. *Quim. Nova*, 2011, 34, 808-811.

[72] Pedro Setti-Perdigao, Maria A. R. Serrano, Otávio A. Flausino, Jr, Vanderlan S. Bolzani, Marília Z. P. Guimaraaes, Newton G. Castro. *Erythrina mulungu* Alkaloids Are Potent Inhibitors of Neuronal Nicotinic Receptor Currents in Mammalian Cells. *PLOS ONE*, 2013, 8, e82726.

In: Alkaloids
Editor: Eduardo Sobarzo-Sánchez
ISBN: 978-1-63482-074-5

Chapter 6

BIOTECHNOLOGICAL APPROACHES FOR THE IMPROVEMENT IN THE PRODUCTION OF TROPANE ALKALOIDS: IMPLICATIONS AND MEDICINAL APPLICATIONS

Alejandra B. Cardillo, Julián Rodríguez Talou and Ana María Giulietti
Microbiología Industrial y Biotecnología, Facultad de Farmacia y Bioquímica, Universidad de Buenos Aires, Buenos Aires, Argentina

ABSTRACT

Hyoscyamine, anisodamine (6β-hydroxyhyoscyamine) and scopolamine are the most important tropane alkaloids from a therapeutical point of view. These alkaloids are well known anticholinergic agents and were historically used in medicine.

Scopolamine is the more valuable alkaloid, having a 10 times higher commercial demand than that of hyoscyamine. Further therapeutical properties were described for anisodamine that includes the treatment of microvascular diseases, glomerulonephritis, rheumatoid arthritis, gastrointestinal colic, eclampsia as well as the control of toxic shock, septic shock, and organophosphorus poisoning. Moreover, anisodamine is not as toxic as atropine, having lesser negative effects on the Central Nervous System than scopolamine.

Since chemical synthesis of tropane alkaloids is difficult and expensive, these compounds are still extracted from plants belonging to several species of *Solanaceae* family including, *Hyoscyamuns niger* L., *Anisodus tanguticus*, *Scopolia tangutica* Maxim, *Atropa belladonna*, and several *Datura* species.

Obtaining these compounds by *in vitro* culture techniques is an interesting alternative since it guarantees a stable and uniform year round supply, independent of weather and soil conditions. Undifferentiated cultures fail to produce tropane alkaloids or produce them at a low level. In order to increase the production of these secondary metabolites several strategies have been assayed.

Some of them were the application of differentiated cultures such as hairy roots obtained by the transformation with *Agrobacterium rhizogenes*, media design, elicitation and the genetic engineering approach. However, the results obtained to date were not applicable for an industrial scale production.

In addition to the advantages mentioned before, plant organ cultures such as hairy roots were useful models in the study of secondary metabolic pathways. From these cultures it was possible the isolation of different key enzymes and its characterization.

Furthermore, the expression of plant genes in microorganisms represents an interesting strategy for the production of plant enzymes and/or the products derived from its activity. The production of the more valuable alkaloids by biocatalytic process es appears to be a promising strategy.

In this chapter we discuss the implications and the results of the different approaches for the improvement of tropane alkaloids production in order to replace the isolation from the natural source.

Keywords: tropane, anticholinergic agents

1. Introduction

The worldwide impact of the natural products isolated from medicinal plants is enormous [1]. It is calculated that a 25% of the prescribed medicines have bioactive compounds produced by plants [2-4]. It is also important to consider that herbal extracts or infusions are used in traditional medicine by 75-80% of the world's population.

In addition, these compounds can be used as models for drug development in the synthesis of improved medicines [5] and as food additive s, cosmetics, pigments and insecticides [6].

Several findings have demonstrated that the ancient human civilizations used plants as medicines [2]. Before the birth of writing, the knowledge on medicinal plants was transmitted from generation to generation, giving them a broad experience about the drugs which could be obtained from their environment [1]. Among the oldest written documents that mention the properties of medicinal plants we can mention the Sumerian tablets, the Pen Tsao, the Evers Papyrus, the Rig Vedá and *De Materia Medica* which is considered the precursor of the actual pharmacopeia [1, 7].

Regarding to the application and knowledge on medicinal plants in America, it is well known that the Pre-Hispanic Cultures (Mayas, Incas, Aztecas) had a wide experience about the properties of plants. *Libellus de medicinalibus indorum herbis*, known as the *Codice Badiano*, is considered the first scientific manuscript of the "New World". It gathered the information about the medicinal plants used by the inhabitants of Mexico [8].

The bioactive compounds isolated from plants and responsible for the medicinal properties mentioned here and different health -improving effects are named secondary metabolites [9]. It is generally believe that they do not play an essential role in plant growth and development. However, they participate in plant adaptation responses and defense functions. It is well known that these compounds play an important role in their survival in the ecosystem because they are involved in the resistance against pests, diseases and other functions such as attraction of pollinators and interactions with organisms of their environment [1, 6, 9].

Frequently, these compounds have complex structures which are difficult to produce chemically and in a cost effective process.

For this reason, many compounds of interest for the pharmaceutical industry are still isolated from the natural producer plants [1]. Even though, many of them are commercially available, they are frequently expensive due to their low abundance in the natural producer plants [9].

The non-rational and unplanned exploitation of the natural resources may result in the loss of important medicinal species. The non-rational exploitation of *Taxus brevifolia* is a clear example. This specie was threatened with the discovery of *paclitaxel* and it became necessary to found another renewable source of this metabolite [1]. The agronomical practices are an alternative to avoid the devastation of medicinal plants. Growing medicinal plants instead of collect them from nature has several advantages because it is possible to control the presence of pathogens, temperature, soil conditions and humidity, among others, in order to ensure the optimum growth and secondary metabolites production.

Tropane alkaloids can be synthesized chemically by 11 chemical steps [10], however, the chemical synthesis is complicated, costly and time consuming. For this reason, the isolation of tropane alkaloids from natural sources is still the strategy of choice and it is performed from plants belonging to several species of *Solanaceae* family [1, 6].

Biotechnological approaches are of increasing interest for the pharmaceutical production of medicinal compounds, including tropane alkaloids, in order to replace the isolation from the natural sources.

Plant *in vitro* culture techniques of undifferentiated and differentiated cultures are an interesting alternative since guarantees a stable and uniform year round supply, independent of weather and soil conditions.

Among the strategies used for increasing the production of medicinal compounds it can be mentioned the media design, elicitation and the genetic engineering approach.

Furthermore, the expression of plant genes in microorganisms represents an interesting strategy for the production of plant enzymes and/or the products derived from its activity. Regarding to tropane alkaloids, the production of the more valuable alkaloids by biocatalytic process es appears to be a promising strategy.

2. Tropane Alkaloids

Tropane alkaloids, including hyoscyamine, its race mic form atropine and scopolamine are anticholinergic agents that have been extensively used as pharmaceuticals [11-13]. These alkaloids were historically and widely used and they are among the oldest drugs applied in medicine [14]. Scopolamine is the more valuable alkaloid, having a 10 times higher commercial demand than that of hyoscyamine and atropine [13, 15-17].

In addition to the anticholinergic action, there were described further therapeutical properties for 6β-hydoxyhyoscyamine (also named anisodamine) that includes the treatment of microvascular diseases, glomerulonephritis, rheumatoid arthritis, gastrointestinal colic, eclampsia, hemorrhagic necrotic enteritis as well as the control of toxic shock, septic shock, and organophosphorus poisoning [18-22]. Also, 6β-hydoxyhyoscyamine is not as toxic as atropine having lesser negative effects on the Central Nervous System than scopolamine [18].

The biosynthetic pathway of tropane alkaloids is not fully understood [23]. In fact, just a few enzymes of the pathway have been isolated and very little is known about the regulation of the biosynthesis of these alkaloids [6].

Hyoscyamine 6β-hydroxylase (H6H, EC 1.14.11.11) is a 2-oxoglutarate-dependent dioxygenase and is the final enzyme of the tropane alkaloid biosynthetic pathway. The H6H catalyzes the conversion of hyoscyamine into 6β-hydoxyhyoscyamine and scopolamine by two sequential reactions [24, 25]. Firstly, it catalyses the 6β-hydroxylation of hyoscyamine and secondly, the formation of the epoxy group leading to scopolamine [13]. Several works have revealed that the epoxidation activity is much lower than the hydroxylation one [11, 26]. These findings are in agreement to those reported in our laboratory about the H6H enzyme isolated from *Brugmansia candida* (syn. *Datura candida)* [27-29]. This South America n native plant is a natural hybrid between *B. aurea* and *B. versicolor* and a tropane alkaloids producer [30, 31].

Putrescine N-methyltransferase (PMT, EC 2.1.1.53) is another key enzyme of the tropane alkaloid pathway and is involved in putrescine N-methylation leading to the first specific precursor of tropane alkaloids [32]. Several reports suggested that PMT is a flux-limiting of the tropane alkaloid pathway [32-35].

According to the well established use of these compounds in medicine, the demand for them is continuous [11-13, 36]. For this reason, much effort has been invested in the development of cost -effective strategies for tropane alkaloids production.

Wu et al. [10] reported the chemical synthesis of scopolamine by 11 chemical steps using 3α-hydroxy-6β-acetiltropine as precursor for the synthesis. Because of the time required for the chemical synthesis of tropane alkaloids and the complications and cost that it represents [10], nowadays, the industrial production of topane alkaloids is carried out from plants belonging to the *Solanaceae* family. Among them we can mention *Hyoscyamuns niger* L., *Anisodus tanguticus*, *Scopolia tangutica* Maxim, *Atropa belladonna*, and several *Datura* species [10, 19-22, 37-40].

It is known that tropane alkaloids are synthesized in root pericicle and traslocated to leaves [41]. For this reason, the investigations on scopolamine production were mainly focused in the hairy root cultures of several *Solanaceous* plants [31, 42-45]. However, the yields obtained to date were not suitable for a commercial production in order to replace the extraction from the natural producer plants [46].

3. TROPANE ALKALOIDS PRODUCTION

As we mentioned above, the isolation of tropane alkaloids is performed from the natural producer plants [1, 6]. Under these conditions, plants in the same stage of development are grown and distributed in a small area facilitating the operations of collection, harvest and even processing of the material [1]. It is worth to point out that the amount of the bioactive compounds is frequently not constant through the different stages of plant development. For this reason it is important to collect medicinal plants in the adequate stage in order to maximize the yield of the desired metabolites.

Extrinsic and intrinsic factors affect not only plant development but also the production of bioactive compounds. Extrinsic factors, such as, climate conditions should be as favorable as possible for the optimum development and metabolite production.

The leaves of the Australian native tree *Duboisia myoporoides* are the commercial source of the pharmacologically useful alkaloid scopolamine. It is known that this plant contains a 3% of hyoscyamine in October (spring) and this alkaloid is almost complete transformed into scopolamine by April (autumn) [1]. For this reason, it is important to identify the optimum harvest time in order to maximize the yield of the desired compound.

There are other species that produce tropane alkaloids, such as *Atropa*, which mainly accumulate hyoscyamine without reaching to the final product scopolamine [14, 26].

The advantage of cultivate medicinal plants rather than collect them from nature is that plant breeding strategies may be combined in order to improve secondary metabolites yield. This advantage is related to the fact that the limits in the production established by the genetic background (intrinsic factors) can not be overcome.

4 Biotechnological Strategies

4.1. *In Vitro* Cultures

Differentiated and undifferentiated plant *in vitro* cultures are an attractive source of secondary metabolites because guarantees that the process will be carried out under defined and controlled conditions and consequently preventing or reducing the variations in the product quality and the alkaloid yield [47].

The production of tropane alkaloids using plant cell culture s has always been an attractive and widely studied option as an alternative production system [46]. However, to date the yields obtained by *in vitro* cultures could not improve that obtained by the whole plant.

4.1.1. Undifferentiated Cultures

Cell suspension cultures are established from callus after plant cell dedifferentiation using growth regulators, such as auxins. The dedifferentiation process it is frequently accompanied by the loss of the ability to produce secondary metabolites. In addition, undifferentiated cells grows slowly (doubling times of 4-6 days) and it is frequently observed instability and genetic heterogeneity [48].

Tropane alkaloids production was analyzed in callus and cell suspension cultures of several plant species such as *Atropa, Datura, Duboisia, Hyoscyamus* and *Scopolia.* Hyoscyamine and scopolamine concentrations were lower in undifferentiated cultures than intact tissues cultures and even in some cases were not detected [49]. Various strategies have been used in order to optimize the production of tropane alkaloids in plant cell culture s. Hashimoto et al. [49, 50] analyzed the influence of different culture conditions on the production of hyoscyamine and scopolamine. However, because of the low productivity and probably the biochemical instability of the undifferentiated cell cultures it was not possible to achieve alkaloids concentrations comparable to that produced by the whole plant.

The biotechnological processes based on undifferentiated cell culture s were not successful according to the low productivity. For this reason, the investigations in order to improve 6β-hydroxyhyoscyamine and scopolamine yields are based mainly on hairy root culture systems [6, 43].

4.1.2. Differentiated Cultures – Hairy Roots

The use of undifferentiated cultures proved to be unsuccessful for tropane alkaloids production. These alkaloids are produced in normal and transformed roots [42, 51]. Several lines of evidence suggest that the differentiation of the tissue is necessary for the synthesis of these metabolites [42, 51]. Different studies suggest that this is probably related to the localization of key biosynthetic enzymes [6]. Among them, Suzuki et al. [52, 53] demonstrated that the *h6h* and *pmt* genes were expressed specifically in root pericycle of *Atropa belladonna* plants. In addition, Nakajima et al. [54, 55] pointed out that Tropinone reductase enzymes were accumulated in lateral roots of *Hyoscyamus niger*.

On the other hand, Rothe et al. [56] postulated that the low yield of tropane alkaloids found in undifferentiated cultures could be related to the hormones added to callus cultures that may inhibit key enzymes involved in alkaloids synthesis [6].

The application of hairy root culture technology is valuable for the production of compounds that are synthesized in roots, including the tropane alkaloids [4]. These cultures are characterized by its genetic and biochemical stability, rapid growth in hormone-free medium as well as the possibility of genetic manipulation [57]. It is generally believe that the secondary metabolites produced by hairy root cultures are the same as those synthesized by non-transformed roots and are produced with the same or even higher yields [6].

Nowadays, the application of *in vitro* cultures, such as hairy roots, is becoming a very interesting system for large scale secondary metabolite production [58-60]. Additionally, recent progress achieved for the scaling -up of hairy root processes makes this technology an attractive tool for industrial processes [61]. According to the potential of these cultures to produce valuable compounds found in roots, the stability of the production and the technological progress in the scale-up of hairy root cultures, this technology has moved to large-scale industrial production of pharmaceuticals becoming a powerful biotechnological tool [61].

Tropane alkaloids are synthesized in root pericycle and translocated to leaves [41]. For this reason, the investigations on scopolamine production were focused on the hairy root cultures of several *Solanaceous* plants [31, 42-45]. However, the yields obtained to date were not suitable for a commercial production in order to replace the extraction from the natural producer plants [46].

In our laboratory, in order to study the influence of plant ecotypes in tropane alkaloids profile, we analyzed and compared the alkaloid production in different organs of Argentinean and Colombian *B. candida* in vitro -germinated seedlings as well as in hairy root clones obtained from both ecotypes [27]. In addition, we scale-up one of the Argentinean hairy root clones obtained in a 1.5L bioreactor [28].

Hairy root cultures were established after transformation with *Agrobacterium rhizogenes* strain LBA 9402 [50]. Figure 1A shows the alkaloid content found in the hairy root cultures belonging to Colombian and Argentinean plants. We reported that the content of hyoscyamine, 6β-hydroxyhyoscyamine and scopolamine was higher in the transformed roots from Argentinean plants and significantly different to the content of the Colombian ones [27].

In all Argentinean clones, the 6β-hydroxyhyoscyamine content predominates over hyoscyamine and scopolamine, while in Colombian clones the main alkaloid was hyoscyamine. In addition, the Argentinean clones produced the highest 6β-hydroxyhyoscyamine concentration. In these cultures, scopolamine constituted about 4% of the three alkaloids measured and 6β-hydroxyhyoscyamine about 68%.

Kursinszki et al. [36] have found that *D. innoxia* and *A. belladonna* hairy root cultures behaved in a similar way to Argentinean *B. candida* hairy roots. In these cultures 6β-hydroxyhyoscyamine was also the predominant alkaloid detected [27, 36].

In addition, we have previously demonstrated that the Argentinean H6H hydroxylase and epoxidase activities were higher and significantly different compared to the Colombian ones (Table 1) [27].

The results reported by Zarate et al. [62] about the clone 19 of *A. baetica* hairy roots were coincident to the results that we found for Colombian hairy root clones. In this case, the scopolamine content predominated over the 6β-hydroxyhyoscyamine one [27].

When we compared the alkaloid content of the *B. candida* hairy roots to that measured in seedlings we found that the profile of the three alkaloids in hairy roots and non transformed Argentinean roots was constant being 6β-hydroxyhyoscyamine > hyoscyamine > scopolamine (Figure 1A and B) [27]. However, in Colombian hairy roots the profile of the mentioned alkaloids was modified from 6β-hydroxyhyoscyamine > hyoscyamine > scopolamine to hyoscyamine > scopolamine >6β-hydroxyhyoscyamine (Figure 1B) [27]. It is worth pointing out that Colombian seedling roots presented the highest content of tropane alkaloids with 6β-hydroxyhyoscyamine being the main one.

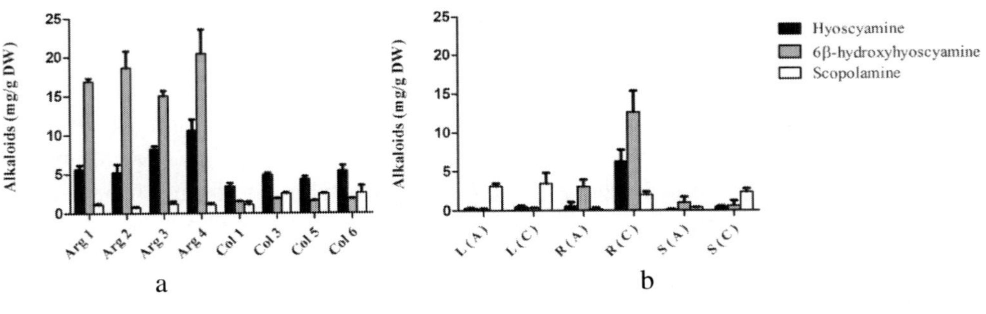

Figure 1. Hyoscyamine, 6β-hydroxyhyoscyamine and scopolamine content in (A) hairy root clones (HR). Arg1-Arg4: HR obtained from Argentinean *B. candida* plants. Col1-Col6: HR obtained from Colombian *B. candida* plants. (B) *B. candida* stem, leaf and root samples from Argentinean and Colombian plants. L: leaves; R: roots; S: steams; A: Argentinean ecotype; C: Colombian ecotype [27].

Table 1. Hydroxylase and epoxidase activities measured in Argentinean and Colombian hairy root ecotypes [27]

	Hydroxylase activity (pKat/mg)	Epoxidase activity (pKat/mg)
Argentinean hairy roots	3.21 ± 1.58	0.22 ± 0.09
Colombian hairy roots	0.36 ± 0.16	0.03 ± 0.01

In addition, we observed that the main alkaloid in leaves was scopolamine and there were not differences between Argentinean and Colombian ecotypes [27]. Our results are in agreement with the hypothesis of scopolamine translocation to the aerial parts [41, 63].

According to the results presented, we suggest that the Colombian seedling roots are an alternative source for 6β-hydroxyhyoscyamine production to the traditional raw material of this alkaloid, *A. tanguticus* [10, 22, 64]. In addition, the high values of 6β-hydroxyhyoscyamine achieved in the Argentinean hairy root culture processes make this *in vitro* culture system an interesting option which deserves further investigation [27].

Also, it is important to consider the advantage of using *in vitro* cultures for producing compounds of interest for the pharmaceutical industry instead of the natural source. Certain compounds, such as tropane alkaloids are accumulated in roots making the hairy root cultures an interesting alternative source for these compounds.

Additionally, many researchers are concentrated on the improvement of large scale cultures of hairy roots which could be very advantageous over the plant production for commercialization [59, 61]. The tropane alkaloid production in bioreactors, instead of their isolation from plants is an interesting strategy because guarantees that the process will be carried out under defined and controlled conditions and consequently preventing or reducing the variations in the product quality and the alkaloid yield [47].

For the reasons mentioned above, we explored the possibility of scaling up the Argentinean *B. candida* hairy roots in a 1.5L stirred tank as an alternative production system of tropane alkaloids [28].

Previous works carried out with stirred tank bioreactors reported that the damage of hairy roots by the impeller induced callus formation and consequently reduced the biomass production [65]. To overcome this problem the bioreactor configuration was modified. A plastic mesh was placed forming a zigzag arrangement around the baffle in order to increase the surface where the roots could be trapped [28].

Table 2 shows the kinetic parameters calculated for the bioreactor and the Erlenmeyer system [28]. It is interesting to note that the growth index found in bioreactor cultures was a 36% higher than that reached in Erlenmeyer flask processes showing an improvement in the liquid-phase culture environment by the application of the modified bioreactor [47]. In this case, 6β-hydroxyhyoscyamine was also the predominant alkaloid (Table 2).

It is frequently mentioned that there is a decrease in productivities when processes are scale-up from Erlenmeyer flasks to bioreactors [66]. However, our studies show that 6β-hydroxyhyoscyamine volumetric productivity was 3-fold higher in the bioreactor system [28].

There were published several reports about the hyoscyamine production by hairy roots grown in bioreactors [65]. As fare as we are aware there is not enough information about the scopolamine and particularly about 6β-hydroxyhyoscyamine production in these systems. Among them, Hilton and Rhodes [67] studied the hyoscyamine production by *D. stramonium* in a modified 14 L stirred tank reactor operated under different conditions in batch and continuous mode. The 35 day culture produced 5.2 mg/g DW and 3.3 mg/g DW of hyoscyamine in Gamborg B5/2 and B5 medium, respectively [67]. *B. candida* hairy roots produced a slightly higher amount of hyoscyamine. Specifically, the process carried out in the modified stirred tank produced 7.0 ± 1.3 mg/g DW of hyoscyamine at the harvest time (Table 2) [28]. Hilton and Rhodes [67] also reported a low release of the alkaloid into the culture medium. The biomass productivity attained in this work was 0.24 g DW/l/d which is very similar to that reported here for *B. candida* hairy root processes (Table 2).

Table 2. Influence of the *B. candida* hairy root culture system on alkaloid yields and productivity [28]

		Erlenmeyer	Bioreactor
GI		2.90	3.93
	H	6.00 ± 2.00	7.01 ± 1.35
Alkaloids	6β	7.80 ± 4.00	10.05 ± 0.76
(mg/g DW)	S	0.50 ± 0.30	1.05 ± 0.27
	H	1.45	3.90
Yp/s	6β	1.89	5.60
(mg/g)	S	0.12	0.58
	T.A	3.47	10.08
	H	6.00	7.01
Yp/x	6β	7.80	10.05
(mg/g DW)	S	0.50	1.05
	T.A	14.30	18.10
	H	0.30 ± 0.10	0.35 ± 0.07
Productivity	6β	0.39 ± 0.20	0.50 ± 0.04
(mg/g DW d)	S	0.02 ± 0.01	0.05 ± 0.01
	T.A	0.71 ± 0.31	0.91 ± 0.11
	H	0.52 ± 0.17	1.44 ± 0.28
Volumetric productivity	6β	0.67 ± 0.34	2.07 ± 0.16
(mg/L d)	S	0.04 ± 0.02	0.22 ± 0.05
	T.A	1.23 ± 0.54	3.72 ± 0.47

H: Hyoscyamine, 6β: 6β-hydroxyhyoscyamine, S: Scopolamine, T.A: Total alkaloids, $Y_{p/x}$: Alkaloid yield referred to biomass formed, $Y_{p/s}$: Alkaloid yield referred to substrate consumed.

The differences regarding to secondary metabolite production could be attributed not only to the plant specie used but also to an oxygen deficiency according to the higher reactor scale [65].

A simple modification introduced in a stirred tank bioreactor allowed us the establishment of *B. candida* hairy root cultures for scopolamine and 6β-hydroxyhyoscyamine production with increased alkaloid concentration compared to the Erlenmeyer flask cultures [28]. It is worth pointing out that these results are potentially applicable to perform the rational scale-up of the process [28].

4.1.3. Other Strategies – Elicitation, Medium Design

An additional strategy for improving alkaloid yield and productivities of *in vitro* cultures is by the appropriate media design [68]. It is important to consider that the biosynthetic pathways that lead to the accumulation of secondary metabolites are complex and sometimes are not fully elucidated [6, 23]. Also, the regulation of these pathways is frequently unknown and these features are needed in order to perform a rational media design [68]. However, it is generally accepted that sucrose is the carbon source preferentially used. The increase of carbon source up to certain levels favors growth and secondary metabolites production. However, further increases may derive in catabolite repression [68].

Concerning to nitrogen source, the basal media contain NO^{3-} or NH^{4+}. Nitrogen is essential for growth, even though, in same cases inhibition of product formation may occur [69-71]. Also, previous works have demonstrated that phosphate limitation stimulates alkaloids production in *Catharantus roseus* [68]. It is well known that phosphate acts as a growth regulator. For this reason, it is desirable to reduce phosphate and nitrogen concentrations in production media.

Amdoun et al. [72] analyzed the effect of B5 medium mineral salts (NO_3^-, $H_2PO_4^-$ and Ca^{2+}) on tropane alkaloids production in *D. stramonium* hairy roots. In addition, they evaluated the occurrence of interaction effect s. The results reported revealed that NO_3^-, $H_2PO_4^-$ and Ca^{2+} and several combinations of these compounds induced alkaloids production [72]. They found that deficiency of NO_3^- reduced alkaloid metabolism even in combination with an elicitor [72].

On the other hand, Pavlov et al. [73] analyzed the effect of NO_3^-, $H_2PO_4^-$ and sucrose on the production of hyoscyamine in *D. stramonium* (diploid and tetraploid) hairy roots. These authors proposed a modified MS medium for an optimal hyoscyamine yield. The optimized MS medium have increased sucrose concentration and decreased NO_3^- concentration compared to standard MS. The results reported by Pavlov et al. [73] showed that it is possible to exploit the biosynthetic potential of *in vitro* cultures by the optimum media composition.

In order to improve yields and productivities of secondary metabolites in plant cell culture s it is essential the selection of the most efficient plant growth regulators (auxins, cytokinins, gibberellic acid and abscicic acid) and the relationship between them [74, 75].

In addition, the supplementation with precursors may improve product formation because the flux through the biosynthetic pathway may be limited by the availability of some intermediates [49]. Methylputrescine, phenylalanine and tropic acid were able to effectively increase hyoscyamine and scopolamine production when they were added from the beginning of the incubation [49]. It is worth to consider that the precursor should be easily available and not expensive. The concentration, time of incorporation and the relationship with the product to be synthesized should be considered.

Another strategy for the improvement of secondary metabolites production is the addition of elicitors (biotic or abiotic) [68].

It is well known that the elicitors are compounds that can induce defense responses on plant cultures deriving in an increase on secondary metabolites production [42]. In addition, it was frequently observed that these compounds were able to release the plant products formed to the medium [42].

However, plant responses are unpredictable because they do not always respond to elicitation techniques.

Regarding to tropane alkaloids, Sevón et al. [76] reported that after elicitation with chitosan the hairy roots of *H. muticus* produced 5-folf increased hyoscyamine concentration than roots control. However, tropane alkaloids production was not increased after elicitation of *D. stramonium* cell culture s by chitosan [77].

On the other hand, Furze et al. [78] observed that using abiotic compounds on *D. stramonium* hairy roots the sesquiterpenoid phytoalexin production was elicitated but no response was observed on tropane alkaloids production [78]. Also, Pitta-Alvarez et al. [42, 45] reported that *B. candida* hairy roots responded positively to different elicitors. Among them it can be mentioned chitosan, acetic acid and citric acid. These compounds promoted the release of alkaloids to the culture medium [42].

In conclusion, for the optimization of secondary metabolites production by media design several considerations should be taken into account. In many cases it is necessary to develop a growth medium and a production one. In the first case, it is necessary to select a basal medium and optimize it for the maximal biomass production.

On the other hand, in the production medium is necessary maintain the carbohydrate concentration of the growth medium, optimize the phosphate and nitrogen content and find the adequate relationship between growth regulator levels [68].

If the biosynthetic pathway it is known, the addition of precursors may stimulate the product formation. Also, the addition of biotic and abiotic elicitors is a strategy for the enhancement of secondary metabolites production.

4.2. Metabolic Engineering

The possibility to introduce genes into plants using the *Agrobacterium* technology as well as particle bombardment has increased the potential for secondary metabolites production in plant cultures by the exploitation and improvement of their own biosynthetic capacities [9, 79]. It is well known that the yield of tropane alkaloids is low for a commercial production. For this reason, is desirable to increase production of these alkaloids in plant species [32]. Numerous works have focused on the genetic engineering of the tropane alkaloids pathway according to the interest in these compounds.

Several genes involved in alkaloids biosynthesis were overexpressed in different plant species [6, 14, 80-82]. It is worth pointing out that organ cultures of *Solanaceous* plants, including hairy roots, have been useful models for the study of tropane alkaloids pathway. From these cultures it was possible the isolation of different key enzymes and its characterization enabling the development of the genetic engineering approach. Among these elucidated biochemical steps we can mention the isolation of the cDNAs encoding PMT from cultures of *A. belladonna* and *H. niger* [9], the characterization of Tropinone reductase enzymes from cultured roots of *H. niger* by Hashimoto et al. [83] and the isolation of cDNAs encoding H6H from cultured roots of *H. niger* [41].

However, the success in the engineering of biochemical steps is unpredictable. In this sense, Sato et al. [34] have revealed that tropane alkaloids production was not increased by the overexpression of PMT in *A. belladonna* hairy roots. However, the production of tropane alkaloids was improved in hairy roots of *D. metel* overexpresing PMT [6]. When this enzyme was overexpressed in hairy roots of *H. muticus*, only hyoscyamine concentration was improved [84].

Several reports have determined that the overexpression of *h6h*gene resulted in the accumulation of higher amounts of tropane alkaloids. Among them, Hashimoto et al. [85] introduced this gene in *A. belladona* hairy roots. The resulting roots contained almost 5-fold scopolamine concentration than wild type hairy roots [85]. In addition, Jouhikainen et al. [14] overexpressed the H6H in *H. muticus.* The results showed that the best hairy root clone produced a 100-fold higher alkaloids concentration than the others [14].

On the other hand, Yun et al. [26] overexpressed the H6H in *A. belladonna* plants by *A. tumefanciens* transformation. In this case, leafs of the transformed plants accumulated almost entirely scopolamine [26].

In addition, a few reports about the overexpression of more than one key enzyme were published. Zhang et al. [13] produced *H. niger* hairy roots overexpressing PMT and H6H. The resulting root clones produced higher concentrations of tropane alkaloids compared to the wild type hairy roots [13].

Additional studies are needed for the improvement of tropane alkaloids production in a scale useful for commercialization. However, the overexpression of several key biosynthetic enzymes is a promising strategy for the improvement of secondary metabolites production.

In our laboratory, we isolate the *h6h*cDNA from immature anthers of *B. candida* flower buttons [29, 86]. Concerning to the sequence analysis, the results are in agreement to previous reports about the H6H from other related species [41, 80]. The 2OG-Fe (II) oxygenase superfamily domain is also conserved in the *h6h* gene from *B. candida* (GenBank: EU530633). This sequence element that is considered to be an iron binding site is also found in other 2-oxoglutarate dependent dioxygenases from plant, bacteria and fungi [87, 88]. In addition, the sequence analysis shows that the *h6h* sequence has similarity to other hydroxylases including those involved in the biosynthesis of ethylene and anthocyanins [29, 89].

It is worth to pointing out that the isolation and characterization of *B. candida* H6H enzyme may constitute a useful tool in order to develop improved plant cell line s for alkaloids production by overexpressing *B. candida h6h*cDNA.

4.3. Combined Systems – Biotransformation

The production of tropane alkaloids by *in vitro* culture techniques is an interesting alternative according to several advantages mentioned previously. However, the results obtained to date were not applicable for an industrial scale production. The low yield of *in vitro* cultures compared to the whole plant production is probably the most important obstacle to overcome in order to reach competitively in the commercial production.

In addition, several plant species that produce tropane alkaloids mainly accumulate hyoscyamine without reaching to the more valuable products [14, 26].

For these reasons, the production of the more valuable alkaloids by biocatalytic process es using genetically modified organisms appears to be a promising strategy.

The expression of plant genes in microorganisms can be an economically attractive and useful strategy for the production of vegetal proteins and/or the products resulting from its activity. Some of the main advantages of microorganisms over plant systems are the broad knowledge on microbial molecular genetics that facilitates its manipulation and the fact that they reach higher concentrations of biomass rapidly [90-95].

For all these features, microorganisms are considered an attractive tool for the production of plant alkaloids by the expression of enzymes involved in alkaloid biosynthesis. In addition, the heterologous expression of plant enzymes in microorganisms allows the detailed analysis of the catalytic mechanisms for an application of plant enzymes in synthetic organic chemistry and also for improving alkaloids production by metabolic engineering in plant cell culture s [90, 96]. The natural activities of wild type strains may be also used for the biotransformation of certain compounds. The modification of medicinal alkaloids by plant or microbial enzymes offers several advantages as reactions are carried out stereospecificaly, they do not operate under extreme conditions of pH, temperature, reducing levels of waste products and eliminating the need to protect and de-protect functional groups [90, 97, 98].

Several studies have reported the production in microorganisms of plant enzymes that are involved in alkaloid biosynthesis [29, 99-102]. In this sense, Unterlinner et al. [100] have cloned and expressed in *Escherichia coli* four cDNAs encoding for different isoforms of the Codeinone reductase NADPH-dependent enzyme isolated from *Papaver somniferum.* In this report has been investigated the substrate specificity of the enzyme and the structural analysis, being the first report about the cloning and expression of genes of the biosynthetic pathway of morphine [90].

Regarding to the H6H enzyme, Liu et al. [103] reported the cloning, expression in *E. coli* and characterization of the enzyme isolated from hairy roots of *A. tanguticus.* This report has focused on the functional studies of the H6H enzyme with the aim to improve alkaloids production in plants by the application of metabolic engineering strategies [103].

Cardillo et al. [29] explored the development of an alternative strategy for the production of the most valuable tropane alkaloids, 6β-hydroxyhyoscyamine and scopolamine, using the H6H from *B. candida* as biocatalyst. This strategy would allow the conversion of hyoscyamine, the more abundant alkaloid found in nature, into the more commercially demanded ones. The host for the heterologous production of the H6H enzyme was *Saccharomyces cerevisiae,* according to several characteristics of this microorganism [29, 91-95]. This yeast is frequently considered the start point in the development of industrial biocatalysts due to the robustness, availability of large scale process technology and regulatory advantages related to the wide use of this microorganism as a host for the expression of therapeutic proteins approved by the FDA [104-108].

Previous works have examined the application of *S. cerevisiae* as a recombinant biocatalyst [102]. Geerlings et al. [102] reported the construction of a *S. cerevisiae* strain harboring two cDNAs coding for Strictosidine synthase and Strictosidine β-glucosidase from the medicinal plant *C. roseus.*

Transgenic yeast expressing the plant enzymes were able to produce indole alkaloids thus providing an interesting alternative for the production of plant naturally occurring alkaloids [102].

According to the features mentioned above, the fact that scopolamine has a 10-times higher commercial demand than that of hyoscyamine [17] and the increased interest in 6β-hydroxyhyoscyamine for medical applications, we constructed transgenic *S. cerevisiae* strains which are able to produce the H6H enzyme from *B. candida* [29].

For this purpose the *h6h*cDNA was amplified from total RNA preparations obtained from immature anthers and hairy roots of the South America n tropane alkaloid producer plant, *B. candida.*

These findings are in agreement with previous reports that have determined the presence of the *h6h*gene in anthers and roots of related plant species [12, 53, 109]. Suzuki et al. [53] analyzed the *h6h* gene expression in several organs of *A. belladona* by RT-PCR and *in situ* hybridization. In this sense, they found that the *h6h*RNA was present in cultures of transformed roots, normal roots and anthers [53]. However, in stem, leaves, pistils and petals were not possible to detect the *h6h*RNA as well as in *B. candida* stems and leaves [29, 86]. The role of the H6H enzyme localized in anthers is a subject of debate [53]. It is well known that the enzyme requires hyoscyamine and other co-factors for the proper catalytic activity [12, 24, 29, 110].

As far as the authors are aware, the occurrence of hyoscyamine and enzyme co-factors was not confirmed in the immature anthers tissue. Nevertheless, hyoscyamine was found in mature pollen grains [53, 111]. Flavonoids are compounds commonly found in this reproductive tissue. Previous studies have demonstrated that they participate in the proper functioning of cells constituting the anther [112, 113].

According to several reports, if scopolamine was produced locally, it could have a similar function to that assigned to flavonoids [53]. On the other hand, the presence of tropane alkaloids in anthers could be attributed to defense functions [96].

In our laboratory, the *h6h*cDNA was cloned into different vectors in order to produce tagged and untagged enzymes. The H6H enzyme was expressed fused at its C-terminus to a V5 epitope and a His-tag (H6H-V5-6His) [29]. The constructions were introduced by chemical transformation in *S. cerevisiae* CEN PK2. In order to explore the different strategies, protein extracts of the induced yeast strains were assayed for the enzyme activity at 30 °C for 2 and 15h [29]. The analysis of the alkaloids was carried out by HPLC with UV detection. The mobile phase used was octanesulfonic acid 0.01M pH3/methanol (65:35), flow rate 1ml/min [29]. The results showed that the tagged and untagged enzymes were able to transform hyoscymine, showing a functional expression of the *h6h*cDNA. However, different products were obtained depending on the enzyme used in the hyoscyamine bioconversion assay (tagged or untagged enzyme). Table 3 shows the percentages of the alkaloids produced in the activity assays.

When the reaction was incubated for 2hs, the tagged protein was able to produce a 15% of 6β-hydroxyhyoscyamine. The other 85% remained as hyoscyamine. On the other hand, the untagged protein produced a 53.7% of the intermediate in the same incubation time.

The reaction carried out with the tagged protein produced approximately a 35.7% of 6β-hydroxyhyoscyamine in 15h of incubation. This time was arbitrarily selected and does not indicate the end of bioconversion. In contrast, the reaction carried out with the untagged enzyme yielded two products. In this case 83.3% of 6β-hydroxyhyoscyamine and 7.6% of scopolamine were produced. Only a 9% of hyoscyamine was not transformed by the untagged enzyme. No product formation was seen in control reactions performed with crude extract of induced wild type strain. Table 4 shows the activity of the tagged and the untagged enzymes produced per mg of total proteins of the crude extract. Comparing these results to previous reports of the H6H, it can be noticed that the activity of these preparations is of the same order of magnitude to results obtained using the H6H purified from hairy root extracts [41, 80, 87].

Table 3. Alkaloids produced from hyoscyamine by biotransformation with H6H and H6H-V5-6His enzymes [29]

	6β-hydroxyhyoscyamine (%)		Scopolamine (%)	
Time (h)	2	15	2	15
H6H	53.7	83.3	-	7,6
H6H-V5-6His	15.0	35.7	-	-
Negative control	-	-	-	-

The percentage values of 6β-hydroxyhyoscyamine and scopolamine are the means of three independent determinations which differs in no more than 10%.

Table 4. Hydroxylase and epoxidase activities in crude protein extract of recombinant *S. cerevisiae* strains, overexpressing the tagged and untagged H6H enzyme from *B. candida* [29]

Crude extract	Hydroxylase activity	Epoxidase activity
	Specific Activity (nKat mg^{-1})	Specific Activity (nKat mg^{-1})
H6H	2.60 ± 0.19	0.24 ± 0.02
H6H-V5-6His	0.89 ± 0.06	ND*

*ND: Not detected.

On the other hand, the H6H enzyme was recently fused at its N-terminus to a cellulose binding domain (CBD -H6H) in order to combine the purification step with the enzyme immobilization on cellulose, a low cost matrix [114]. Several studies have revealed that the immobilization on a cellulose matrix is a highly effective technique for protein immobilization. In addition, immobilization to this low cost matrix can be carried out under mild conditions which are desirable for industrial catalysis [114, 115]. The untagged H6H and the CBD-H6H were able to convert hyoscyamine into 6β-hydroxyhyoscyamine and scopolamine while H6H-V5-6His only produced 6β-hydroxyhyoscyamine [29]. Surprisingly, the cellulose binding domain did not negatively affect the catalytic activity of the H6H as the V5 epitope and the Histidine tag affected it (results not published). These facts are encouraging for the development of a biocatalytic process using immobilized enzymes.

CONCLUSION

The biotechnological approaches have great potential in order to produce effectively a wide variety of chemical compounds, including tropane alkaloids. However, further studies are needed to reach a complete knowledge about tropane alkaloids biosynthesis and the regulation mechanisms. We can conclude that hairy root cultures of *Solanaceous* plants are an interesting system for tropane alkaloids production. These cultures are stable and it is possible to scale-up this system. In addition, hairy root secondary metabolites yield can be improved by several strategies such as elicitation, media design and by the addition of precursors.

Also, the isolation of key enzymes of the biosynthetic pathway allows the application of metabolic engineering approaches in order to improved alkaloids production.

It is worth pointing out that the isolation and characterization of H6H enzyme constitutes a useful tool in order to develop biocatalytic process es for the production of the most valuable alkaloids from a medical and economical point of view.

ACKNOWLEDGMENTS

This work was supported by CONICET (Consejo Nacional de Investigaciones Científicas y Técnicas, Argentina), UBA (Universidad de Buenos Aires) and ANPCyT (Agencia Nacional de Promoción Científica y Tecnológica, Argentina). ABC, JRT and AMG are researchers from CONICET.

The authors wish to thank Dr. Sabrina B. Cardillo for her critical comments and suggestions during the writing of this chapter.

REFERENCES

[1] Samuelsson, G., ed. (1999) *Drugs of natural origin.* Gunnar Samuelsson and Apotekarsocieteten - Swedish Pharmaceutical Society, Swedish Pharmaceutical Press, Sweden.

[2] Rates, S. M. (2001) Plants as source of drugs. *Toxicon.* 39, 603-613.

[3] Kumar, P. (2004) Valuation of medicinal plants for pharmaceutical uses. *Curr. Sci.* 86, 930-937.

[4] Georgiev, M. I., Pavlov, A. I. and Bley, T. (2007) Hairy root type plant in vitro systems as sources of bioactive substances. *Appl. Microbiol. Biotechnol.* 74, 1175-1185.

[5] Zarate, R., el Jaber-Vazdekis, N., Cequier-Sanchez, E., Gutierrez-Nicolas, F., and Ravelo, A. G. (2008) Biotechnology for the production of plant natural products. *Studies in Natural Products Chemistry.* 34, 309-392.

[6] Palazon, J., Navarro-Ocana, A., Hernandez-Vazquez, L., and Mirjalili, M. H. (2008) Application of metabolic engineering to the production of scopolamine. *Molecules.* 13, 1722-1742.

[7] Barquero, A. A. (2007) Plantas sanadoras: pasado, presente y futuro. *Revista QuímicaViva.* 6, 53-69.

[8] De la Cruz, M. (1991) *Libellus de medicinalibus indorum herbis, manuscrito azteca de 1552,* según traducción latina de Juan Badiano., Mexico DC.

[9] Zhang, L., Kai, G. Y., Lu, B., Zhang, H. M., Tang, K. X., Jiang, J. H., and Chen, W. S. (2005) Metabolic Engineering of Tropane Alkaloid Biosynthesis in Plants. *Journal of Integrative Plant Biology.* 47, 136-143.

[10] Wu, Y. F., Lü, W., Lu, Q., and Zhang, W. S. (2005) Asymmetric Synthesis of Anisodine. *Chinese Chemical Letters.* 16, 1287-1289.

[11] Palazón, J., Moyano, E., Cusidó, R. M., Bonfill, M., Oksman-Caldentey, K., and Piñol, M. T. (2003) Alkaloid production in Duboisia hybrid hairy roots and plants overexpressing the h6h gene. *Plant Sci.* 165, 1289-1295.

[12] Hashimoto, T. and Yamada, Y. (1986) Hyoscyamine 6beta-Hydroxylase, a 2-Oxoglutarate-Dependent Dioxygenase, in Alkaloid-Producing Root Cultures. *Plant Physiol.* 81, 619-625.

[13] Zhang, L., Ding, R., Chai, Y., Bonfill, M., Moyano, E., Oksman-Caldentey, K. M., Xu, T., Pi, Y., Wang, Z., Zhang, H., Kai, G., Liao, Z., Sun, X., and Tang, K. (2004) Engineering tropane biosynthetic pathway in Hyoscyamus niger hairy root cultures. *Proc. Natl. Acad. Sci. US* 101, 6786-6791.

[14] Jouhikainen, K., Lindgren, L., Jokelainen, T., Hiltunen, R., Teeri, T. H., and Oksman-Caldentey, K. M. (1999) Enhancement of scopolamine production in Hyoscyamus muticus L. hairy root cultures by genetic engineering. *Planta* 208, 545-551.

[15] Hashimoto, T. and Yamada, Y. (1986) Hyoscyamine 6β-hydroxylase, a 2-oxoglutarate-dependent dioxygenase, in alkaloid-producing root cultures. *Plant Physiol.* 81, 619-625.

[16] Yamada, Y., Yun, D. J. and Hashimoto, T. (1994) Genetic Engineering of Medicinal Plants for Tropane Alkaloid Production. In: *Studies in Plant Sci. Advances in Plant Biotechnology*. pp. 83-93, D.D.Y. Ryu and S. Furusaki (eds.).
[17] Hakkinen, S. T., Moyano, E., Cusido, R. M., Palazon, J., Pinol, M. T., and Oksman-Caldentey, K. M. (2005) Enhanced secretion of tropane alkaloids in Nicotiana tabacum hairy roots expressing heterologous hyoscyamine-6beta-hydroxylase. *J. Exp. Bot.* 56, 2611-2618.
[18] Poupko, J. M., Baskin, S. I. and Moore, E. (2006) The pharmacological properties of anisodamine. *J. Appl. Toxicol.* 27, 116-121.
[19] Wang, T. N., Yang, H. J., Gu-Ling, Li, J. Y., and Zheng, X. X. (2005) Advanced Measurement and Quantitative Appraise of Anisodamine on Calcium Triggered in Cardiac Myocyte. In: *Engineering in Medicine and Biology 27th Annual Conference* ed.)^eds.). pp. 7710-7713, Shanghai, China.
[20] Wang, P. Y., Chen, J. W. and Hwang, F. (1993) Anisodamine causes acyl chain interdigitation in phosphatidylglycerol. *FEBS Lett.* 332, 193-196.
[21] Wang, H., Lu, Y. and Chen, H. Z. (2003) Differentiating effects of anisodamine on cognitive amelioration and peripheral muscarinic side effects induced by pilocarpine in mice. *Neurosci. Lett.* 344, 173-176.
[22] Sheng, C. Y., Gao, W. Y., Guo, Z. R., and He, L. X. (1997) Anisodamine restores bowel circulation in burn shock. *Burns*. 23, 142-146.
[23] Humphrey, A. J. and O'Hagan, D. (2001) Tropane alkaloid biosynthesis. A century old problem unresolved. *Nat. Prod. Rep.* 18, 494-502.
[24] Hashimoto, T. and Yamada, Y. (1987) Purification and characterization of hyoscyamine 6 beta-hydroxylase from root cultures of Hyoscyamus niger L. Hydroxylase and epoxidase activities in the enzyme preparation. *Eur. J. Biochem.* 164, 277-285.
[25] Hashimoto, T., Matsuda, J. and Yamada, Y. (1993) Two-step epoxidation of hyoscyamine to scopolamine is catalyzed by bifunctional hyoscyamine 6 beta-hydroxylase. *FEBS Lett.* 329, 35-39.
[26] Yun, D. J., Hashimoto, T. and Yamada, Y. (1992) Metabolic engineering of medicinal plants: transgenic Atropa belladonna with an improved alkaloid composition. *Proc. Natl. Acad. Sci. US* 89, 11799-11803.
[27] Cardillo, A. B., Otalvaro Alvarez, A. M., Calabro Lopez, A., Velasquez Lozano, M. E., Rodriguez Talou, J., and Giulietti, A. M. (2010) Anisodamine production from natural sources: seedlings and hairy root cultures of argentinean and colombian Brugmansia candida plants. *Planta Med.* 76, 402-405.
[28] Cardillo, A. B., Otálvaro, A. M., Busto, V. D., Rodriguez Talou, J., Velásquez, M. E., and Giulietti, A. M. (2010) Scopolamine, anisodamine and hyoscyamine production by *Brugmansia candida* hairy root cultures in bioreactors. *Process Biochemistry*. 45, 1577-1581.
[29] Cardillo, A. B., Rodriguez Talou, J. and Giulietti, A. M. (2008) Expression of Brugmansia candida Hyoscyamine 6beta-Hydroxylase gene in Saccharomyces cerevisiae and its potential use as biocatalyst. *Microb. Cell Fact.* 7, 17-23.
[30] Roses, O. E., Miño, J. and Villamil, E. C. (1988) Acción farmacodinámica de las flores de *Brugmansia candida. Fitoterapia*. 59, 120-127.

[31] Giulietti, A. M., Parr, A. J. and Rhodes, M. J. (1993) Tropane alkaloid production in transformed root cultures of Brugmansia candida. *Planta Med.* 59, 428-431.

[32] Moyano, E., Fornale, S., Palazon, J., Cusido, R. M., Bagni, N., and Pinol, M. T. (2002) Alkaloid production in Duboisia hybrid hairy root cultures overexpressing the pmt gene. *Phytochemistry.* 59, 697-702.

[33] Robins, R. J., Bent, E. G. and Rhodes, M. J. C. (1991) Studies on the biosynthesis of tropane alkaloids by *Datura stramonium* L. transformed root cultures. *Planta.* 185, 385-390.

[34] Sato, F., Hashimoto, T., Hachiya, A., Tamura, K., Choi, K. B., Morishige, T., Fujimoto, H., and Yamada, Y. (2001) Metabolic engineering of plant alkaloid biosynthesis. *PNAS.* 98, 367-372.

[35] Hashimoto, T., Yukimune, Y. and Yamada, Y. (1989) Putrescine and putrescine N-methyltransferase in the biosynthesis of tropane alkaloids in cultured roots of *Hyocyamus albus* I. Biochemical studies. *Planta.* 178, 123-130.

[36] Kursinszki, L., Hank, H., Laszlo, I., and Szoke, E. (2005) Simultaneous analysis of hyoscyamine, scopolamine, 6beta-hydroxyhyoscyamine and apoatropine in Solanaceous hairy roots by reversed-phase high-performance liquid chromatography. *J. Chromatogr. A.* 1091, 32-39.

[37] Norby, F. L. and Ren, J. (2002) Anisodamine inhibits cardiac contraction and intracellular Ca(2+) transients in isolated adult rat ventricular myocytes. *Eur. J. Pharmacol.* 439, 21-25.

[38] Yang, L. M., Xie, Y. F., Chen, H. Z., and Lu, Y. (2007) Diastereomeric and enantiomeric high-performance liquid chromatographic separation of synthetic anisodamine. *J. Pharm. Biomed. Anal.* 43, 905-909.

[39] Lee, K. T., Yamakawa, T., Kodama, T., and Shimomura, K. (1998) Effects of chemicals on alkaloid production by transformed roots of Belladonna. *Phytochemistry.* 49, 2343-2347.

[40] Li, J., Chun, Y. and Jua, H. (2007) Simultaneous electrochemiluminiscence detection of anisodamine, atropine and scopolamine in Flos daturae by capillary electrophoresis using b-cyclodextrin as additive. *Electroanalysis.* 19, 1569-1574.

[41] Matsuda, J., Okabe, S., Hashimoto, T., and Yamada, Y. (1991) Molecular cloning of hyoscyamine 6 beta-hydroxylase, a 2-oxoglutarate-dependent dioxygenase, from cultured roots of Hyoscyamus niger. *J. Biol. Chem.* 266, 9460-9464.

[42] Pitta-Alvarez, S. I. and Giulietti, A. M. (1999) Influence of chitosan, acetic acid and citric acid on growth and tropane alkaloid production in transformed roots of Brugmansia candida. Effect of medium pH and growth phase. *Plant Cell Tissue and Organ Culture.* 59, 31-38.

[43] Pitta-Alvarez, S. A. and Giulietti, A. M. (1995) Advantage and limitations in the use of hairy roots cultures for the production of tropane alkaloids: use of anti-auxins in the maintenance of normal root morphology. *In Vitro Cell. Dev. Biol. Plant.* 31, 215-220.

[44] Hashimoto, T., Yun, D. J. and Yamada, Y. (1993) Production of tropane alkaloids in genetically engineered root cultures. *Phytochemistry.* 32 713-718.

[45] Pitta-Alvarez, S. I., Marconi, P. L. and Giulietti, A. M. (2003) Comparison of the influence of different elicitors on hyoscyamine and scopolamine content in hairy root cultures of *Brugmansia candida. In Vitro Cell. Dev. Biol. Plant.* 39, 640-644.

[46] Oksman-Caldentey, K. M. (2007) Tropane and nicotine alkaloid biosynthesis -novel approaches towards biotechnological production of plant-derived pharmaceuticals. *Curr. Pharm. Biotechnol.* 8, 203-210.

[47] Eibl, R. and Eibl, D. (2008) Design of bioreactors suitable for plant cell and tissue cultures. *Phytochem. Rev.* 7, 593-598.

[48] Giulietti, A. M. and Ertola, R. J. (1997) Biotechnological strategies for production of plants and secondary metabolites of pharmaceutical interest. In: *Second word congress on medicinal and aromatic plants for human welfare WOCMAP-2* (Giberti, G., ed.)^eds.). pp. 269-280, ISHS, Mendoza, Argentina.

[49] Hashimoto, T. and Yamada, Y. (1987) Effects of Culture Conditions on Tropane Alkaloid Formation in Hyoscyamus niger Suspension Cultures. *Agric. Biol. Chem.* 51, 2769-2774.

[50] Hamill, J. D., Parr, A. J., Rhodes, M. J. C., Robins, R. J., and Walton, N. J. (1987) New Routes to Plant Secondary Products. *Bio/Technology* 5, 800-804.

[51] Hashimoto, T. and Yamada, Y. (1983) Scopolamine Production in Suspension Cultures and Redifferentiated Roots of Hyoscyamus niger. *Planta Med.* 47, 195-199.

[52] Suzuki, K., Yamada, Y. and Hashimoto, T. (1999) Expression of Atropa belladonna putrescine N-methyltransferase gene in root pericycle. *Plant Cell Physiol.* 40, 289-297.

[53] Suzuki, K., Yun, D., Chen, X., Yamada, Y., and Hashimoto, T. (1999) An *Atropa belladonna* hyoscyamine 6 beta-hydroxylase gene is differentially expressed in the root pericycle and anthers. *Plant Molecular Biology.* 40, 141-152.

[54] Nakajima, K. and Hashimoto, T. (1999) Two tropinone reductases, that catalyze opposite stereospecific reductions in tropane alkaloid biosynthesis, are localized in plant root with different cell-specific patterns. *Plant Cell Physiol.* 40, 1099-1107.

[55] Nakajima, K., Oshita, Y., Kaya, M., Yamada, Y., and Hashimoto, T. (1999) Structures and expression patterns of two tropinone reductase genes from Hyoscyamus niger. *Biosci. Biotechnol. Biochem.* 63, 1756-1764.

[56] Rothe, G., Hachiya, A., Yamada, Y., Hashimoto, T., and Drager, B. (2003) Alkaloids in plants and root cultures of Atropa belladonna overexpressing putrescine N-methyltransferase. *J. Exp. Bot.* 54, 2065-2070.

[57] Robins, R. J., Walton, N. J., Parr, A. J., Aird, E. H., Rhodes, J. C., and Hamill, J. D. (1994) Progress in the genetic engineering of the pyridine and tropane alkaloid biosynthetic pathways of Solanaceous plants. In: *Genetic Engineering of Plant Secondary Metabolism.* pp. 1-33, Plenum Press, New York.

[58] Diwan, R. and Malpathak, N. (2008) Novel technique for scaling up of micropropagated Ruta graveolens shoots using liquid culture systems: a step towards commercialization. *N. Biotechnol.* 25, 85-91.

[59] Liu, C. (2008) Novel Plant Reactors for Pharmaceutical Production. In: *World Congress on In Vitro Biology.* ed.)^eds.), NewsRelease, University communications office, Arkansas State University.

[60] Cardillo, A. B., Maria Otalvaro Alvarez, A., Calabro Lopez, A., Enrique Velasquez Lozano, M., Rodriguez Talou, J., and Maria Giulietti, A. (2009) Anisodamine Production from Natural Sources: Seedlings and Hairy Root Cultures of Argentinean and Colombian Brugmansia candida Plants. *Planta Med.*

[61] Guillon, S., Tremouillaux-Guiller, J., Pati, P. K., Rideau, M., and Gantet, P. (2006) Hairy root research: recent scenario and exciting prospects. *Curr. Opin. Plant Biol*. 9, 341-346.

[62] Zarate, R., el Jaber-Vazdekis, N., Medina, B., and Ravelo, A. G. (2006) Tailoring tropane alkaloid accumulation in transgenic hairy roots of Atropa baetica by over-expressing the gene encoding hyoscyamine 6beta-hydroxylase. *Biotechnol. Lett*. 28, 1271-1277.

[63] Hashimoto, T., Hayashi, A., Amano, Y., Kohno, J., Iwanari, H., Usuda, S., and Yamada, Y. (1991) Hyoscyamine 6 beta-hydroxylase, an enzyme involved in tropane alkaloid biosynthesis, is localized at the pericycle of the root. *J. Biol. Chem*. 266, 4648-4653.

[64] Zhang, H. M., Ou, Z. L. and Yamamoto, T. (2001) Anisodamine inhibits shiga toxin type 2-mediated tumor necrosis factor -alpha production in vitro and in vivo. *Exp. Biol. Med.* (Maywood). 226, 597-604.

[65] Kim, Y., Wyslouzil, B. E. and Weathers, P. J. (2002) Secondary metabolism of hairy root cultures in bioreactors. *In Vitro Cell. Dev. Biol. Plant*. 38, 1-10.

[66] Palazón, J., Mallol, A., Eibl, R., Lettenbauer, C., Cusidó, R. M., and Piñol, M. T. (2003) Growth and ginsenoside production in hairy root cultures of Panax ginseng using a novel bioreactor. *Planta Med*. 69, 344-349.

[67] Hilton, M. G. and Rhodes, M. J. (1990) Growth and hyoscyamine production of 'hairy root' cultures of Datura stramonium in a modified stirred tank reactor. *Appl. Microbiol. Biotechnol.* 33, 132-138.

[68] Ertola, R. J., Giulietti, A. M. and Castillo, F. J. (1994) Design, formulation, and optimization of media. In: *Bioreactor System Design* (Asenjo, J. A., ed.). pp. 89-137, M. Dekker.

[69] Nigra, H. M., A., A. M. and Giulietti, A. M. (1990) Effect of carbon and nitrogen source on growth and solasodine production in batch suspension cultures of *Solanum eleagnifolium Cav. Plant Cell Tissue and Organ Culture*. 21, 55-60.

[70] Nakagawa, K., Konagai, A., Fukui, H., and Tabata, M. (1984) Release and crystallization of berberine in the liquid medium of *Thalicitum minus* cells suspension cultures. *Plant Cell Rep*. 3, 254-257.

[71] Hilton, M. G. and Rhodes, M. J. C. (1994) The effect of varying levels of Gamborg's B5 salts and temperature on the accumulation of starch and hyoscyamine in batch cultures of transformed roots of Datura stramonium. *Plant Cell, Tissue and Organ Culture*. 38, 45-51.

[72] Amdoun, R., Khelifi, L., Khelifi-Slaoui, M., Amroune, S., Benyoussef, E. H., Vu Thi, D., Assaf-Ducrocq, C., and Gontier, E. (2009) Influence of minerals and elicitation on *Datura stramonium L.* tropane alkaloid production: Modelization of the in vitro biochemical response. *Plant Science* 177, 81-87.

[73] Pavlov, A. I., Georgiev, V., Marchev, A., and Berkov, S. (2009) Nutrient medium optimization for hyoscyamine production in diploid and tetraploid *Datura stramonium* L. hairy root cultures. *World Journal of Microbiology and Biotechnology*. 25, 2239-2245.

[74] Nigra, H. M., Alvarez, M. A. and Giulietti, A. M. (1989) The influence of auxins, light and cell differentiation on solasodine production by *Solanum eleagnifolium* Cav. calli. *Plant Cell Rep*. 8, 230-233.

[75] Rhodes, J. C., Parr, A., Giulietti, A. M., and Aird, E. H. (1994) Influence of exogenous hormones on the growth and secondary metabolite formation in transformed root cultures. *Plant Cell Tissue and Organ Culture*. 38, 143-151.

[76] Sevón, N., Hiltunen, R. and Oksman-Caldentey, K. M. (1992) Chitosan increases hyoscyamine content in hairy root cultures of *Hyoscyamus muticus*. *Pharm. Pharmacol. Lett*. 2, 96-99.

[77] Ballica, R., Ryu, D. D. and Kado, C. I. (1993) Tropane alkaloid production in Datura stramonium suspension cultures: elicitor and precursor effects. *Biotechnol. Bioeng*. 41, 1075-1081.

[78] Furze, J. M., Rhodes, M. J. C., Parr, A. J., Robins, R. J., Withehead, I. M., and Threlfall, D. R. (1991) Abiotic factors elicit sesquiterpenoid phytoalexin production but not alkaloid production in transformed root cultures of *Datura stramonium*. *Plant Cell Reports*. 10, 111-114.

[79] Verpoorte, R., van der Heijden, R. and Memelink, J. (2000) Engineering the plant cell factory for secondary metabolite production. *Transgenic Res*. 9, 323-343; discussion 321.

[80] Liu, T., Zhu, P., Cheng, K. D., Meng, C., and Zhu, H. X. (2005) Molecular cloning and expression of putrescine N-methyltransferase from the hairy roots of Anisodus tanguticus. *Planta Med*. 71, 987-989.

[81] Bulgakov, V. P. (2008) Functions of rol genes in plant secondary metabolism. *Biotechnol. Adv*. 26, 318-324.

[82] Palazon, J., Cusido, R. M., Roig, C., and Piñol, M. T. (1997) Effect of rol genes from *Agrobacterium rhizogenes* TL-DNA on nicotine production in tobacco root cultures. *Plant Physiol. Biochem*. 35, 155-162.

[83] Hashimoto, T., Nakajima, K., Ongena, G., and Yamada, Y. (1992) Two Tropinone Reductases with Distinct Stereospecificities from Cultured Roots of Hyoscyamus niger. *Plant Physiol*. 100, 836-845.

[84] Moyano, E., Jouhikainen, K., Tammela, P., Palazon, J., Cusido, R. M., Pinol, M. T., Teeri, T. H., and Oksman-Caldentey, K. M. (2003) Effect of pmt gene overexpression on tropane alkaloid production in transformed root cultures of Datura metel and Hyoscyamus muticus. *J. Exp. Bot*. 54, 203-211.

[85] Hashimoto, T., Yun, D. and Yamada, Y. (1993) Production of tropane alkaloids in genetically engineered root cultures. *Phytochemistry*. 32, 713-718.

[86] Cardillo, A. B., Giulietti, A. M. and Marconi, P. L. (2006) Analysis and sequencing of h6hmRNA, last enzyme in the tropane alkaloids pathway from anthers and hairy root cultures of Brugmansia candida (Solanaceae). *Electronic Journal of Biotechnology*. 9, 195-198.

[87] Matsuda, J., Hashimoto, T. and Yamada, Y. (1997) Analysis of active-site residues in hyoscyamine 6β-hydroxylase. *Plant Biotech*. 14, 51-57.

[88] Schofield, C. J. and Zhang, Z. (1999) Structural and mechanistic studies on 2-oxoglutarate-dependent oxygenases and related enzymes. *Curr. Opin. Struct. Biol*. 9, 722-731.

[89] Robins, R. J., Walton, N. J., Parr, A. J., Aird, E. H., Rhodes, J. C., and Hamill, J. D. (1993) Progress in the genetic engineering of the pyridine and tropane alkaloid biosynthetic pathways of Solanaceous plants. In: *Recent advances in phytochemistry, genetic engineering of plant secondary metabolism*. pp. 1-33 Plenum Press NY.

[90] Rathbone, D. A. and Bruce, N. C. (2002) Microbial transformation of alkaloids. *Curr. Opin. Microbiol.* 5, 274-281.
[91] Chemler, J. A. and Koffas, M. A. G. (2008) Metabolic engineering for plant natural product biosynthesis in microbes. *Curr. Opinion in Biotechnology*. 19, 597-605.
[92] Becker, J. V. W., Armstrong, G. O., van der Merwe, M. J., Lambrechts, M. G., Vivier, M. A., and Pretorius, I. S. (2003) Metabolic engineering of Saccharomyces cerevisiae for the synthesis of the wine-related antioxidant resveratrol. *FEMS Yeast Research*. 4, 79-85.
[93] Engels, B., Dahmb, P. and Jennewein, S. (2008) Metabolic engineering of taxadiene biosynthesis in yeast as a first step towards Taxol (Paclitaxel) production. *Metabolic Engineering*. 10, 201-206.
[94] Zeng, Q., Qiu, F. and Yuan, L. (2007) Production of artemisinin by genetically-modified microbes. *Biotechnol. Lett.*
[95] Limem, I., Guedon, E., Hehn, A., Bourgaud, F., Ghedira, L. C., Engasser, J. M., and Ghoul, M. (2008) Production of phenylpropanoid compounds by recombinant microorganisms expressing plant-specific biosynthesis genes. *Process Biochemistry*. 43, 463-479.
[96] Kutchan, T. M. (1995) Alkaloid Biosynthesis -The Basis for Metabolic Engineering of Medicinal Plants. *Plant Cell*. 7, 1059-1070.
[97] Straathof, A. J., Panke, S. and Schmid, A. (2002) The production of fine chemicals by biotransformations. *Curr. Opin. Biotechnol.* 13, 548-556.
[98] Leuenberger, H. G. W. (1990) Biotransformation-A useful tool in organic chemistry. *Pure and Appl. Chem*. 62, 753-768.
[99] Pauli, H. H. and Kutchan, T. M. (1998) Molecular cloning and functional heterologous expression of two alleles encoding (S)-N-methylcoclaurine 3'-hydroxylase (CYP80B1), a new methyl jasmonate-inducible cytochrome P-450-dependent mono-oxygenase of benzylisoquinoline alkaloid biosynthesis. *Plant J*. 13, 793-801.
[100] Unterlinner, B., Lenz, R. and Kutchan, T. M. (1999) Molecular cloning and functional expression of codeinone reductase: the penultimate enzyme in morphine biosynthesis in the opium poppy Papaver somniferum. *Plant J*. 18, 465-475.
[101] Warzecha, H., Gerasimenko, I., Kutchan, T. M., and Stockigt, J. (2000) Molecular cloning and functional bacterial expression of a plant glucosidase specifically involved in alkaloid biosynthesis. *Phytochemistry*. 54, 657-666.
[102] Geerlings, A., Redondo, F. J., Contin, A., Memelink, J., van der Heijden, R., and Verpoorte, R. (2001) Biotransformation of tryptamine and secologanin into plant terpenoid indole alkaloids by transgenic yeast. *Appl. Microbiol. Biotechnol.* 56, 420-424.
[103] Liu, T., Zhu, P., Cheng, K. D., Meng, C., and He, H. X. (2005) Molecular cloning, expression and characterization of hyoscyamine 6-beta-hydroxylase from hairy roots of *Anisodus tanguticus*. *Planta Med.* 71, 249-253.
[104] Gerngross, T. U. (2004) Advances in the production of human therapeutic proteins in yeasts and filamentous fungi. *Nature Biotechnology*. 22, 1409-1414.
[105] Cereghino, G. P. and Cregg, J. M. (1999) Applications of yeast in biotechnology: protein production and genetic analysis. *Curr. Opin. Biotechnol*. 10, 422-427.
[106] Ostergaard, S., Olsson, L. and Nielsen, J. (2000) Metabolic engineering of Saccharomyces cerevisiae. *Microbiol. Mol. Biol. Rev*. 64, 34-50.

[107] Özcan, N. (2001) Heterologous Expression of Genes in the Yeast Saccharomyces cerevisiae. *Turk. J. Agric. For.* 25, 45-49.

[108] ApoLife. (2006) Company Update: ApoLife Touts Yeast-Based Expression System. Company Cites Platform´s Advantages from Discovery through Commercial Manufacturing. In: *GEN Genetic Engineering and Biotechnology News* ed.)^eds.)

[109] Ebrahimzadeh, H., Teimoori, A. and Lohrasebi, T. (2003) Hyoscyamine 6 b-hydroxylase gene isolation from in vitro cultured roots of *Hyoscyamus niger* L. and *Hyoscyamus tenuicaulis* Schonbeck-Temesy. *DARU*. 11, 1-4.

[110] Prescott, A. G. and Lloyd, M. D. (2000) The iron (II) and 2-oxoacid-dependent dioxygenases and their role in metabolism. *Nat. Prod. Rep*. 17, 367-383.

[111] Detzel, A. and Wink, M. (1993) Attraction, deterrence or intoxication of bees (*Apis mellifera*) by plant allelochemicals. *Chemoecology*. 4, 8-18.

[112] Ylstra, B., Busscher, J., Franken, J., Hollman, P. C. H., Mol, J. N. M., and van Tunen, A. J. (1994) Flavonol and fertilization in Petunia hybrida: localization and mode of action during pollen tube growth. *Plant Cell* 6, 201-212.

[113] Mo, Y., Nagel, C. and Taylor, L. P. (1992) Biochemical complementation of chalcone synthase mutants defines a role for flavonols in functional pollen. *Proc. Natl. Acad. Sci.* 89, 7213-7217.

[114] Engel, S., Vyazmensky, M., Berkovich, D., Barak, Z., Merchuk, J., and Chipman, D. M. (2005) Column flow reactor using acetohydroxyacid synthase I from Escherichia coli as catalyst in continuous synthesis of R-phenylacetyl carbinol. *Biotechnol. Bioeng*. 89, 733-740.

[115] Richins, R. D., Mulchandani, A. and Chen, W. (2000) Expression, immobilization, and enzymatic characterization of cellulose -binding domain-organophosphorus hydrolase fusion enzymes. *Biotechnol. Bioeng*. 69, 591-596.

In: Alkaloids
Editor: Eduardo Sobarzo-Sánchez

ISBN: 978-1-63482-074-5

Chapter 7

BIOACTIVE ALKALOIDS OF MARINE ORIGIN

Anna Aiello, Concetta Imperatore, Filomena D'Aniello and Marialuisa Menna*
The NeaNat Group, Dipartimento di Farmacia,
Università degli Studi di Napoli "Federico II", Napoli, Italy

ABSTRACT

The marine organisms, mainly invertebrates such as sponges and ascidians, are a rich source of varied natural alkaloids with unique chemical scaffolds and diverse bioactivities. From a medicinal chemistry perspective, the marine alkaloids' fascinating structures as well as their broad spectrum of pharmacological activity make them promising lead compounds for various druggable targets. The recent results in the field of marine alkaloids are thus here described. The structures and biological activities of selected typical classes of marine alkaloids are briefly illustrated, with the emphasis on the role of these natural products as leads for drug discovery.

Keywords: alkaloids, bioactive natural products, sponges, ascidians, algae

1. INTRODUCTION

Marine organisms comprise approximately half of the total biodiversity on the earth and the marine ecosystem is now recognized as a great source of bioactive secondary metabolites many of which exhibiting structural/chemical features not found in terrestrial natural products. In recent yeras, a significant number of novel molecules with potent pharmacological properties have been discovered from the marine organisms, including alkaloids which probably constitute the largest class of natural products synthesized by marine organisms. Marine alkaloid s have become increasingly important since 1990s,

* Corresponding author: Marialuisa Menna. The NeaNat Group, Dipartimento di Farmacia, Università degli Studi di Napoli "Federico II", Via D. Montesano 49, 80131, Napoli, Italy. E-mail: mlmenna@unina.it, tel.: +39-081-678-518, fax: +39-081-678-552.

representing approximately 14% of all reported marine natural products [1]. They are usually present in an organism as a mixture consisting of several major and a few minor compounds of the same biosynthetic origin and differing only in functional groups. This group of compounds has apparently evolved as a defense mechanism against predators and, thus, alkaloids are often highly potent and toxic molecules [2]. Sessile marine invertebrates, such as sponges, mollusks, and tunicates, mostly lacking morphological defense, have proven to be an outstanding source of bioactive alkaloids and show the largest chemical diversity in this class of compounds. Many kinds of alkaloids with extraordinary structures and significant biological activities have been isolated from these organisms; their unique structural properties have made them attractive in terms of natural product synthesis and, mainly, pharmaceutical research. The number of marine alkaloids in preclinical trials is quite large [3], and the list of medical alkaloids is currently extending with the increasing interest and attention from the academy and industry. Some modifications are added based on the original template of marine alkaloids, beyond that, a series of compounds are synthesized using marine alkaloids as leads.

In this chapter, the structures, the source, and bioactivity of selected classes of typical marine alkaloids are described. The emphasis is on compounds that have emerged as novel leads, such as lamellarins (pyrroles), cephalostatins/ritterazines (pyrazines) and ecteinascidins (isoquinolines), and manzamines (β-carbolines) to give relevance to the role of marine alkaloids as an important source of leads for drug discovery.

2. Pyrrole-Imidazole Alkaloids

The pyrrole-imidazole alkaloids family comprises hundreds of secondary metabolites originating from marine sponges exclusively.

Those natural products, whose architectural complexity goes from simple, achiral, monomeric oroidin (**1**, Figure 1) to the breath-taking 16-stereocenter-containing tetrameric stylissadines (**76-77**, Figure 11), have been mainly isolated from various species of *Agelasidae*, *Axinellidae*, *Dyctionellidae* and *Hymeniacidonidae* families of sponges [4]. The systematic recurrence of these alkaloids in these families of sponges allowed to speculate their taxon-specificity and then to consider these secondary metabolites as chemical markers for phylogenetically related sponges [5]. Most of these alkaloids share the key building block pyrrole-imidazole; oroidin (**1**), the first member of this group to be isolated [6], is the underlying structural motif of this family of alkaloids.

The anti-predatory role of these alkaloids can be considered their most important biological function. It has been determined that Caribbean reef sponges of the genus *Agelas* are chemically defended from fish predation by bromopyrrole alkaloids [7]. However, bromopyrrole alkaloids are important not only for their ecological role and for chemotaxonomic considerations, but also for the interesting pharmacological activities they have been shown to possess, including antibacterial, antifungal, antimalarial, anticoagulant, antiprotozoal, antiviral, antihistaminic, anticancer as well as anti-inflammatory effects [8].

Pyrrole-imidazole alkaloids structures can vary with regard to (a) oxidation, reduction or hydration of the 2-amino -4(5)-vinylimidazole unit, (b) dimerization, and (c) cyclization. The pyrrole-2-carboxamide moiety can be non-, mono-, or dibrominated exclusively in the 2- and 3-positions (numbering according figure 1).

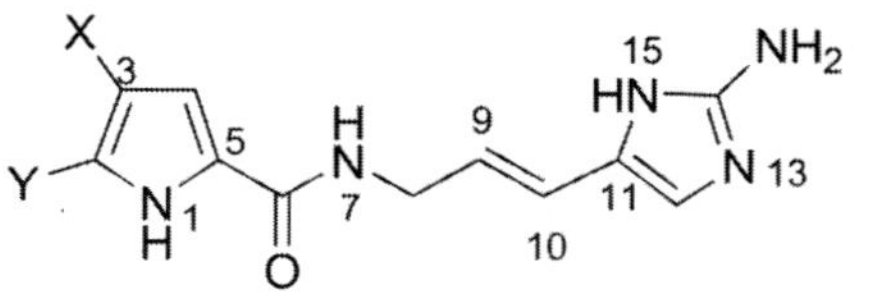

1 X = Y = Br
2 X = Br Y = H
3 X = Y = H

Figure 1. Structures of oroidin (**1**), hymenidin (**2**), and clathrodin (**3**).

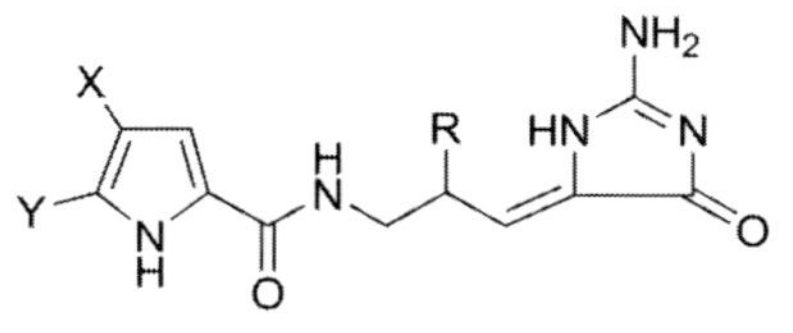

X = Y = Br, R=H **4**
X = Br Y = H, R=H **5**
X = Y = Br, R=OH **6**
X = Br Y = H, R=OH **7** (dispacamide D/mukanadin A)
X = Br, Y = H , R = OMe **8**

Figure 2. Structures of dispacamides (**4-7**) and 9-methoxy dispacamide B (**8**).

Bromination of the pyrrole 4-position or of the imidazole part has never been observed. A great molecular diversity is observed in this group of alkaloids; extremely varied chemical architectures are reported, ranging from linear oroidin-like monomers to complex polyciclic and tetrameric molecules. In this chapter illustrative structures of this alkaloids family are reported, divided into the following groups:

a) Oroidin-like linear monomers. These structures contain the skeleton of oroidin without any further C-C or C-N bond formation. Hymenidin (**2**, Figure 2), isolated from the Okinawan sponge *Hymeniacidon sp.* [9] is the 2-debromo derivative of oroidin, while clathrodin (**3**), isolated from the Caribbean sea sponge *Agelas clathrodes* [10], is its 2,3-debromo derivative.

The degree of bromination of the pyrrole moiety has been shown to affect the biological properties of these compounds. Both oroidin and hymenidin have been found to reduce voltage dependent calcium elevation in PC12 cells, and the potency of the tested alkaloids increases with the number of bromine atoms associated with the pyrrole ring. Analogously, the degree of bromination also affect the feeding deterrent properties, hymenidin being less potent than oroidin [11]. Oroidin, hymenidin and clathrodin showed marked antiserotonergic and anticholinergic activities [9, 12]. Clathrodin is a sodium channel neurotoxin, which acts influencing channel ionic conductance [13].

In dispacamides A-D (**4-7**, Figure 2), isolated from four Caribean *Agelas* sponges (*A. conifera, A. longissima, A. clathrodes, A. dispar*) [14, 15], the 2-aminoimidazole moiety is oxidized to an alkylidene glycocyamidine. All dispacamides exhibited a remarkable antihistaminic activity on the guinea pig ileum [14]. They were shown to produce a reversible non-competitive antagonistic effect, specific toward histamine receptors; dispacamide A was the most potent compound in the series, while the activity of dispacamides C and D was mild, when compared to dispacamides A and B. Thus, the insertion of a hydroxyl group in the alkyl chain causes a marked reduction of the antihistaminic activity, indicating the importance of the central segment for the pharmacological activity.

In has been hypothesized that this portion of the molecule could interact with hydrophobic groups in the receptor zone, so that the insertion of a polar group achieves the effect of making this interaction worse [15]. The 9-methoxy derivative of dispacamide B (**8**), isolated as a racemate from the Mediterranean sponge *Axinella verrucosa* [16], displayed neuroprotective activity against the agonist serotonin in vitro, with a potential to treat psychosis, different phobia and mood fluctuation disorders.

In mukanadins A (**7**) and B (**9**), from the Okinawan sponge *Agelas nakamurai* [17] (actually, the structure of mukanadin A corresponds to that of dispacamide D), mukanadin D (**10**), from the Jamaican sponge *Didiscus oleata* [18], and compound **11**, from *Axinella verrucosa* [16], the 2-aminoimidazole ring of dispacamides is replaced by a hydantoin moiety. Midpacamide (**12**) from an unidentified marine sponge collected in the Marshall Islands [19] and mauritiamide A (**13**) from the sponge *Agelas mauritiana* [20], possess the *N*-methylpyrrole-2-carboxamide partial structure linked to a *N*-methylated hydantoyn and aminoimidazolone moieties, respectively. Also for compound **11** a neuroprotective activity has evidenced, exerted through a potent glutamate antagonist effect [16]. Mauritiamide A was the first member of the bromopyrrole alkaloids class to include a taurine moiety. Successively, this uncommon feature has been found attached to a 2-aminoimidazole ring in tauroacidin A (**14**) and its debromoderivative tauroacidin B (**15**), isolated from an Okinawan *Hymeniacidon* sp. [21], in taurodispacamide (**16**), isolated from the Mediterranean sponge *Agelas oroides* [22], and its debromoderivative **17**, isolated from *Axinella verrucosa* [16]. Different pharmacological activities have been evidenced for compounds **14**-**17**, including inhibitory activity against EGF receptor kinase and *c-erbB*-2 kinase (Tauroacidins A and B) [21], antihistaminic activity (taurodispacamide, **16**) [22] and neuro-protective (compound **17**) [16]. The cytotoxic slagenins A-C (**18-20**), isolated from the Okinawan sponge *Agelas nakamurai* [23] and possessing the unique tetrahydrofuro[2,3-*d*]imidazolidin-2-one moiety also can be included in this section, although they are formally tricyclic, because the third ring is not formed by an intramolecular C-C or C-N cyclization of the key oroidin building block but through the participation of a hydroxy group at C-9.

b) Polyciclic oroidin derivatives. This gorup of bromopyrrole alkaloids includes all those molecules whose cyclized skeleton can be formally derived through the formation of one (or more) C-C or C-N bonds within the oroidin framework (see Figure 1). Consequently, we have found very useful to classify these molecules according to the oroidin atoms involved in the formal cyclization. Several alkaloids belonging to this class have been shown to possess interesting pharmacological activity, particularly as antitumor agent s (agelastatins and palau'amine) or inhibitors of pro-inflammatory cytokines (hymenialdisins).

The hymenin family alkaloids (compounds **21-34**, Figure 4) can be considered C4/C10 derivatives. Hymenin (**21**) has been isolated as an α-adrenoceptor blocking agent from the sponge *Hymeniacidon* sp. [24, 25]. In addition, **24** exhibited also a good antibacterial activity against *Bacillus subtilis* and *Escherichia coli* [24].

Stevensine (= odiline) (**23**) has been isolated by Faulkner et al. from an unidentified marine sponge [26] and subsequently found in *Pseudaxynissa cantharella* [27], *Axinella verrucosa* [16], and in *Stylissa carteri* [28] along with 2-Debromostevensine (**24**).

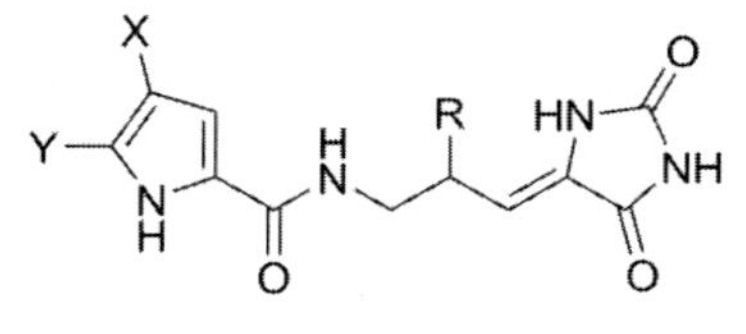

9 X = Br, Y = R = H
10 X = Y = Br, R = H
11 X = Br, Y = H, R = OH

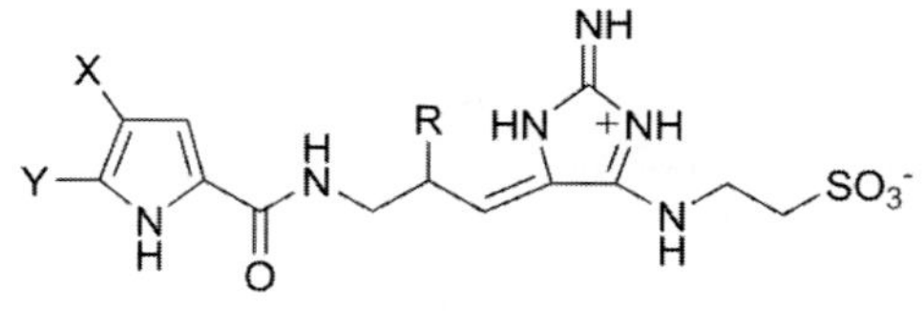

14 X = Y = Br, R = OH (9S/9R = 6:4)
15 X = Br, Y = H, R = OH (9S/9R = 1:1)
16 X = Y = Br, R = H
17 X = Br, Y = R = H

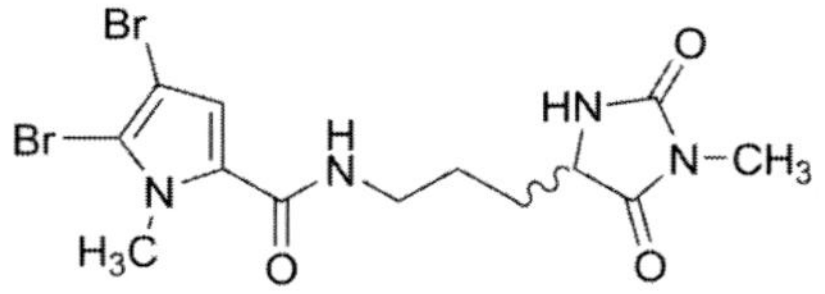

12

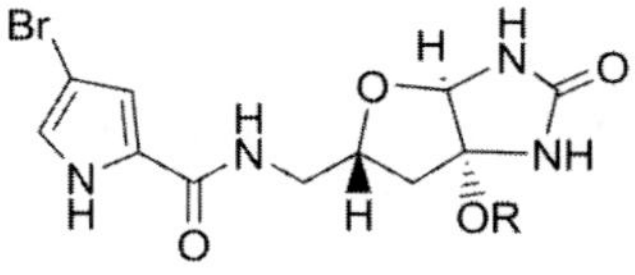

18 R = H
19 R = CH_3

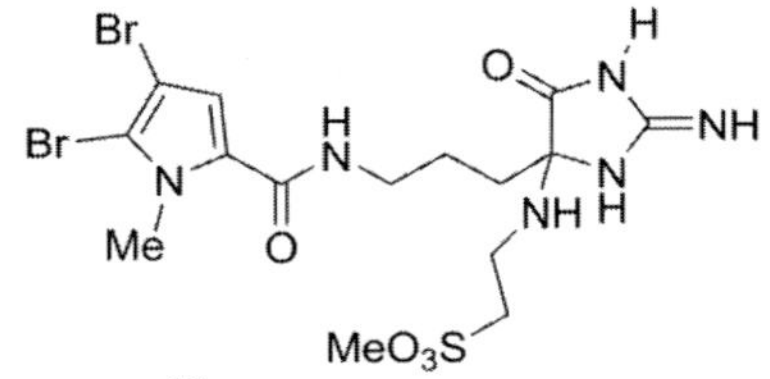

13

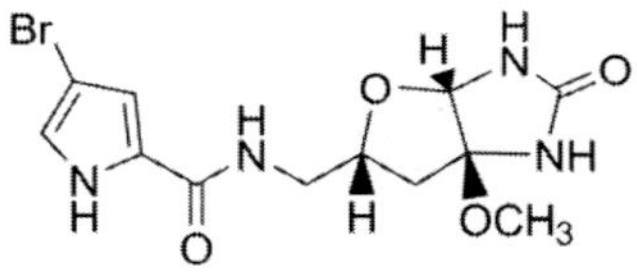

20

Figure 3. Structures of of compounds **9-20.**

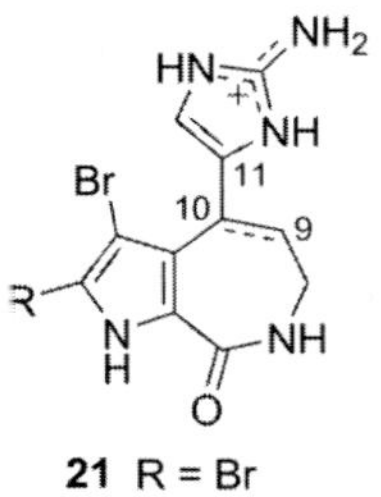

21 R = Br
22 R = H
23 R = Br, Δ9
24 R = H, Δ9

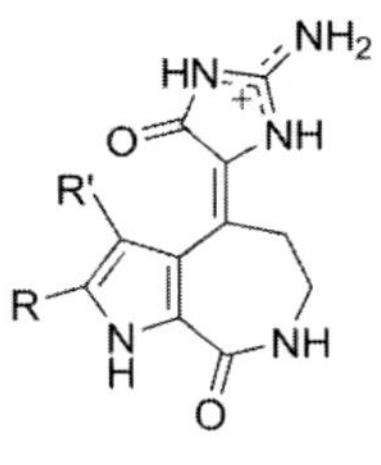

25 R = R' = H
26 R = Br R' = H
27 R = R' = Br

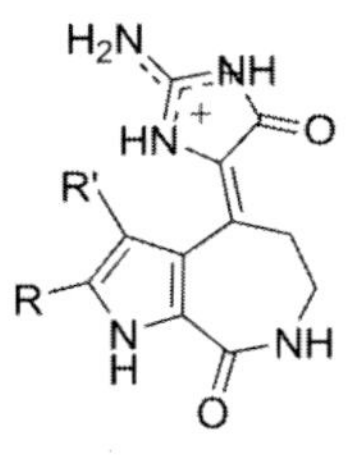

28 R = R' = H
29 R = Br R' = H
30 R = R' = Br

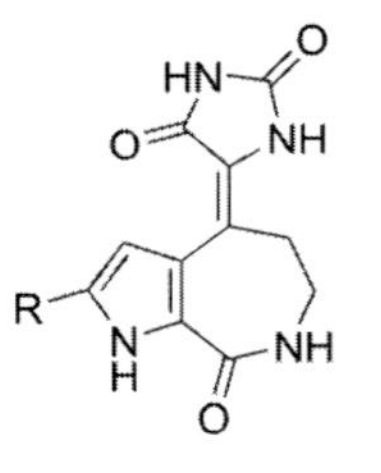

31 R = H
32 R = Br

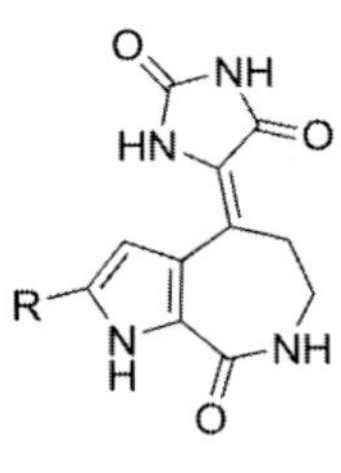

33 R = H
34 R = Br

Figure 4. Polyciclic C4/C10 derivatives: the hymenin family alkaloids **21-34**.

Stevensine is one of the main responsible of chemical defence against predatory fishes for the reef sponge *Axinella corrugata* [29].

Hymenialdisins (**25-30**) are hymenin analogues possessing a double bond $\Delta^{10,11}$ and a carbonyl group at C-12 which have been isolated from *Phakellia flabellata* [30], *Axinella verrucosa* [16], *Acanthella aurantiaca* [31], *Hymeniacidon aldis* [32], *Stylissa carteri* [28], and *Pseudaxynissa cantharella* [27]. Both (*E*) (**25-27**) and (*Z*) (**28-30**) configurations of the double bond $\Delta^{10,11}$, connecting the azepine and the imidazole rings, have been found in the hymenialdisin class, most likely deriving from interconversion of double bond geometry. Hymenialdisins have been proved proved to be nanomolar inhibitors of G_2 DNA damage checkpoint and of the protein kinases Chk1 and Chk2 [33], mitogen -activated protein kinase 1 (MEK -1) [34], and of other kinases [35], In addition, they inhibit several pro-inflammatory cytokines (interleukin-2, IL-6, IL-8, nitric oxide) through inhibition of the NF-*k*B signalling pathway [36, 37], This activity is potentially useful in the treatment of rheumatoid arthritis and osteoarthritis, a pathology for which specific pharmaceutical agent are strongly needed. Structure-activity relationships for MEK-1 inhibitory activity have been established: i) a change in the geometry of the double bond is not important; ii) the activity is strictly dependent from the bromination of the pyrrole ring; iii) the presence of an aminoimidazolone ring is essential for the activity. Interestingly, axinohydantoins (**31**-**32**) and spongiacidins (**33**-**34**), de-aminated analogues of hymenialdisins, do not share the interesting bioactivities shown by hymenialdisins, evidencing the key role played by the guanidinium group in the interaction with the targets have also been found [38, 39].

Cyclooroidin (**35**), from *Agelas oroides* [22], and its analogues agesamide (**36**) from *Agelas* sp. [40] and oxocyclostylidol (**37**) from *Stylissa caribica* [41] are N1/C9 derivatives (Figure 5). Cyclooroidin (**35**) could be envisaged as the precursor of the non-imidazole bromopyrrole alkaloids longamides (see below).

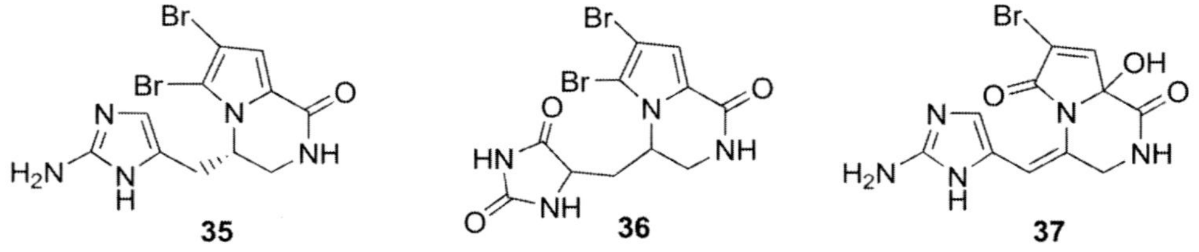

Figure 5. Polyciclic N1/C9 derivatives: cyclooroidin (**35**) and its analogues (**36**-**37**).

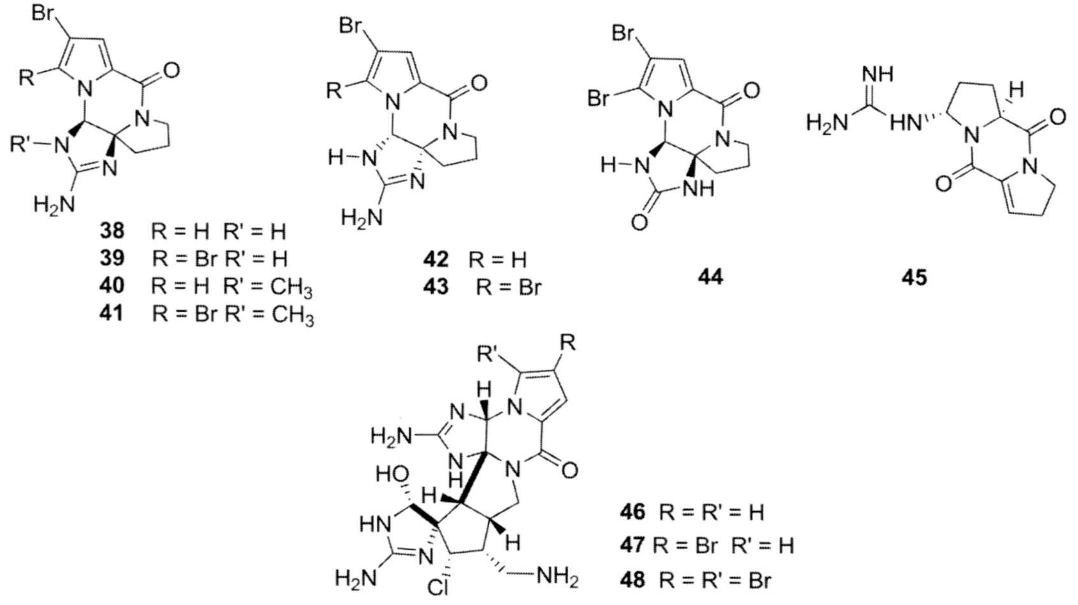

Figure 6. Polyciclic N7/C11 + N1/C12 derivatives: the phakellins (**38**- **45**) and palau'amines (**46**-**48**) family alkaloids.

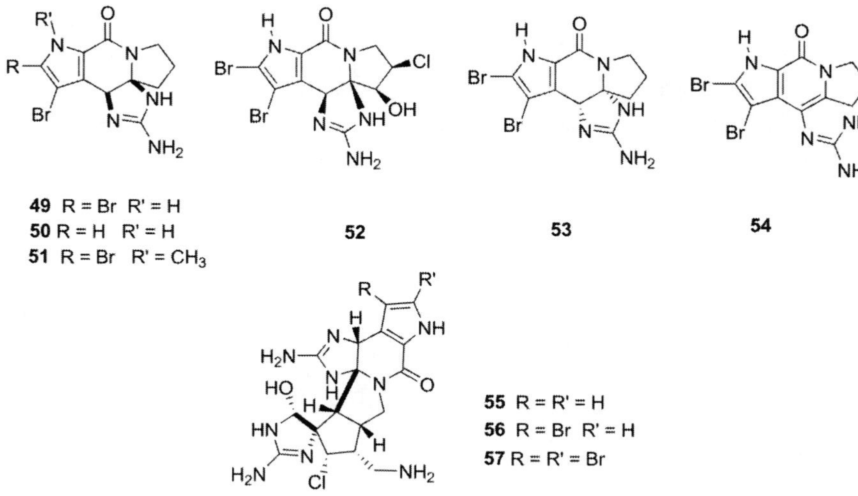

Figure 7. Polyciclic N7/C11 + C4/C12 derivatives: the isophakellins (**49** - **54**) and Styloguanidines (**55**-**57**) family alkaloids.

N7/C11 + N1/C12 derivatives are the phakellins family alkaloids (**38-44**), tetracyclic derivatives in which both the pyrrole and the amidic nitrogen atoms are involved in the formation of a linkage with carbon atoms of the imidazole ring [27, 42-45].

Phakellistatin (**44**) exhibited potent cell growth inhibitory activity against a variety of human cancer cell line s (ED_{50} from 0.3 to 4.0 μM) [45]. Interestingly, a series of similar cyclo(Pro,Pro) diketopiperazines, named verpacamides (e.g., **45**), have been very recently isolated from the marine sponge *Axinella vaceleti* [46]. It was proposed that these molecules could be the biogenetic precursors of the $C_{11}N_5$ skeleton, common to all the alkaloids of this class, through an intramolecular oxidative rearrangement, via a dioxetanone intermediate.

The group of N7/C11 + N1/C12 derivatives includes also palau'amines, complex metabolites possessing the phakellin skeleton conjugated to another heterocyclic moiety. Palau'amine (**46**), from the sponge *Stylotella aurantium* [47], is a strongly cytotoxic and immunosuppressive agent; a good antibacterial activity against *Staphylococcus aureus* and *Bacillus subtilis* was also reported [47]. Its monobrominated (**47**) and dibrominated (**48**) analogues resulted to be much less bioactive [48].

Molecules isomeric with those of the previous group are N7/C11 + C4/C12 derivatives, differing in the linkage of the imidazole carbon C-12 with C-4 in place of N-1. They constitute the isophakellins family alkaloids (**49-54**, Figure 7) [49-53].

This group of cyclic oroidin derivatives parallels the above N7/C11 + N1/C12 class in the presence of complex derivatives in which the isophakellin skeleton appears to be conjugated with an aminoimidazolyl propene unit. These isomers of palau'amines are called styloguanidines (**55-57**) [54]. Styloguanidines were shown to be potent inhibitors of chitinase, an enzyme involved in the ecdysis of many insects and crustaceans, whose inhibitors are supposed to control the settlement of barnacles and, therefore, could have a potential application as antifouling agents.

The only one member of a N1/C12 + N7/C12 derivatives groupo is dibromoagelaspongin (**58**, Figure 8), isolated from an *Agelas* sp. [55]. This molecule is strictly related to dibromophakellin (**39**), but in this case the nitrogen atoms *N*-1 and *N*-7 link the same carbon of the imidazole ring, namely C-12, and, consequently, the pyrrole-condensed ring is five-membered and not six-membered. It has been proposed propose a biogenetic relationship between the two molecules. The N1/C9 + C8/C12 derivatives group includes only agelastatins A-D (**62-65**, Figure 8) [56-58].

Agelastatins are highly bioactive derivatives, possessing nanomolar activity against several cancer cell line s, although the mechanism of this potent action has not yet been elucidated [57].

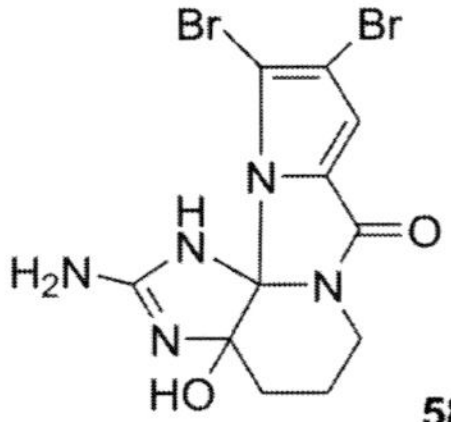

58

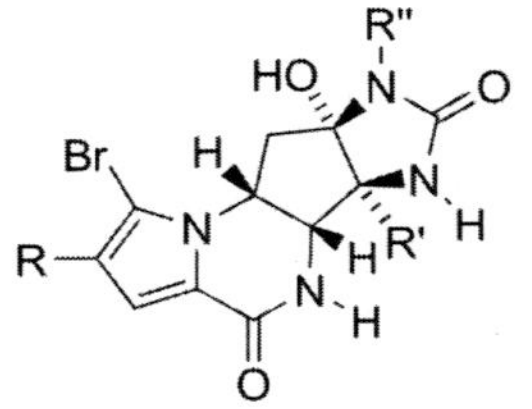

59 R = H R' = H R" =CH_3
60 R = Br R' = H R" = CH_3
61 R = H R' = OH R" = CH_3
62 R = R' = R" = H

Figure 8. Polyciclic N1/C12 + N7/C12 and N1/C9 + C8/C12 derivatives: dibromoagelaspongin (**58**) and agelastatins A-D (**59-62**).

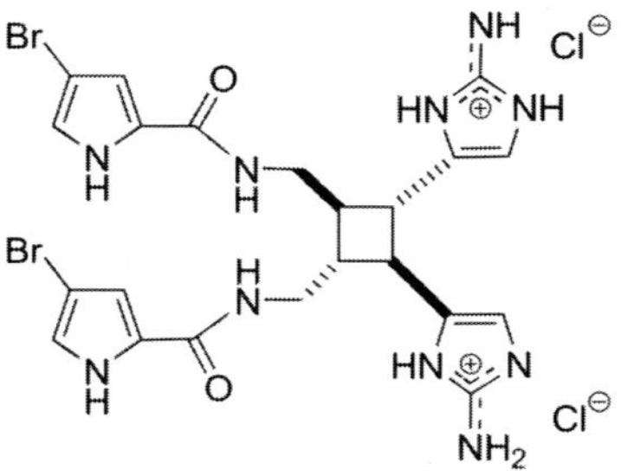

63

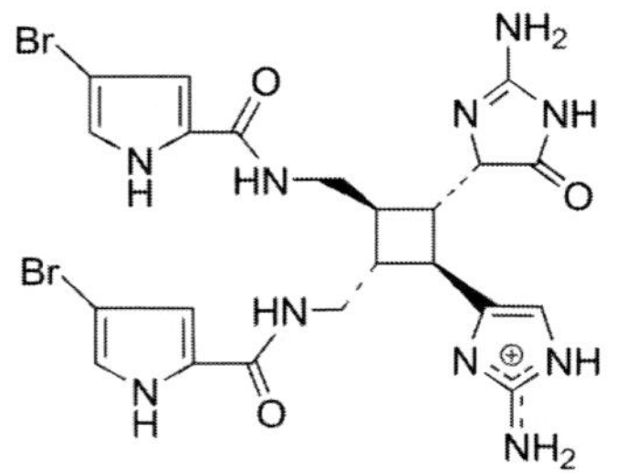

64

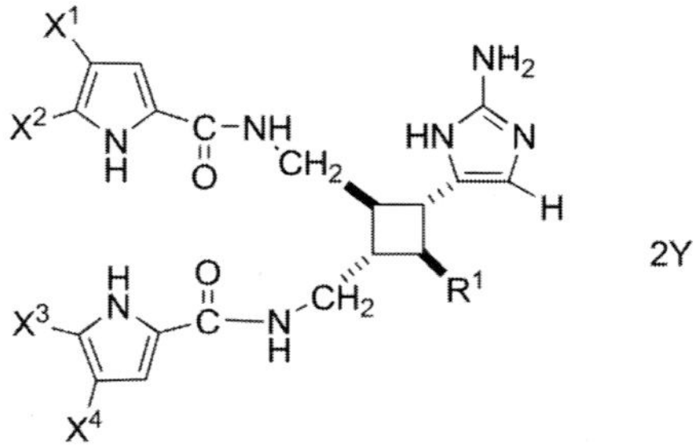

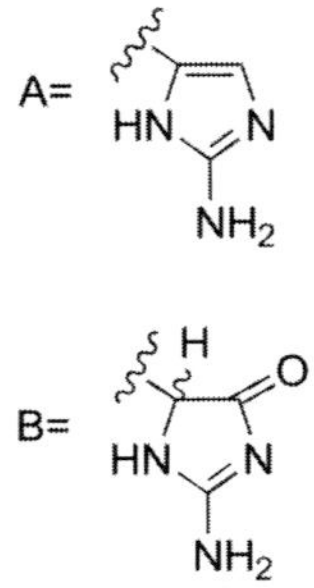

	X^1	X^2	X^3	X^4	R^1	Y
65	Br	H	H	H	A	HOAc
66	Br	Br	Br	Br	A	HOAc
67	H	H	H	Br	B	HOAc
68	Br	Br	H	Br	A	HOAc
69	H	H	H	H	A	HCl

Figure 9. Oroidin-like dimers: sceptrin (**63**) and its analogues **64-69**.

In addition, agelastatin A (**59**) inhibits glycogen synthase kinase-3β (GSK-3β), an activity that could be useful in the treatment of Alzheimer's disease, but, since GSK-3β is also related to the production of tumorigenic promoters involved in melanoma and colon cancer, its inhibition can have a significance also in the cancer prevention and therapy [59].

c) Oroidin-like dimers and tetramers. Dimeric bromopyrrole alkaloids have been almost exclusively found as secondary metabolites of sponges belonging to the genus *Agelas.* The isolation of sceptrin (**63**) from from *Agelas sceptrum* in 1981 [60] marked the first report of a dimer of the pyrrole imidazole alkaloid family. It is a symmetrical dimer of the 2-debromo derivative of oroidin and exhibits a broad range of biological activities, including bacteriostatic [60-62], fungicidal [63], antiviral [64], antimuscarinic [12], and antihistaminic [15] properties. Sceptrin was shown to be the first and most potent non-peptide somatostatin inhibitor in the submicromolar range [65]. Several sceptrin analogues have been reported displaying diverse biological activity. Oxysceptrin (**64**), from *Agelas conifera* [66], was an actimyosin-ATPase activator [67]. Numerous antiviral and antibacterial sceptrin/oxysceptrin derivatives (**65-69**) have been isolated [64, 68, 69]. Dibromosceptrin (**66**) was shown to reduce voltage dependent calcium elevation in PC12 cells [70].

The complex fused-hexacyclic skeleton of konbu'acidin A (**70**), containing two guanidine units is related to palau'amine (**46**) and styloguanidines (**55-57**). This molecule was isolated from an Okinawan *Hymeniacidon* sp. and exhibited inhibitory activity against cyclin dependent kinase 4 (cdk4) [71].

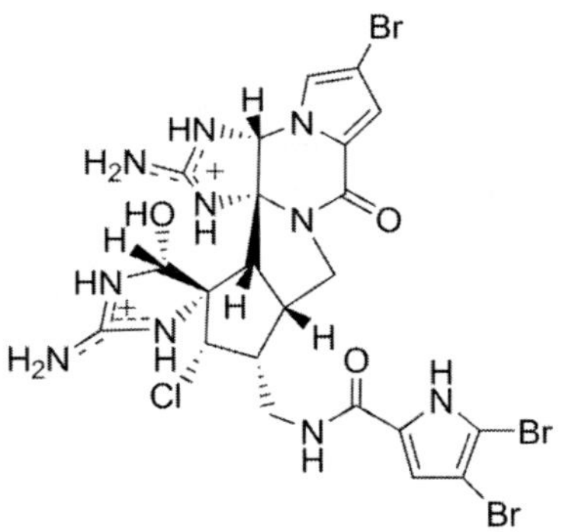

70

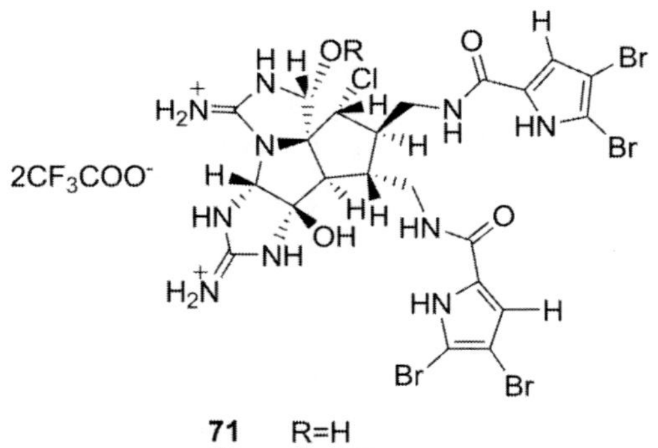

71 R=H
73 $R=CH_3$

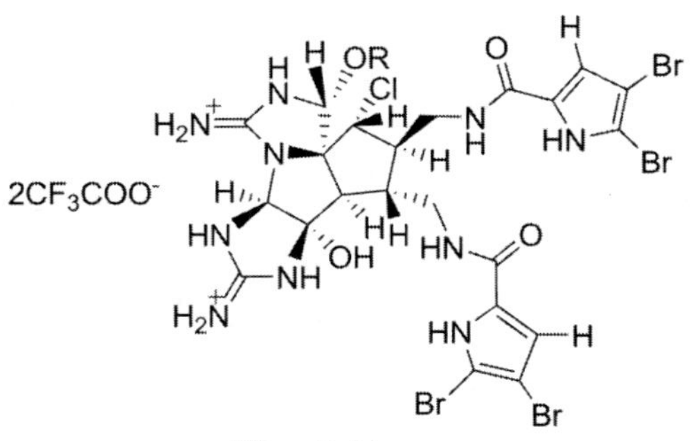

72 R=H
74 $R=CH_3$

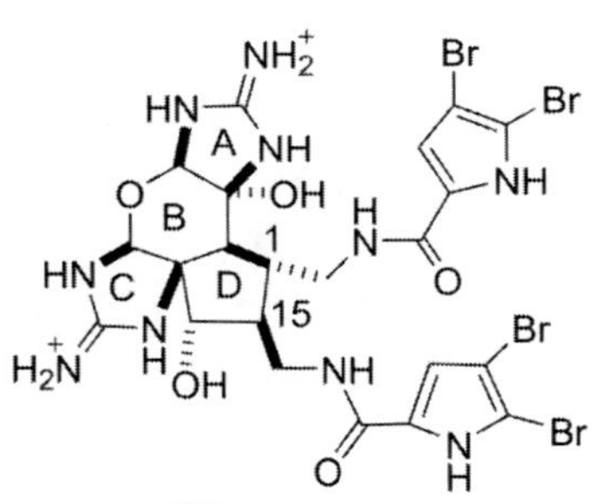

75

Figure 10. Oroidin-like dimers: **konbu'acidin A** (**70**), axinellamines A-D (**71-74**), and massadine (**75**).

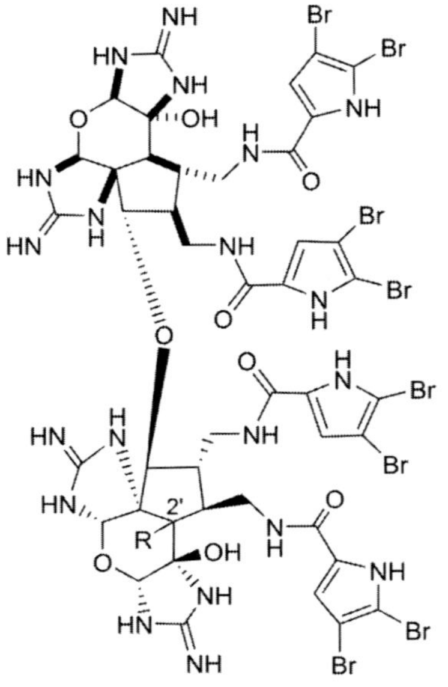

76 R = β-H2'
77 R = α-H2'

Figure 11. Oroidin-like tetramers: stylissadines A and B (**76** and **77**).

Four imidazo-azolo-imidazole alkaloids, axinellamines A-D (**71**-**74**), were isolated from an Australian *Axinella* sp.; compounds **72-74** had bactericidal activity against *Helicobacter pylori,* a Gram-negative bacterium associated with pepticular and gastric cancer [72]. Massadine (**75**), a highly ox ygenated congener of dimeric oroidin derivatives was isolated from the marine sponge *Stylissa* cfr. *massa* as an inhibitor of geranylgeranyltransferase type I (GGTase I) [73]. In the proposed structure two cyclic guanidine faced each other in the endo positions of ring B, which adopted a boat conformation.

Stylissadines (**76** and **77**) are tetrameric molecules. Structurally, they are ether-linked dimers of massadine and they are the largest and, with their 16 stereogenic centers, the most complex structures within the oroidin family of alkaloids discovered so far. Both have biological activity as antagonists of the P2X7 receptor involved in inflammatory disease s, but their high molecular weight and structural complexity render them hard to develop as drugs [74].

d) Others. The pyrrole-imidazole alkaloids class includes several members whose structures hardly fit any of the aforementioned classification criteria. Examples are clathramides A-D (**78-81**), imidazole containing bromopyrrole alkaloids but, since in this case the imidazole ring is part of an histidine -like moiety, their skeleton appears to be different from that of oroidin. They were isolated from *A. clathrodes* [75], and *A. dispar* [76] as mild antifungal agents. Laughine (**82**), from *gelas wiedenmayeri* [153], and the related compound **83**, from *Eurypon laughlini* [77], are a pyrrole-imidazole alkaloids with a guanidine function instead of the aminoimidazole (Figure 12).

Ageladine A (**84**), from the sponge *Agelas nakamurai*, is the first isolated pyrrole-imidazole alkaloid containing a 2-aminoimidazolopyridine moiety [78].

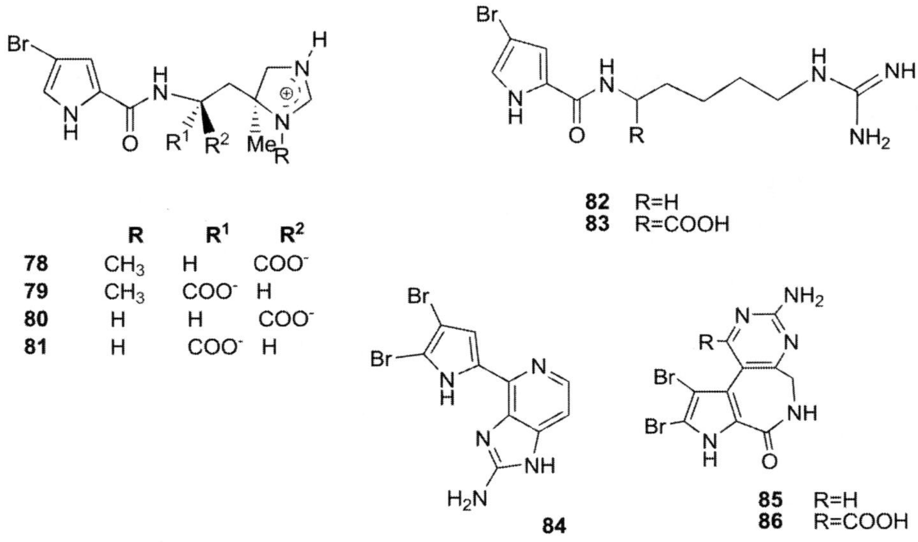

Figure 12. Other pyrrole-imidazole alkaloids: clathramides (**78-81**), laughines (**82-83**) ageladine A (**84**), and latonduines (**85-86**).

It is reported to inhibit several matrix metalloproteinase s (MMPs -1, -2, -8, -9, -12 and -13); as it has been demonstrated that **84** does not chelate zinc ion like all known MMP -inhibitors do, the natural product probably operates with a completely different mechanism of action. Ageladine A showed also antiangiogenic effects. Latonduines A (**85**) and B (**86**), have been isolated from the Indonesian sponge *Stylissa carteri* [79]. Their skeleton cannot be derived from the $C_{11}N_5$ building block of the oroidins; it has been proposed that ornithine is the biogenetic precursor to the aminopyrimidine fragment of the latonduines (Figure 12).

3. Lamellarins

The lamellarins form a group of more than 50 highly condensed DOPA- derived pyrrole alkaloids which have attracted researchers ' interest due to both their structural originality and significant biological properties [80-82]. Lamellarins were originally isolated from the prosobranch mollusk *Lamellaria* sp. in 1985 [83]. They were later extracted mostly from ascidians belonging to *Didemnum* genus [84-89] and rarely from sponges [90, 91]. The invertebrates likely represent the original producer of lamellarins because these organisms are presumed to be the dietary source of the Lamellaria mollusks. The lamellarins have a pyrrole ring as a core component of their skeleton. They fall into two structural groups, depending on whether the central pyrrole ring is fused (Group I) or unfused (Group II) to adjacent aromatic rings. Group I could be further divided into two subgroups, Ia, including compounds possessing an olefin at C5/C6, and Ib, with compounds in which this olefin is saturated (Figure 13). Each group includes derivatives in which phenolic hydroxyl groups are substituted by methoxy, sulfate or acetate functions. Selected structures of lamellarins (compounds **87-106**) are reported in Figure 14.

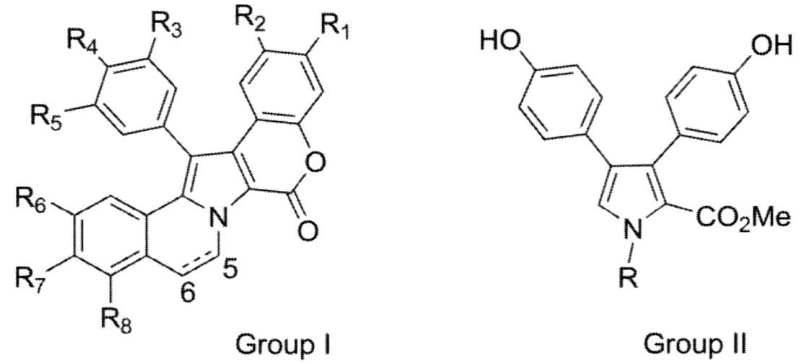

Figure 13. Core structures of lamellarins.

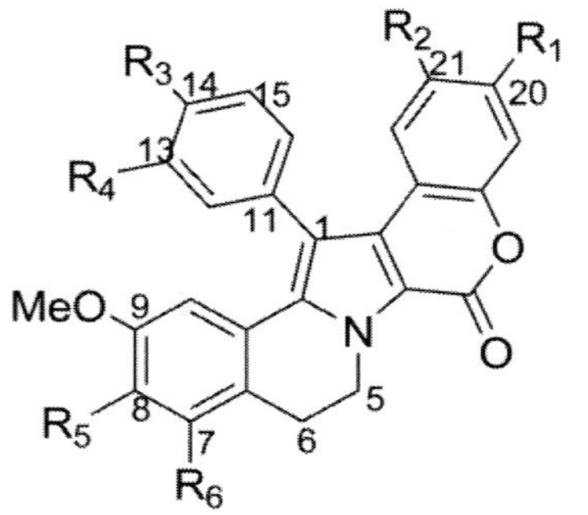

N.	Lamellarin	R1	R2	R3	R4	R5	R6	R7
87	H	OH	OH	OH	OH	OH	OH	H
88	M	OH	OMe	OH	OMe	OMe	OMe	OH
89	N	OH	OMe	OMe	OH	OMe	OH	H
90	N tri-acetate	OAc	OMe	OMe	OAc	OMe	OAc	H
91	X	OH	OMe	OMe	OH	OMe	OMe	OH
92	ε	OH	OMe	OMe	OMe	OMe	OMe	OH
93	α	OH	OMe	OMe	OH	OMe	OMe	H
94	α 20-sulfate	OSO_3Na	OMe	OMe	OH	OMe	OMe	H
95	B 20-sulfate	OSO_3Na	OMe	OH	OMe	OMe	OMe	OMe
96	α13,20-disulfate	OSO_3Na	OMe	OMe	OSO_3Na	OMe	OMe	H

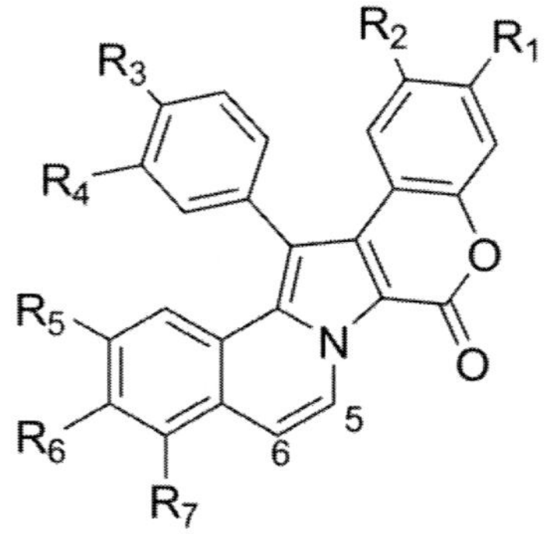

N.	Lamellarin	R1	R2	R3	R4	R5	R6
97	E	OH	OMe	OMe	OH	OMe	OH
98	F	OH	OMe	OMe	OMe	OMe	OH
99	G	OMe	OH	OMe	OH	OH	H
100	C 20-sulfate	OSO_3Na	OMe	OH	OMe	OMe	OMe
101	L 20-sulfate	OSO_3Na	OMe	OMe	OH	OH	H
102	G 8-sulfate	OMe	OH	OMe	OH	OSO_3Na	H
103	I	OH	OMe	OMe	OMe	OMe	OMe
104	J	OH	OMe	OMe	OMe	OH	H
105	K	OH	OMe	OH	OMe	OMe	OH
106	T	OH	OMe	OMe	OH	OMe	OMe

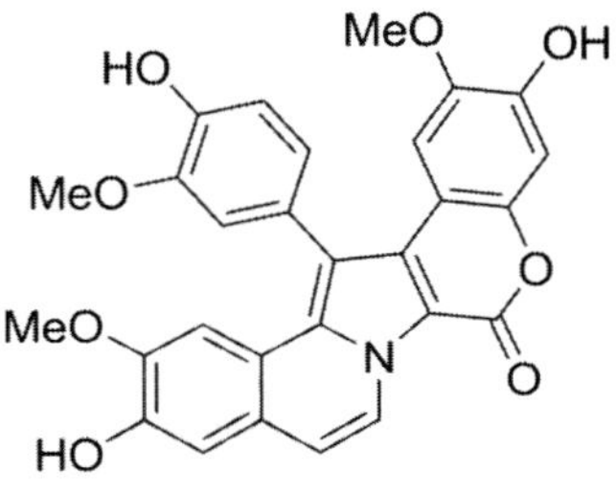

107

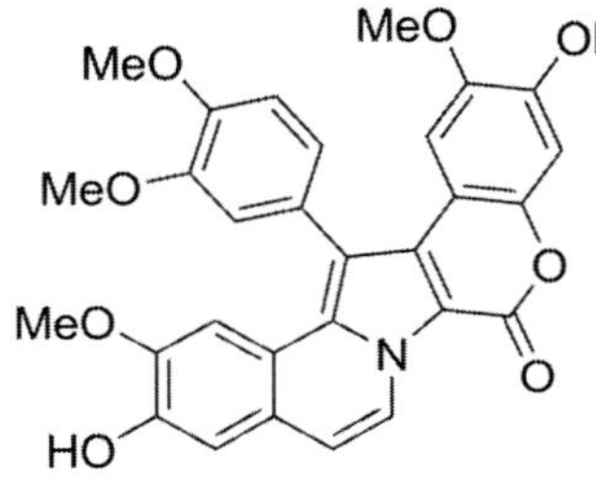

108

Figure 14. Selected structures of lamellarins.

The most common and remarkable property of the lamellarins is their capacity to inhibit the proliferation of cancer cells [92]. They are among the most promising anti-cancer drug candidates derived from marine invertebrates. The majority of lamellarins are considerably cytotoxic, with IC_{50} (or LD_{50}) values in the nanomolar to micromolar range, depending on the experimental condition s and the nature of the compounds. A noticeable exception is that of the sulphated lamellarins, which are not cytotoxic presumably due to reduced cell uptake. Lamellarins D (**107**), K (**105**) and M (**88**), are the most potent compounds in the series [80], but lamellarin N (**51**) and dehydrolamellarin J (**108**) are also promising candidates.

Following the discovery of the potent anti-proliferative and pro-apoptotic activities of lamellarins, their biological activities have been extensively studied, in particular their capacities to interfere with topoisomerase (TOPO) I and mitochondria, both contributing to their potent cytotoxicity. Although all the aspects of the lamellarins mechanism of action are not completely known, it has been demonstrated that they are potent inhibitors of topoisomerase (TOPO) I, they interact with DNA and they target mitochondria directly and induce the release of cytochrome C and apoptosis -inducing factor (AIF) [92-95]. A nuclear TOPO I-independent, direct effect on mitochondria has been also demonstrated; studies on the apoptotic pathway induced by lamellarin D revealed a very early mitochondrial dysfunction, including a reduction of mitochondrial membrane potential and the release of cytochrome c and AIF from the mitochondria to the cytosol [94, 95]. Although the exact mechanism implicated in the targeting of mitochondria is not fully understood, a direct induction of mitochondrial permeability transition (MPT), which leads to swelling and cytochrome c release, has been identified as the mechanism behind the pro-apoptotic function of these molecules [96]. All the aforementioned findings open the path to new cancer therapy strategies that target mitochondria, thereby reinforcing the potential value of mitochondriophilic drugs for this indication.

Some lamellarins also function as multi-drug resistance (MDR) reversal drugs [97, 98]. Lamellarins I (**103**) and K (**105**), and T (**106**) have been reported to act as MDR modulators. It has been demonstrated that lamellarin I (**103**) reversed MDR by directly inhibiting the P-glycoprotein-mediated drug efflux at nontoxic doses and that the potency of lamellarin I as an MDR modulator was 9-16-fold higher than that of approved anticancer agent verapamil, resensitizing the resistant malignant cells to frontline therapeutics [97].

A few members of the lamellarin family revealed HIV -1 integrase inhibition activity, although this pharmacological activity has so far not been thoroughly explored. Examples are lamellarin α 20-sulfate (**94**) which acts at two different steps of the catalytic cycle, the terminal cleavage and the strand transfer, and inhibits the viral replication in cultured cells at non-toxic doses [87]. The sulfate group plays a critical role since the desulfated analogue lamellarin α (**93**) is inactive. Lamellarin H (**87**) and lamellarin α 13,20-disulfate (**96**) were also studied for their HIV-1 integrase inhibition. Lamellarin α 13,20-disulfate **96** was a less effective inhibitor of HIV-1 integrase than was lamellarin α -20 sulfate **94** while lamellarin H (**87**) is a more potent inhibitor of HIV-1 integrase but lacks the specificity required to be clinically useful [99].

Other activities are reported for lamellarin alkaloids, including antioxidant activity [88] and inhibition of protein kinases [100]; however, these are certainly not the most advantageous properties of this class of molecules given their significant cytotoxicity which makes them an original source of structures for the design of drugs active against cancer. In addition, the key discovery that TOPO I was a major target for lamellarin D has opened the door to the determination of structure –function relationships and the rational design of lamellarin analogs of pharmaceutical interest. Elegant synthetic routes to these compounds or their key intermediates have been developed, which afforded sufficient quantities of structurally simplified lamellarin analogs for SAR and other biological studies [81]. SAR studies have been directed at derivatives of lamellarin D also by using their mechanism-based (i.e., TOPO I inhibition) activities in relation to their cytotoxicities and precise structure-activity relationships have been delineated. The structural elements that affect the lamellarin activities are illustrated in Figure 15.

Recognition of the biological potential of this large class of alkaloids can be found in recent reports of lamellarins entering into preclinical development for the treatment of multidrug-resistant tumors. More than 300 derivatives have been synthesized by chemists at PharmaMar (Spain) and drug candidates have been selected for the extended preclinical and toxicological studies required prior to human clinical trials [101].

Seeveral lamellarin-related alkaloids have been isolated also from ascidians species. Examples are lukianols (**109–110**) [102], polycitones (**111–112**), polycitrins (**113–114**) [103, 104], and ningalins (**115–118**) [105], (Figure 16). The structures of the lukianols A and B contain a pyrrolooxazinone moiety; lukianol A (**109**) inhibited DNA synthesis in L1210 lymphocytic leukemia cell line s with less effect on RNA and protein synthesis, demonstrating a therapeutic profile very similar to current clinically used anticancer agents [102]. Ningalins A–D (**115–118**) are condensed aromatic systems with the unifying theme that all appear derived from the condensation of two, three, four, and five DOPA precursors, respectively.

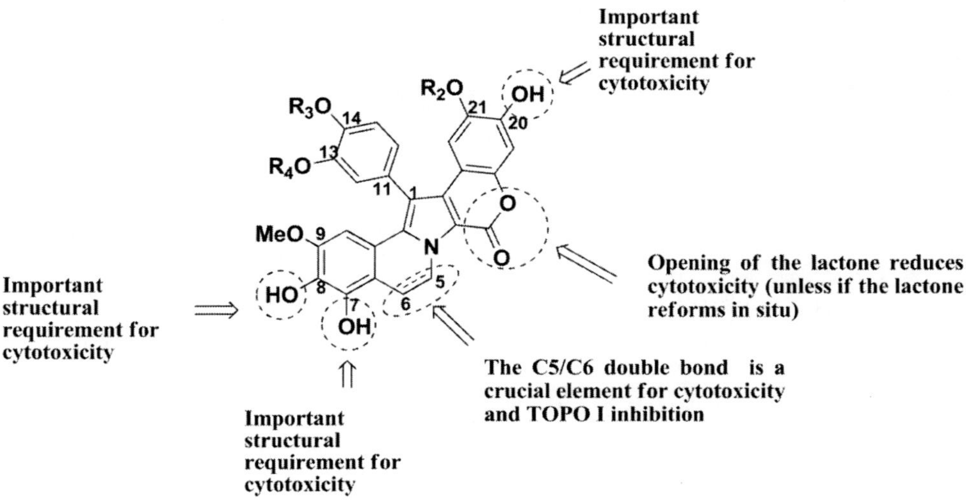

Figure 15. Structural elements that affect lamellarin biological activities.

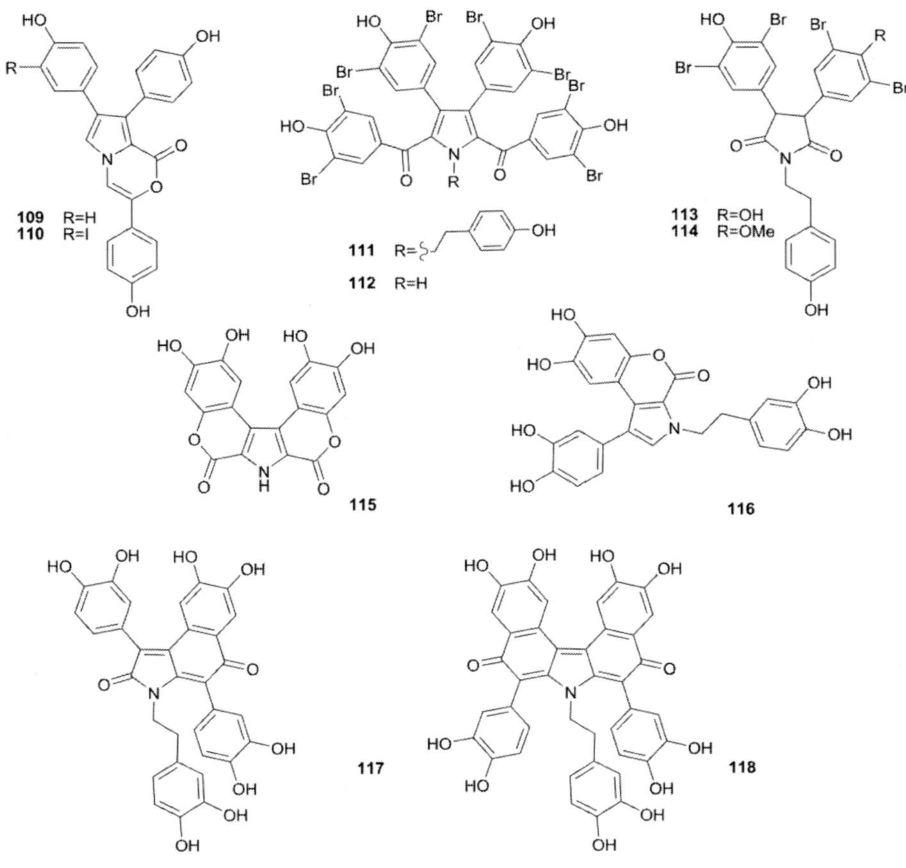

Figure 16. Structures of alkaloids related to lamellarins.

Although a potential role for these alkaloids in metal sequestration has not been demonstrated they are structurally related to other metal binding ascidian-derived *o*-catechols; it is thus conceivable that they participate in the metal chelating phenomena characteristic of this class of marine invertebrates [102, 106].

4. Ritterazines and Cephalostatins

The ritterazines are highly ox ygenated dimeric steroidal alkaloids isolated by Fusetani and co-workers from 1994 to 1997 from the colonial tunicate *Ritterella tokioka* (family Policlinidae) [107, 108].

Together with cephalostatins, isolated from the marine worm *Cephalodiscus gilchristi*, they make up a unique family of 45 trisdecacyclic bis-steroidal pyrazines (Figures 1 and 2) that display extremely potent cytotoxicity against human tumors with unique cell selectivity and apoptotic response [Reviewed in 109, 110].

The outstanding antineoplastic potency together with the new and challenging molecular architecture made these marine invertebrate alkaloids a promise as anticancer leads.

Illustrative examples of cephalostatin and ritterazine structures are reported in figure 17 and 18, respectively. The ritterazines and cephalostatines share many common structural features. Both consist of two highly ox ygenated C27 steroid units fused via a pyrazine ring at C-2 and C-3; both chains of the steroid units usually form either 5/5 or 5/6 spiroketals. While cephalostatins in general are more oxygenated on the right side, the ritterazines have the more oxygenated left side.

Hydroxyl groups are present at C-12, C-17, C-23, C-26, C-12′, and C-23′ in the cephalostatins, whereas C-12, C-7′, C-12′, C-17′, and C-25′ are hydroxylated in the ritterazines. Cephalostatins 5 (**123**) and 6 are unusual in that they both contain an aromatic C ring. While naturally occurring and synthetic steroids with aromatic A rings are well-known, steroids bearing an aromatic C ring are quite rare. Cephalostatin 12 (**127**) is the only symmetric costituent (right unit = left unit). Ritterazines N (**131**) -S, having two non-polar steroidal units, were much less active than ritterazine B. Ritterazines T (**132**)-Y are related to ritterazine A (**128**) and B (**129**), composed of polar and non-polar steroidal units, lacking the C7' and C17' hydroxyl groups. Ritterazine Y (**134**) differs from ritterazine B in the absence of the two hydroxyl groups whereas ritterazines W and X have the 5/5 spiroketal terminus in the polar steroidal unit instead of the 5/6 spiroketal terminus. In ritterazine T, and in ritterazine V, both steroidal units are rearranged.

Cephalostatin 1 (**119**) is among the most powerful anticancer agents ever tested, displaying subnanomolar-to-picomolar cytotoxicity against much of the National Cancer Institute's (NCI) 60-cell line panel, with femtomolar activity against the P388 cell line and in the Purdue Cell Culture Laboratory (PCCL) human tumor panel [109, 110].

Most of the remaining cephalostatins, were potent much the same. A dramatic reduction in PS cell growth inhibition was displayed by cephalostatins 5 and 6, suggesting that preservation of structural integrity in the right-hand side unit, including C/D ring stereochemistry, is very important to realizing powerful cytotoxicity. Cephalostatin 12 also dispalyed significantly reduced inhibitory activity in comparison to other cephalostatins, suggesting that asymmetry is necessary for optimum cytotoxicity [109].

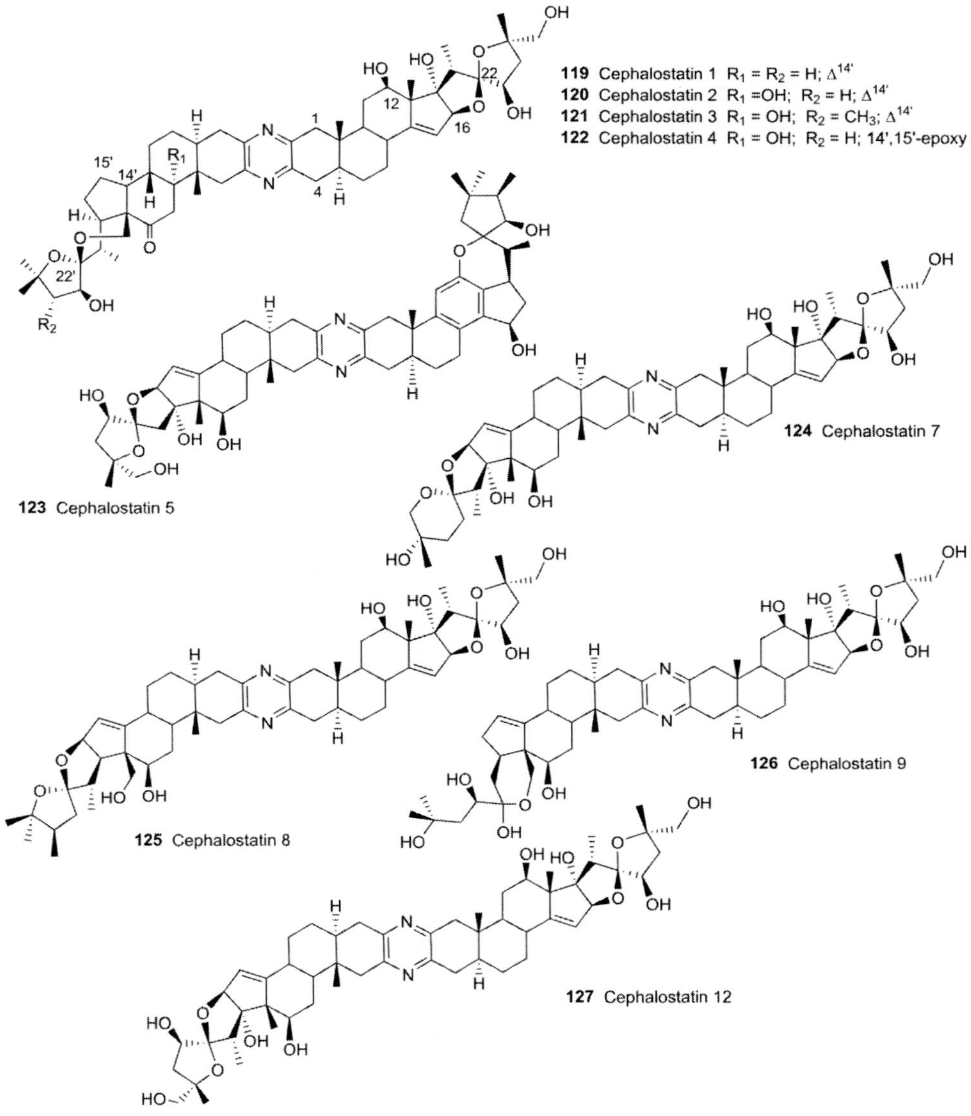

Figure 17. Selected cephalostatin structures.

Ritterazines are less potent than cephalostatins, with P388 inhibitory activities ranging from 3.6 μM for ritterazine W (**133**) to 0.17 nM for ritterazine B (**129**), (see Table 1) the most active constituent of *R. tokioka*, which, not surprisingly, contains nearly the same right-hand side steroid unit (**129**: no 17*R*-hydroxy moiety) as the most active cephalostatins (**119-122**, **124-126** see Table 1).

The 45 members of the cephalostatin/ritterazine family isolated to date, together with a growing number of analogues and related monosteroidal antineoplastics provide the basis for elucidating some structure -activity relationships (SAR) of these potent cytotoxins and for discovering the minimum pharmacophore required to maintain potent cancer cell growth inhibitory behavior.

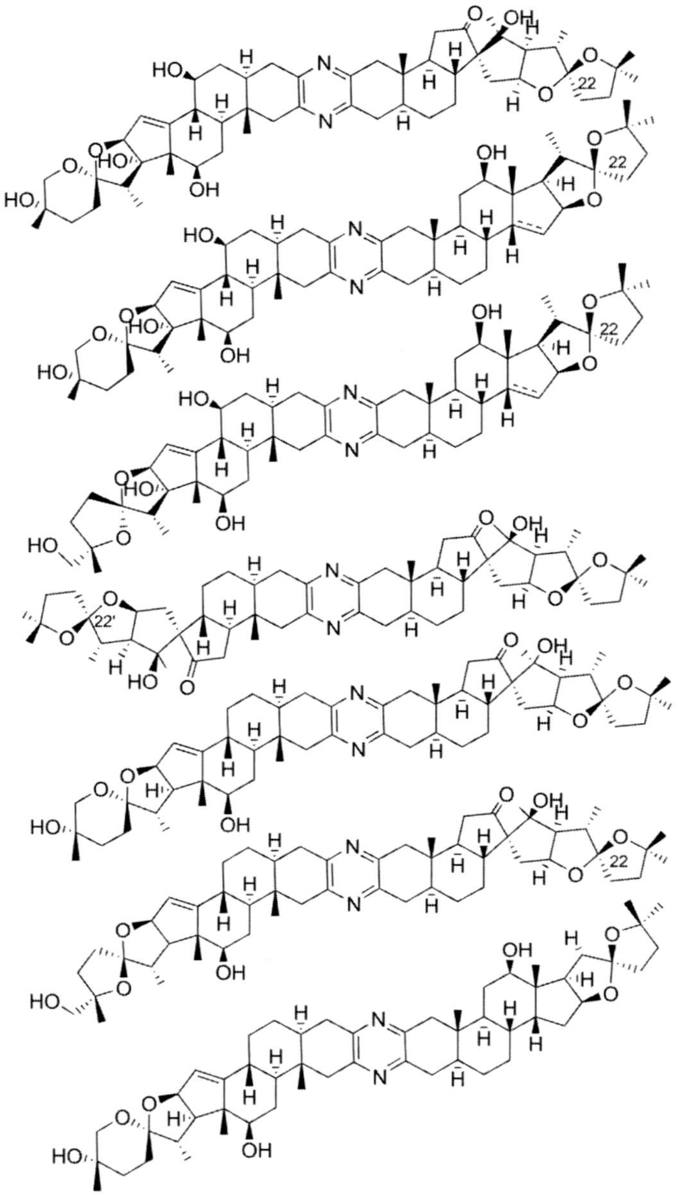

128 Ritterazine A

129 Ritterazine B

130 Ritterazine C

131 Ritterazine N

132 Ritterazine T

133 Ritterazine W

134 Ritterazine Y

Figure 18. Selected ritterazine structures.

These studies suggest that four features conspire to provide active *in vitro* materials: (i) a molecular dipole consisting of covalently linked lipophilic "nonpolar" and hydroxylated "polar" domains, with a molecular length of ~30Å; (ii) a spiroketal or other latent precursor of an oxacarbenium ion; (iii) one or more homoallylic oxygen arrays; and (iv) a 17-OH function. The pyrazine ring, though present in most examples, is absent in several subnanomolar active monosteroids. Questions regarding the necessity, location, and molecular function for the latter two features remain, but both are present in the most potent natural and analogue examples, whereas one or more of these distinctive units are missing in structures with notably inferior *in vitro* activity [110].

Table 1. Murine P388 lymphocytic leukemia inhibitory activity of ritterazines and cephalostatins

Ritterazine	(ED_{50} nM)	Cephalostatin	(ED_{50} nM)
A	14.2	**1**	10^{-4}-10^{-6}
B	0.17	**2**	10^{-4}-10^{-6}
C	102.3	**3**	10^{-4}-10^{-6}
D	17.5	**4**	10^{-4}-10^{-6}
E	3.8	**5**	42.5
F	0.81	**6**	2.3
G	0.81	**7**	1- < 0.1
H	17.8	**8**	1- < 0.1
I	15.3	**9**	1- < 0.1
J	14.0	**10**	3.2
K	10.4	**11**	2.7
L	11.1	**12**	76.2
M	16.7	**13**	47.9
N	522	**14**	4.4
O	2383	**15**	26.2
P	819	**16**	< 1.1
Q	657	**17**	4.4
R	2461	**18**	4.6
S	539	**19**	7.9
T	522		
U	2341		
V	2341		
W	3631		
X	3404		
Y	4.0		
Z	2200		

A COMPARE pattern-recognition analysis gave correlation coefficient s of ~0.9 between cephalostatin 1 and ritterazine B in NCI-10 cell line s, suggesting that ritterazines act by the same mechanism as does cephalostatins which, however, is presently largely unknown [111]. The fingerprint of their activity in the NCI 60-tumor panel is quite different from known anticancer agents, likely indicating a new mechanism of action. bis-Steroidal pyrazine alkaloids do not contain functional groups commonly associated with cytotoxicity such as alkylation sites, Michael acceptors, intercalators, or redox-active quinones. Early speculation on the mode of action of these compounds centered around the likelihood of cell membrane penetration due to their steroidal nature and dimensions or, alternatively, the possibility that the compounds serve as a spatially defined set of hydrogen -bond donors /acceptors for enzyme binding [112]. Pro-apoptotic properties have been demonstrated for ritterazine B (**2**) although apoptosis induced by this compound appeared to be independent of the caspase pathway [111]. A recent biological study clearly demonstrated that cephalostatin 1 (**119**) evokes a new cytochrome c-independent apoptosis signaling pathway. This is in contrast to most of the well-known anticancer drug s, which act in a cytochrome c-dependent route.

The lack of cytochrome c release by **119** indicates that it induces apoptosis in cancer cells via caspase-9 activation without formation of an apoptosome (complex of cytochrome c, procaspase-9, and the cytosolic factor Apaf-1). As part of this unique apoptosis mechanism, **119** has been found to selectively release mitochondrial Smac (second mitochondria -derived activator of caspases), necessary for caspase-9 and caspase-2 activations. Furthermore, it produces an endoplasmic reticulum stress response, inducing caspase-4, which activates the caspase-9 route to apoptosis. The new pathway is marked by structural changes in the mitochondria [113-115]. These results show that ritterazine/cephalostatine alkaloids family might be potent inducer of apoptosis acting via a novel antimitotic mechanism implying that they could be used to treat drug-resistant cancers.

Despites the number and value of synthetic contributions to the cephalostatins and ritterazines [109, 110], owing to the complexity of the targets, only very small amounts of these substances were produced and in very low overall yields, not suitable to supply sufficient samples at reasonable cost for extended biological evaluation. The availability of these compounds from their only known natural sources, the marine worm *C. gilchristi* and the marine tunicate *Ritterella tokioka*, is still extremely limited; while the isolation yields of the ritterazines are slightly better than the cephalostatins, they also are too low to supply clinical trials. As a result, *in vivo* anticancer evaluation of these very promising natural products and subsequent preclinical development has been greatly restricted. Nevertheless, the occurrence in different phyla of the cephalostatins and ritterazines, closely related in structure metabolites, raises questions as to their true producer and may indicate a microbial origin for this family of compounds [109, 110].

The current good possibility that the cephalostatin/ritterazine alkaloids inhibit cancer cell growth by affecting a novel molecular target (s), the ongoing total synthetic and SAR challenges, the possibility of locating a marine microorganism source actually responsible for their biosynthesis, and clinical development prospects suggest the bis steroidal alkaloids field will become increasingly productive and useful.

5. Tetrahydroisoquinoline Alkaloids: Ecteinascidins

Ecteinascidins, isolated mainly from the Caribbean ascidian *Ecteinascidia turbinata*, are probably the most useful anticancer agents found to date in a marine source. The lead compound, trabectedin (ET-743, **136**), is regarded as a successful story of modern marine drug research: it is indeed the first representative of a marine natural product to receive marketing authorization for the treatment of patients with advanced or metastatic soft tissue sarcomas (STS) and relapsed platinum -sensitive ovarian cancer under the brand name Yondelis® [116, 117].

The first description and structural characterization of six new chemical entities called ecteinascidins, ET 729 (**135**), ET 743 (**136**), ET 745 (**137**), ET 759A (**138**), ET 759B (**139**), and ET 770 (**140**), was reported by the Rinehart group in 1990 of which ET-743 was the most abundant representative [118]. Simultaneously, Wright and coworkers described compounds **135** and **136** [119], but the unequivocal assignment of the absolute stereochemistry was achieved only when the X-ray crystal structure s of the natural N^{12}-oxide of ET-743 (**141**) and a synthetic O-methyl analogue of N^{12}-formyl ET-729 (**142**) were solved [120].

Successively, a number of additional new members of this class of molecules have been isolated, such as compounds **143** and **144** [121], **145–148** [122], **149** [123], and **150–153** [124] (Figure 19).

Apart from the biological activity, ecteinascidins exhibit an extraordinary three-dimensional molecular architecture; their unique structure consists of a monobridged pentacyclic skeleton composed of two fused tetrahydroisoquinoline rings (subunits A and B) linked to a 10-membered lactone bridge through a benzylic sulfide linkage. Most ecteinascidins have an additional tetrahydroisoquinoline or tetrahydro-β-carboline ring (subunit C) attached to the rest of the structure through a spiro-ring (Figure 19).

The carbon and nitrogen framework of units A-B of the ecteinascidins is the same as that of the saframycins, safracins, and renieramicins families antitumor agent s isolated from bacteria and sponges [125]. It has been proposed that A-B units could be formed by condensation of two DOPA-derived building blocks, and the tetrahydroisoquinoline ring in unit B is closed by condensation with a serine -(or glycine -) derived aldehyde as in the case of the related saframycins.

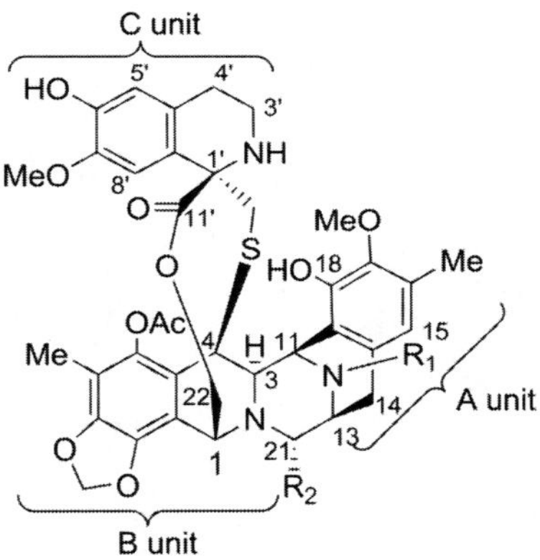

135 ET729 R_1=H; R_2=OH
136 ET743 R_1=Me; R_2=OH
137 ET745 R_1=Me; R_2=H
138 ET759A R_1=Me; R_2= =O (lactam)
139 ET759B R_1=Me; R_2=OH, S-oxide
140 ET770 R_1=Me; R_2=CN
141 ET 743 N12-oxide R_1=Me; R_2= OH, N-oxide
142 ET 729 N12-formyl R_1=CHO; R_2= OH
146 ET 815 R_1=Me; R_2=CH(CHO)$_2$
147 ET 731 R_1= R_2=H
148 ET 745B R_1=H; R_2=OH, S-oxide
149 ET 786 R_1=Me; R_2=CN, S-oxide

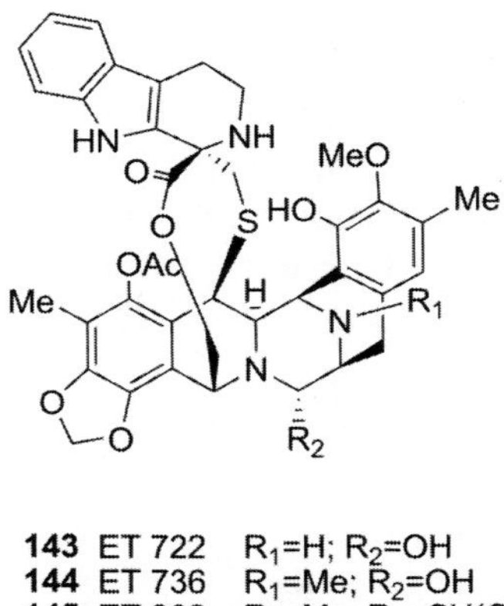

143 ET 722 R_1=H; R_2=OH
144 ET 736 R_1=Me; R_2=OH
145 ET 808 R_1=Me; R_2=CH(CHO)$_2$

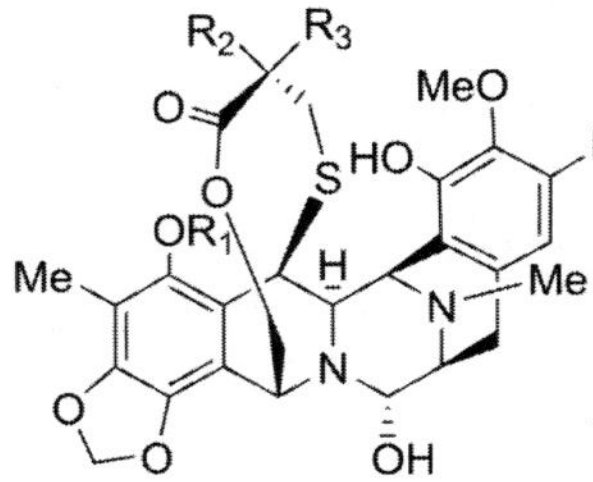

150 ET 594

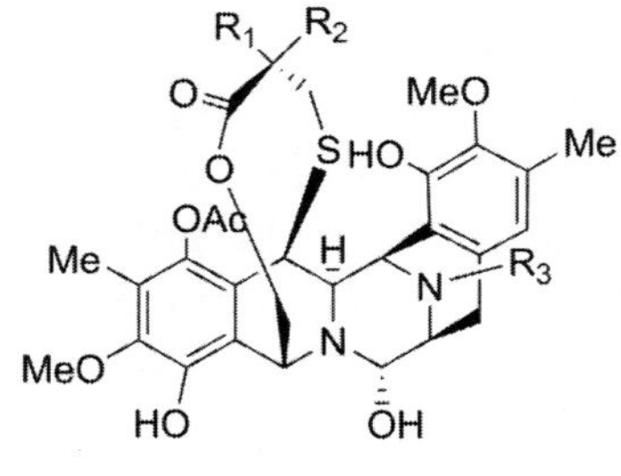

151 ET 583 R_1=H; R_2=NH$_2$; R_3=H
152 ET 597 R_1=H; R_2=NH$_2$; R_3=Me
153 ET 596 R_1= R_2= =O; R_3=Me

Figure 19. Structures of ecteinascidins.

S-Adenosylmethionine is the likely source of methyl group s at C-6, O-7, C-16, O-17, and N-1.2 [124]. Initial structure –activity relationships of natural members of ecteinascidins were established based on results from an *in vitro* cytotoxic screening assay against murine leukemia cells L1210. These data revealed that the carbinolamine of C-21 is crucial for optimal bioactivity and the presence of the aromatic C subunit was found to be important for potency. Later, these compounds were shown to possess strong *in vivo* antitumor effects in various mice models bearing P388 lymphoma, B16 melanoma, M5076 ovarian sarcoma lewis and lX-1 human lung carcinoma, and MX-1 human mammary carcinoma xenografts [121], ET-743 (trabectedin, **136**) was selected for further development; the classical NCI60 human tumor cell line s anticancer drug screen revealed that the drug was several orders of magnitude more potent than other anticancer agents with prominent growth inhibition in the pM to low nM dose range, whereas no correlation with any other known drugs was detected by COMPARE analysis, indicating that tracectedin has a unique mode of action.

Following studies revealed it as the first of a new class of DNA binding agents with a complex, transcription -targeted mechanism of action. In has been demonstrated that units A and B bind covalently and reversibly to the DNA minor groove, with preference for GC-rich triplets; subsequently, trabectedin forms covalent adducts with the N2-position through its carbinolamine moiety. This induces a DNA bending towards the major groove.

The third unit C, being unbound, could interact with nuclear proteins. The transcriptional activation of inducible genes is inhibited, depending on the presence of the DNA repair systems, particularly the transcription coupled nucleotide excision repair (TC-NER) system [82, 126, 127]. Trabectedin was tested in a great variety and number of models against tumors of murine origin and human sensitive and resistant xenografts and showed a broad spectrum of antineoplastic activity.

It was developed by the Spanish pharmaceutical company PharmaMar under the trade name Yondelis® and approved for the treatment of refractory soft-tissue sarcomas by the European Commission in July 2007; the currently ongoing Phase III trials along with the already existing clinical evidence may provide enough data for the Food and Drug Administration for an approval in the US [128]. In November 2009 Yondelis® has received its second marketing authorization from the European Commission for the treatment of relapsed platinum -sensitive ovarian cancer in combination with DOXIL®/Caelyx®. Phase II trials with Yondelis® are also being carried out for breast cancer and for pediatric tumors. Another Phase III trial is on going for soft tissue sarcoma in first line of treatment [129].

As the usual in marine natural products development, obtaining sufficient amounts of ET743 has been a significant challenge due to its restricted natural availability (1 g from 1 ton of tunicate). Methods to produce ecteinascidins by in-the-sea culture of *E. turbinata* have been evolved [130] and much synthetic effort has been directed towards the synthesis of ecteinascidins [82]. Currently, ET-743 for clinical application is produced by a semi-synthetic process in 17 chemical steps developed by Pharma Mar starting from cyanosafracin B, an antibiotic produced by the fermentation of *Pseudomonas fluorescens* [131].

This approach provides access also to other natural members of the ecteinascidins family, such as ET-729 (**135**), ET-745 (**138**), ET-759B (**139**), ET-736 (**144**), and ET-594 (**150**) [132]. The desirability of finding simpler and more stable analogs of ET 743 has led to design new members of the class; among them, phtalascidin (PT 650), **154**, Figure 20 displayed an antitumor profile and activity comparable to ET 743 and was more readily synthesized and more stable than ET 743 [133].

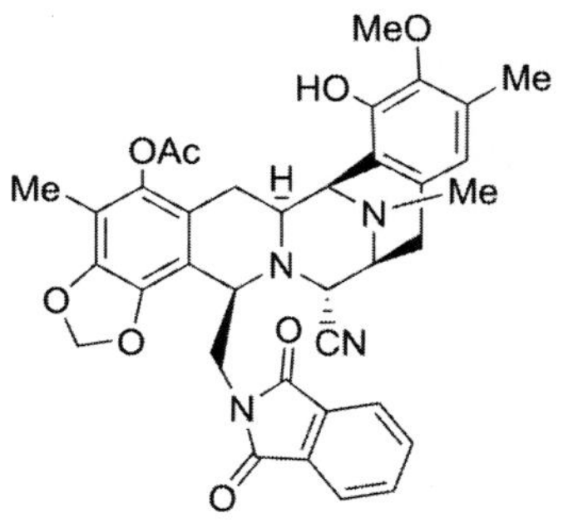

154 PT 650

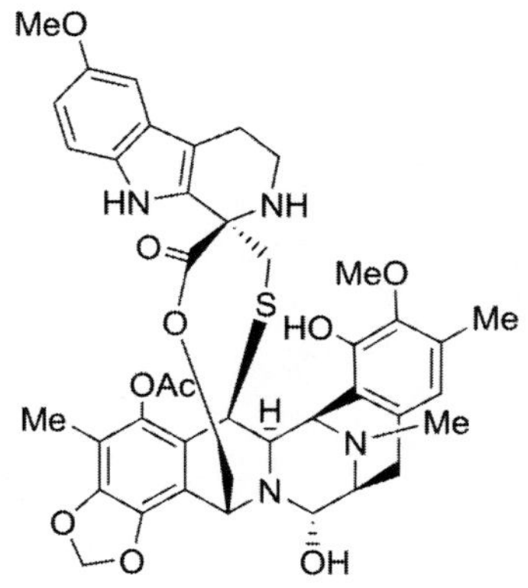

155 PM 01183

Figure 20. Structures of ET 743 synthetic analogs phtalascidin (PT 650, **154**) and lurbinectedin (PM 01183, **155**).

PM01183 (lurbinectedin, **155**, Figure 20) is a further very promising synthetic analog of ET 743; it is structurally similar to ET 743 except for the C subunit, where the tetrahydroisoquinoline present in ET 743 is replaced by a tetrahydro β-carboline in PM01183 [117]. This structural variation is accompanied by important modification of pharmacokinetic and pharmacodinamic properties in cancer patients [135, 136].

Like ET743, PM01183 covalently binds to the minor groove of the DNA forming DNA adducts which give rise to double strand breaks and perturbations of the cell cycle inducing cell death. In preclinical studies, the compound displays a potent cytotoxic activity against tumor cell line s of different origin.

It is currently in Phase II clinical trials for relapsed ovarian, lung, breast and pancreatic cancer. PM01183 is also undergoing Phase I development in combination with other chemotherapies and in haematological tumours. PharmaMar on August 20, 2012 receives FDA orphan drug design ation (ODD) for PM01183 for the treatment of ovarian cancer [126, 134, 135].

6. Guanidine Alkaloids

Marine sponges are recognized as the major source bioactive guanidine alkaloids. Considering the number of the isolated compounds, only selected and illustrative examples have been included in this chapter to provide an overview of marine guanidine alkaloids.

The reported molecules have been divided with respect to their chemical architecture, into the following main groups: a) modified creatinine guanidine derivatives; b) aromatic guanidine alkaloids; c) amino acid and peptide guanidines; d) poliketide-derivative guanidines. Guanidine alkaloids featuring a 2-aminoimidazole moiety have been discussed in Section 2 of this chapter.

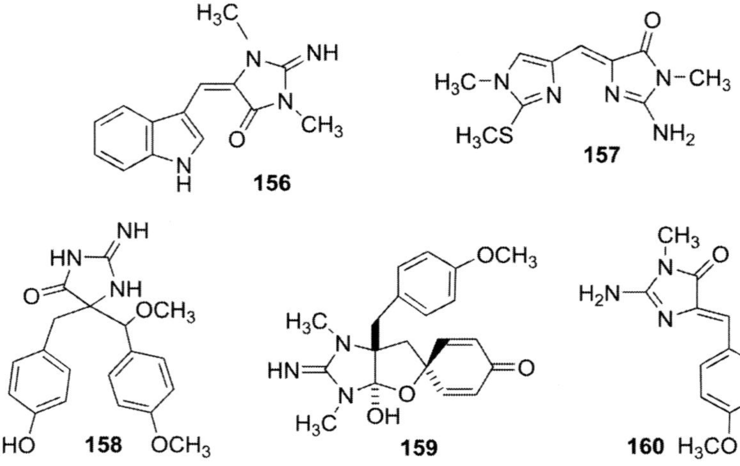

Figure 21. Selected structures of creatinine -guanidine alkaloids.

a) Modified creatinine guanidine derivative. The first member of this class is aplisinopsin (**156**) which has been found, along with many analogs, in several species of marine sponges, including *Thorecta* sp. [136], *Verongia* (= Aplysina) *spengelli* [137], *Smenospongia aurea* [138], *Thorectandra* sp. [139], from the anthozoan *Astroides calycularis* [140], from the scleratinian coral *Tubastrea aurea* [141], from *T. coccinea* and its nudibranch predator *Prestilla melanobranchia* [142], and from the scleratinian coral *Dendrophyllia* sp. [143]. Biological activities of aplisinopsin and naturally occurring derivatives include *in vivo* antineoplastic activity in mice P388 leukemia cells [137], cytotoxic activity against murine lymphoma L2110 and epidermoid carcinoma KB cells [144], and inhibition of neuronal nitric oxide synthase [145]. Aplisinopsin was showed to be a potent reversible inhibitor of monoamine oxidase; unfortunately, it was also hepatotoxic [146, 147]. Further examples of this group are corallistine (**157**), from the sponge *Corallister fulvodesmus* [148], calcaridine A (**158**), (-)-spiroleucettadine (**159**) and leucettamine (**160**), from *Leucetta* spp. [149-151]. Compounds **159** and **160** displayed antibacterial activity against a series of human pathogenic bacteria (Figure 21).

b) Aromatic guanidine alkaloids. This group includes molecules featuring the guanidine moiety linked or included into aromatic and/or heteroaromatic monocilyc or plyciclic systems. Representative examples are the pyridopyrrolopyrimidine alkaloid 3'-methyltetrahydrovariolin (**161**), isolated from the sponge *Kirkpatrickia varialosa* along with other related variolins, which displayed cytotoxic and antifungal properties [152, 153]. A number of indole derivatives have been isolated from ascidians, such as compound **162** from *Dendrodoa grossularia* [154] and the β-carboline alkaloids trypargine (**163**), tryparginine (**164**), and 1-carboxy-trypargine (**165**) from *Eudistoma* sp. [155]. The quinoline alkaloids distomadines A (**166**) and B (**167**) have been isolated from the ascidian *Pseudodistoma aureum* (Figure 22) [156].

c) Amino acid and peptide guanidines. Marine invertebrates are a rich source of guanidine-bearing amino acid and peptide derivatives. Their structres range from simple amino acid derivatives, such as the diketopiperazine cyclo(-l-Arg-deydrotyrosine) (**168**) [157], and the linear dipeptide *N*-(*p*-hydroxybenzoyl)-l-arginine (**169**), isolated from the sponge *Geodia baretti* [158] to complex cyclic pentapetide scffolds.

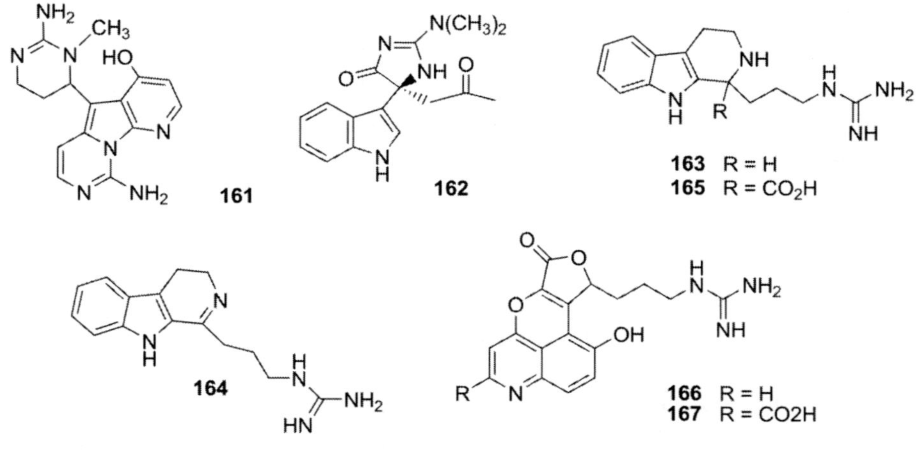

Figure 22. Selected structures of aromatic guanidine alkaloids.

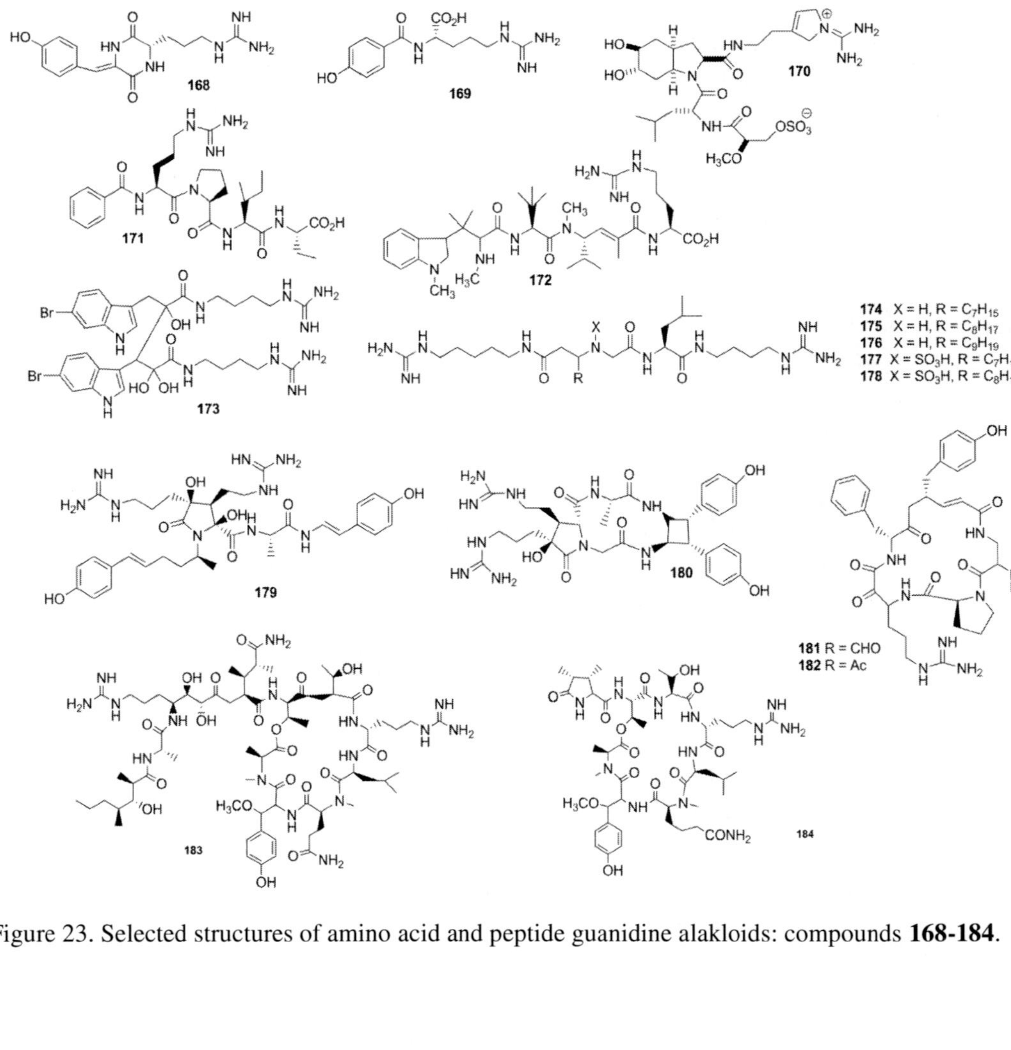

Figure 23. Selected structures of amino acid and peptide guanidine alakloids: compounds **168-184**.

Dysinosins are alkaloid tripeptides with very unusual α-substituted acid residues, including 5,6-dyhydroxyoctahydroindole-2-carboxylic acid and 3-sulfonic-2-methoxy propionic acid, as well as the basic moiety 2-aminoethyl-(1-N-amidino-Δ-3-pyrroline). Among them, dysinosin A (**170**), isolated from an identified sponge belonging to the family Dysideidae, inhibited factor VIIa with a K_i of 108 nM and thrombin with a K_i of 452 nM [159].

Several guanidine-modified tetrapeptides have been isolated from marine invertebrates, in particular from sponges. Examples are nazumamides (**171**, nazunamide A), from *Theonella* sp. [160, 161] and criamides (**172**, criamide B), from *Cymbastela* sp [162]. Criamide B showed a potent cytotoxic activity against human cancer cell line s such as murine leukemia P388 (ED_{50} = 7.3 ng/mL), glioblastome/astrocytoma U373 (ED_{50} = 0.27 μg/mL), ovarian carcinoma HEY (ED_{50} = 0.19 ng/mL) and human lung A549 P388 (ED_{50} = 0.29 ng/mL). Bis-guanidine derivatives are eusynstyelamide (**173**), from the ascidian *Eusynstyela misakiensis* [163] and minalemines (**174-178**), a group of modified tripeptides bearing two homoagmatine moieties isolated from the ascidian *Didemnum rodriguesi* [164].

Very unusual modified tetrapeptides have been isolated from the sponge *Anchinoe tenacior*, among them anchinopetolide A (**179**) and cycloanchinopeptolide C (**180**) [165, 166]. The cyclotheonamides are unusul pentapeptides isolated from sponges of the genus *Theonella*. The first two members of this series, cyclotheonamides A (**181**) and B (**182**) were isolated from the sponge *Theonella* sp. as thrombin inhibitors and contained two new amino acid s, β–ketohomoarginine and (E)-4-amino-5-(4-hydroxyphenil)pent-2-enoic acid [167]. Then, cyclotheonamides C-E have been isolated from *Theonella swinhoei* [168]. Further cyclotheonamides were isolated from *Theonella spp*. as serine protease inhibitors, the activity being due to the presence of the β–keto group in the β-ketohomoarginine residue [169], and from an *Ircinia* sponge as human tryptase inhibitors [170]. Callipeltins A (**183**), B (**184**) and C (the linear peptide corresponding to callipeltin A) have been isolated from the sponge *Callipelta* sp. The structures of callipeltins A and C include a new guanidine amino acid, (2*R*, 3*R*,4*S*)-7-guanidino-2,3-dihydroxyheptanoic acid (AGDHE) (Figure 23) [171-175].

Callipeltins displayed a wide range of bioactivities; callipeltins A-C were cytotoxic *in vitro* against several human carcinoma cells, callipeltin A being the most active acting also against CEM4 lymphocytic cell line s infected with HIV -1 [171]. Callipeltins A and C exhibited also antifungal activity against *Candida albicans*. Callipeltin A induces a powerful and selective inhibition of the Na+ /ca2+ exchanger in cardiac sarcolemmal vescicles [176]. Polydiscamide A (**185**), a depsitetradecapeptide isolated from the sponge *Discodermia* sp., inhibited the growth of *Bacillus subtilis* (MIC = 3.1 μ/mL) and the proliferation of the cultured human lung cancer A549 cell line12 (IC_{50} = 0.7 μ/mL) *in vitro* [177].

Closely related are the discodermins, the first polypeptides isolated from *Discodermia kiiensis*. Discodermin A (**186**) promotes permeabilization of the plasma membrane of A10 cells to the nonpermeable flourescent probes ethidium homodimer-1 (MW = 857), calcein (MW = 623), as well as permeabilization of vascular tissuecells to Ca^{2+} and ATP [178-183]. Neopetrosiamides A (**187**) (Figure 24) and B are rather complex linear polypeptides, differing from each other by the sterochemistry of the sulfoxide group. Both peptides were active in an amoeboid invasion assay at 6 μg/mL [184].

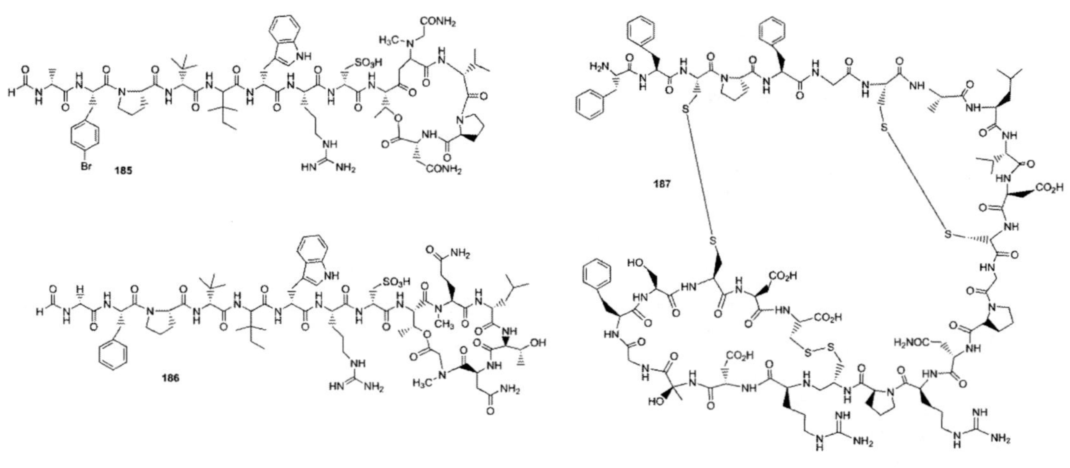

Figure 24. Selected structures of amino acid and peptide guanidine alkaloids: compounds **185-187**.

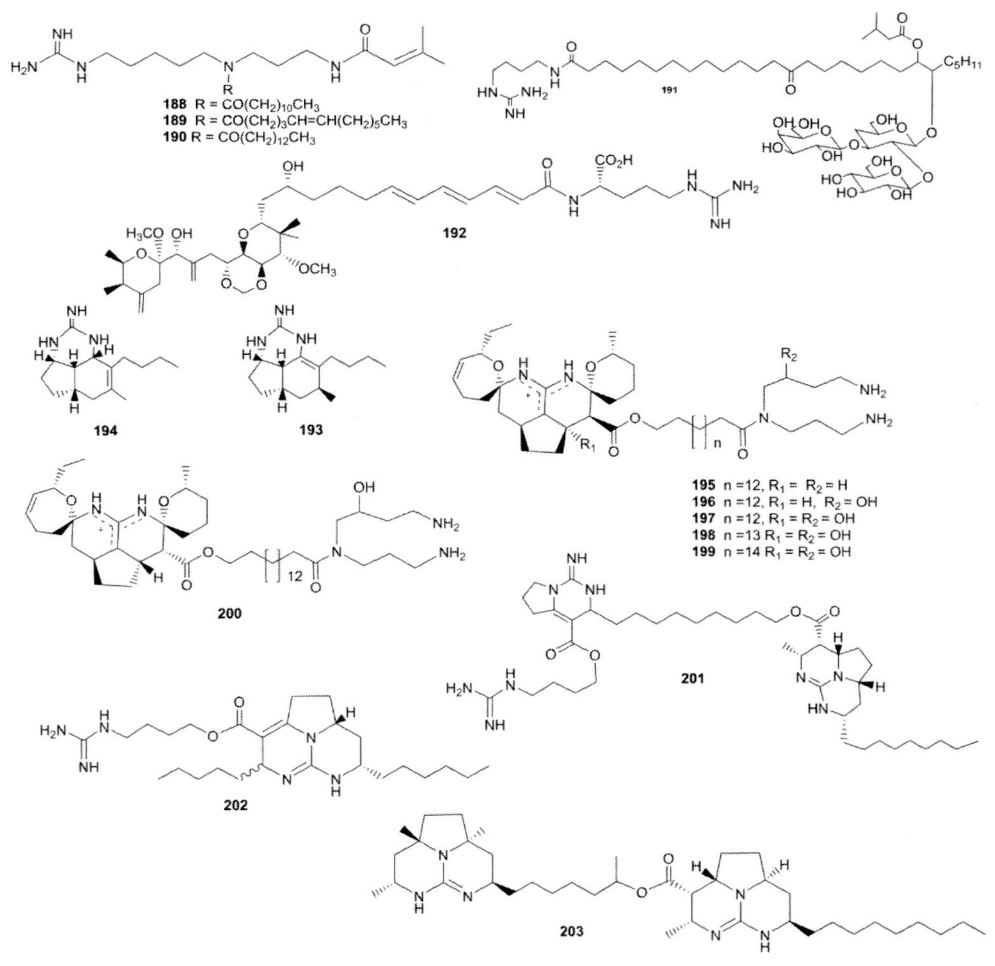

Figure 25. Selected structures of polyketide-derived guanidines alakloids.

d) Polyketide-derived guanidines. This is the largest and varied group of marine guanidine alkaloids. It includes simple linear sctructures, such as those of the antiviral and antimicrobial acardines (**188-190**), from the sponge *Acarnus erithacus* [185]. Phloedictyne C2 (**191**), from the sponge *Phloeodictyon* sp. [186] is an example of glycosilated guanidine alkaloid. In this group, several unique complex polyciclic structures can be found. Onnamide A (**192**) [187] and its analogs [188-189] have been isolated from marine sponges of the genus *Theonella*. The mechanism-of-action of onnamide A cytotoxicity is related to the inhibition of protein synthesis and the induced activation of p38 mitogen -activated protein kinase and c-jun NH_2 -terminal protein kinase (JNK) [190]. Pyilocaulin (**193**) and isoptilocaulin (**194**), from the Caribbean sponge *Ptilocaulis* aff. *spiculifer*, displayed cytotoxic and antimicrobial activity [191].

Ptilomycalin A (**195**), isolated from the Caribbean sponge *Ptilocaulis spiculifer* and from the Red Sea *Hemimycale* sp., possesses a complex pentacyclic guanidine/fattyacid/polyamine skeleton and shows cytotoxic, antifungal, and antiviral activity [192-196]. Related ptilomycalin A derivatives are crambescidins 800 (**196**), 816 (**197**), 830 (**198**), and 844 (**199**), as well as isocrambescidin 800 (**200**), from the sponge *Crambe crambe* [197-199]. Crambescidin 816 (**197**) was active against HCT-16 human colon carcinoma cells (IC_{50} = 0.24 μγ/mL). The activity of crambescidin 816 on voltage-sensitive Ca^{2+} channels was also tested in a neuroblastoma X glioma cell line (NG 108-1 5) and acethylcoline-induced concentration of isolated guinea pig ileum were also evalueted [198].

Crambescidin 816 exters a potent Ca^{2+} antagonist activity (IC_{50} = 1.5 x 10^{-4} μM), more potent than nifepidine (IC_{50} = 1.2 μM), a known selective blocker of L-type Ca^{2+} channels. The biogenetycally related batzelladines have been isolated from several marine sponges belonging to the order Poescilosclerida. The first series of such compounds were batzelladine A (**201**, major homolog), B, C (**202**, major homolog), D, and E, isolated from a sponge of the genus *Batzella* as mixture of homologs [200]. Batzelladine A inhibited the binding of the gp120 domain of HIV -envelope gp 160 glycoprotein to the CD4 receptor on the surface of the human T cell [200]. A second series of batzelladine alkaloids isolated from another *Batzella* sp. included the F (**203**) and G-I members (Figure 25) [201].

7. Indole Alkaloids

A variety of marine sources including sponges, tunicates, and red alga have been shown to generate indole alkaloids, which represent the largest number and most complicated of the marine alkaloids (1/ 4 of total alkaloids) [202]. In this chapter they have been listed based on the number of indole unit present in their frameworks.

a) Monoindoles. A number of 3-substituted indoles have been isolated from ascidians. Examples are 6-bromotryptamine (**204**) isolated from *Didemnum candidum* [203], conicamin (**205**), from *Aplidium conicum*, with histamine -antagonistic activity [204], citorellamine (**206**), from *Polycitorella mariae*, the first indole disulfide dihydrochloride from a marine organism [205, 206], and the indolyl-3-glyoxylic acid derivatives polyandrocarpamides A–C (**207–209**) [207].

Dendrodoa grossularia was the source of several indole alkaloids with different heterocyclic moieties linked at the 3-position (compounds **210–213**) [154, 108-210].

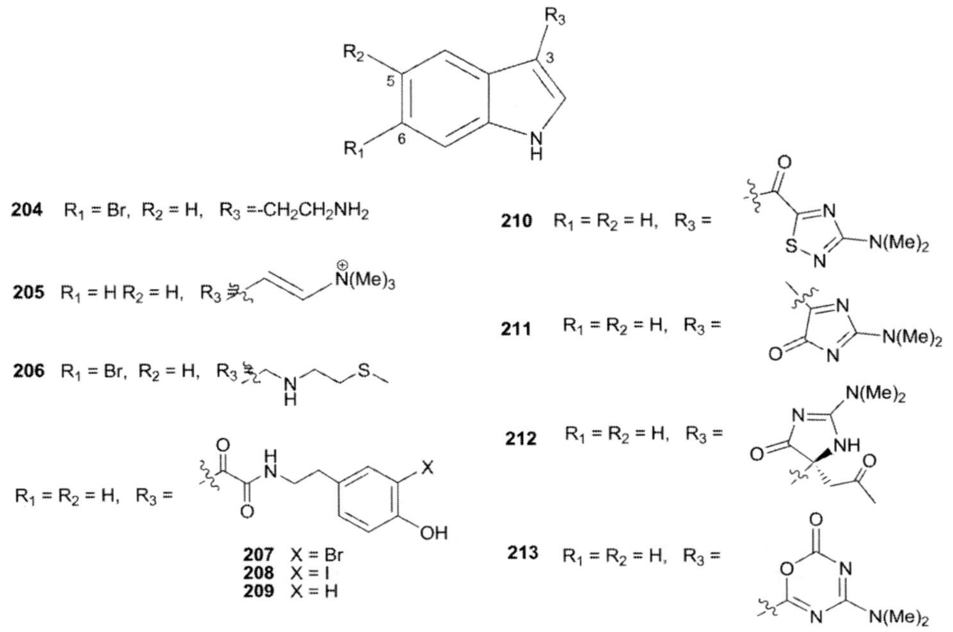

Figure 26. 3-Substituted indoles.

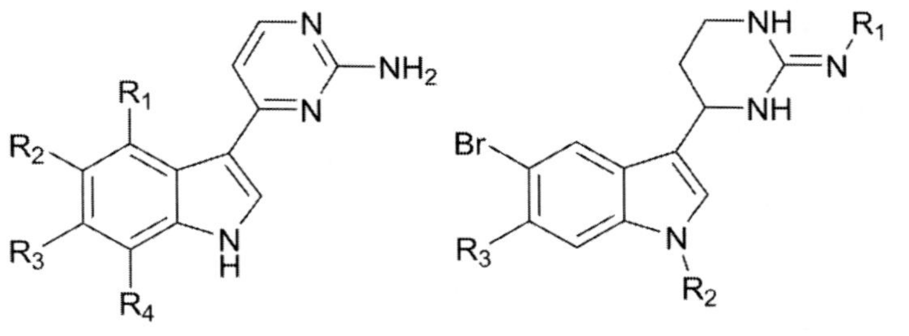

214	R_1=OH; R_2=R_3=R_4=H	**221**	R_1= R_2=R_3=H
215	R_1=OH; R_2=R_4=H; R_3=Br	**222**	R_1=Ac; R_2=R_3=H
216	R_1=R_3=R_4=H; R_2=Br	**223**	R_1=R_3=H; R_2=OMe
217	R_1= R_2=R_4=H; R_3=Br	**224**	R_1=Ac; R_2=OMe; R_3=H
218	R_1=OH; R_2=R_3=H; R_4=Br	**225**	R_1=H; R_2=OMe; R_3=Br
219	R_1=R_4=H; R_2=R_3=Br	**226**	R_1=Ac; R_2=OMe; R_3=Br
220	R_1=R_2=R_3=R_4=H		

Figure 27. 3-(2-Aminopyrimidine)-indoles (meridianins A–G, 2**14–220**) and their reduced forms (aplicyanins A-F, **221–226**).

Among them, alboinon (**213**) contained an oxadiazinone system, found in nature for the first time (Figure 26) [210]. Meridianins A–G (**214–220**), from *Aplidium meridianum* [211, 212], are brominated and/or hydroxylated 3-(2-aminopyrimidine)-indoles differing in the bromine and /or hydroxyl substitution (Figure 27). They constitute a new family of protein kinase inhibitors, inhibiting various protein kinases such as cyclin-dependent kinases, glycogen synthase kinase-3, cyclic nucleotide-dependent kinases and casein kinase 1; they also prevent proliferation and induce apoptosis probably due to their ability to enter cells and to interfere with the activity of kinases important for cell division and death [213].

Structurally related to meridianins are aplicyanins A–F (**221–226**), from *A. cyaneum,* containing a 6-tetrahydropyrimidine substituent at C-3; these alkaloids can be considered reduced forms of the relevant meridianins and, thus, their biogenetic precursors (Figure 27) [214].

Kottamides A–D (**227–230**), from *Pycnoclavella kottae* [215], are imidazol-4-one containing alkaloids; their biogenesis could involve stereospecific imidazolone ring formation from modified Trp-Val-Ile and Trp-Ile-Ala tripeptide precursors (Figure 28) [216]. Kottamide E (**231**) represent the first report of a natural product bearing a 4-amino -1,2-dithiolane-4-carboxylic acid residue [217]. Wakayin (**232**), from *Clavelina* sp., contains an uncommon pyrroloiminoquinone moiety; it is one of the first camptothecin (CPT)-like TOPO I inhibitors isolated from a marine organism, the indole ring linked to the bispyrroleiminoquinone core playing a significant role in this activity (Figure 28) [216].

Four predator-deterrent alkaloids possessing a novel indole-maleimide-imidazole carbon skeleton, didemnimides A–D (**233–236**), were isolated from *Didemnum conchyliatum* [218]. Successively, didemnimide E (**237**) was isolated from *Didemnum granulatum* together with isogranulatimide (**238**) [219] which is the cyclization product of didemnimide A (**233**) formed via a C-2 indole condensation with the imidazole nitrogen. Isogranulatimide (**238**) was reported as the first non-cytotoxic, specific G2 cell cycle checkpoint inhibitor, and the mechanism of this antitumor action was elucidated at molecular level (Figure 29) [219, 222].

Several 3-substituted indoles have been isolated from the red alga *Martensia fragilis (M. denticulata)* such as: fragilamide (**239**), martensines (martensin A, **240**), martefragin A (**241**), and denticins (denticin A, **242**) [223-227]. Fragilamide showed strong antioxidant activity [227] whereas martefragin A (**241**) showed inhibitory activity on NADPH- depending lipid peroxidation in rat liver microsomes [223-226] (Figure 30).

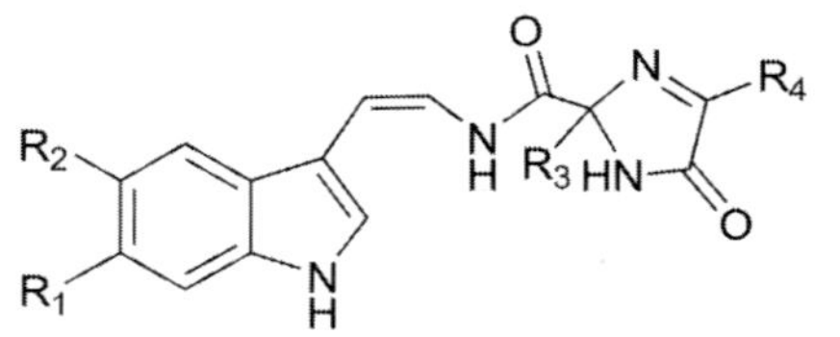

227 $R_1=R_2=Br$; R_3=2-propyl; R_4=2-butyl
228 $R_1=Br$; $R_2=H$; R_3=2-propyl; R_4=2-butyl
229 $R_1=H$; $R_2=Br$; R_3=2-propyl; R_4=2-butyl
230 $R_1=R_2=Br$; R_3=2-Butyl; R_4=Methyl

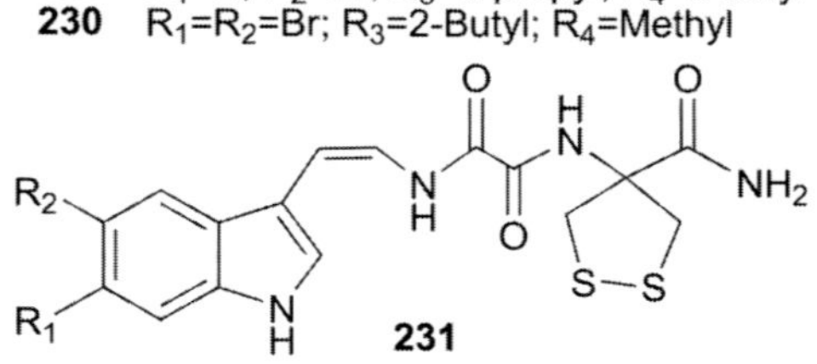

231

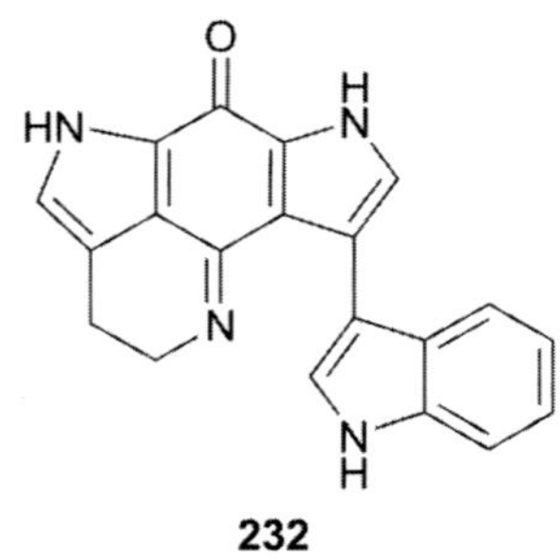

232

Figure 28. Structures of kottamides A–E (**227–231**) and wakayin (**232**).

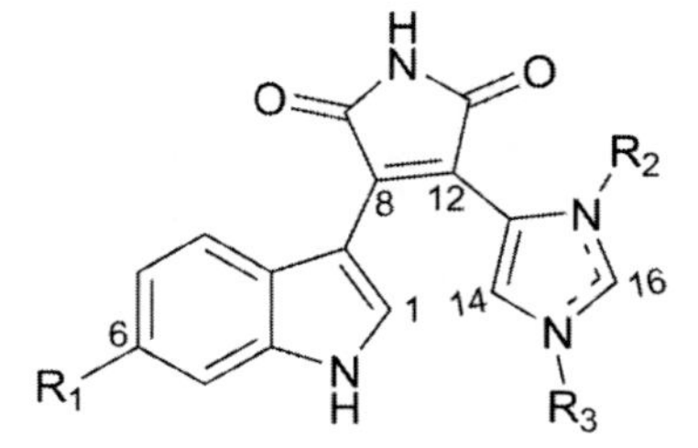

233 $R_1=R_2=H$
234 $R_1=Br$; $R_2=H$
235 $R_1=H$; $R_2=Me$
236 $R_1=Br$; $R_2=Me$
237 $R_1=H$; $R_3=Me$

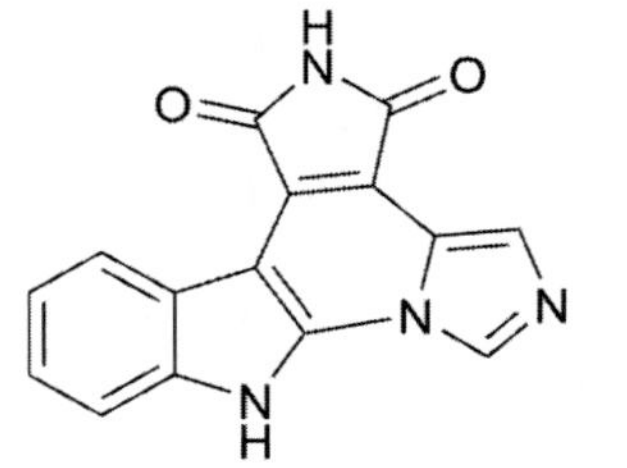

238

Figure 29. Indole-maleimide-imidazole alkaloids.

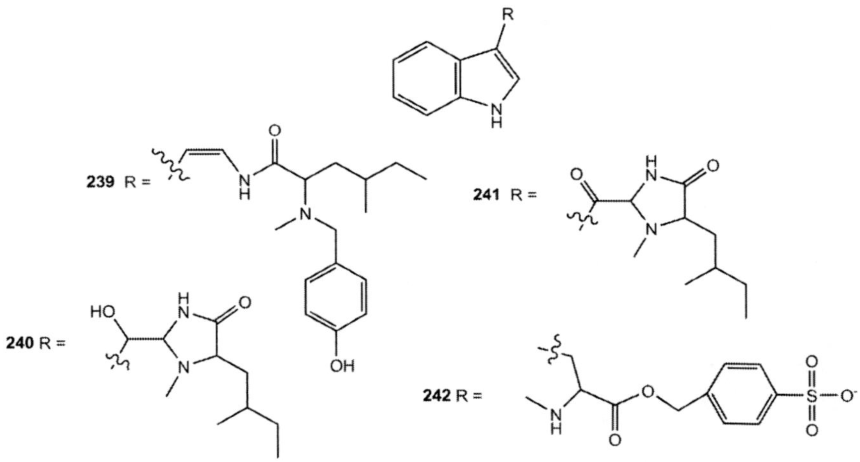

Figure 30. Selected structures of 3-substituted indoles from the red alga *Martensia fragilis*.

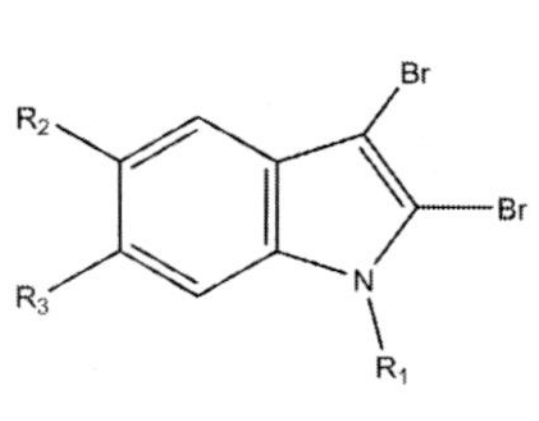

243 R_1=Me, R_2=H, R_3=Br
244 R_1=Me, R_2=Br, R_3=H
245 R_1=H, R_2= R_3= Br
246 R_1=Me, R_2=R_3=Br

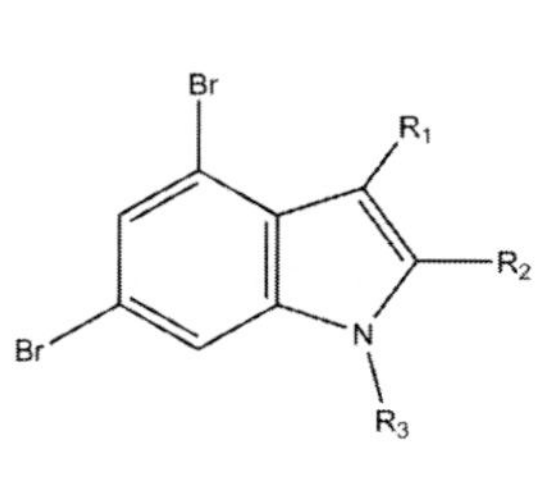

247 R_1=R_3=H, R_2=Br
248 R_1=R_2=Br, R_3=H

Figure 31. Selected Structures of halogenated indole alkaloids alkaloids isolated from *Laurencia* species.

Halogenated indole alkaloids alkaloids have been isolated only in marine organisms, especially from red algae; also sulfur -containing bromoalkaloids were extracted from red algae many bromoindoles and *N*-methylbromoindoles have been isolated from *Laurencia* species; representative examples are 2,3,6-tribromo-1-methyl indole (**243**), 2,3,5-tribromo-1-methyl indole (**244**), 2,3,5,6-tetrabromo-1*H*-indole (**245**), 2,3,5,6-tetrabromo-1-methyl indole (**246**), 2,4,6- tribromo-1*H*-indole (**247**) and 2,3,4,6-tetrabromo-1*H*-indole (**248**) from *L. brongniartii, L. similis*, and *L. decumbens* [228-231] (Figure 31).

Discodermindole (**249**) [232] and 6-hydroxydiscodermindole (**250**) [233], isolated from the Bahamas sponge *Discodermia polydiscus*, as well as trachycladindoles A-G (**251-257**), [234] from the South Australian sponge *Trachycladus laevispirulifer*, belong to the rather limited class of indole alkaloids containing a 2-aminoimidazoline substituent at position 3 (Figure 32). In addition they all, with the exception of trachycladindole G, have one or more bromine atoms in their structure, which is common for many secondary metabolites of marine origin. All compounds of this group have cytotoxic properties [235, 231-234] and are of interest in the search for new anticancer drug s [236].

b) Bisindoles. Staurosporine and its derivatives (**258-266**) (Figure 33) have been isolated from various actinomycete strains as well as from some marine organisms, including ascidians [237].

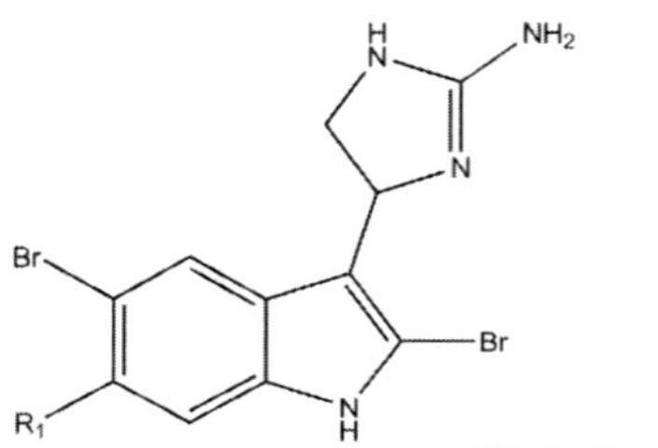

249 R_1=H
250 R_1=OH

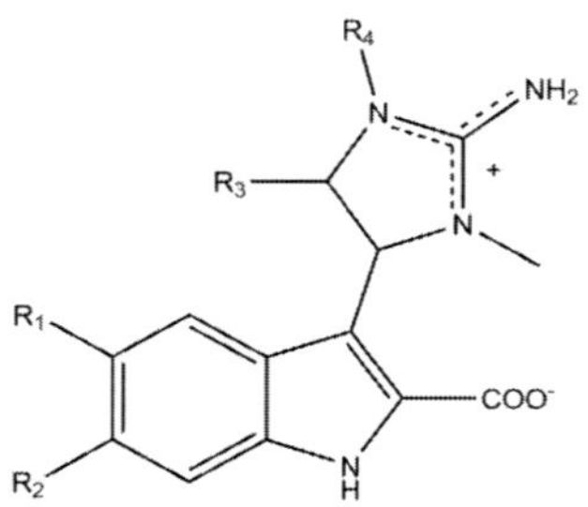

251 R_1=Br, R_2=R_3=R_4=H
252 R_1=Br, R_2=R_3=H, R_4=Me
253 R_1=Br, R_2=OH, R_3=R_4=H
254 R_1=Br, R_2=OH, R_3=H, R_4=Me
255 R_1=Br, R_2=H, R_3=OH, R_4=Me
256 R_1=Br, R_2= R_3=OH, R_4=Me
257 R_1= R_2= R_3= R_4=H

Figure 32. Structures of discodermindole (**249**), 6-hydroxydiscodermindole (**250**), and trachycladindoles A-G (**251**-**257**).

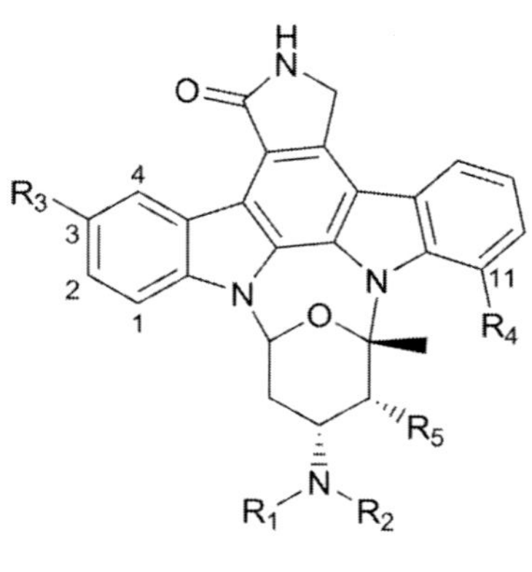

258 R_1=H, R_2=Me, R_3=OH, R_4=H, R_5=OMe
259 R_1=H, R_2=Me, R_3=H, R_4=OH, R_5=OMe
260 R_1=H, R_2=Me, R_3=OH, R_4=H, R_5=OH
261 R_1=H, R_2=H, R_3=H, R_4=OH, R_5=OMe
262 R_1=Me, R_2=Me, R_3=OH, R_4=H, R_5=OMe
263 R_1=H R_2=H, R_3=OH, R_4=H, R_5=OMe
264 R_1=H, R_2=H, R_3=H, R_4=H, R_5=OH

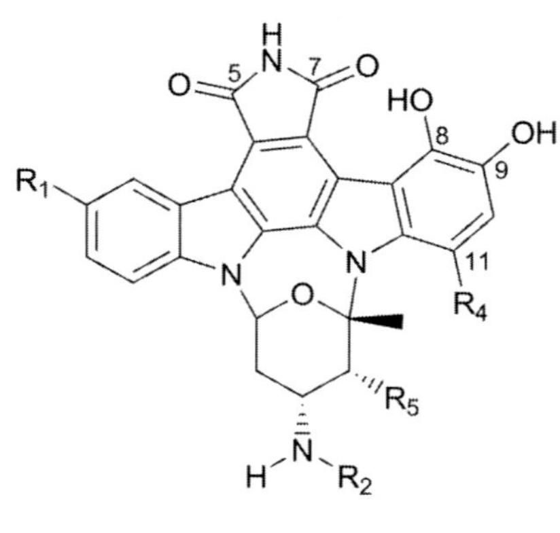

265 R_1=OH, R_2=Me
266 R_1=R_2=H

Figure 33. Staurosporine derivatives **258–266**.

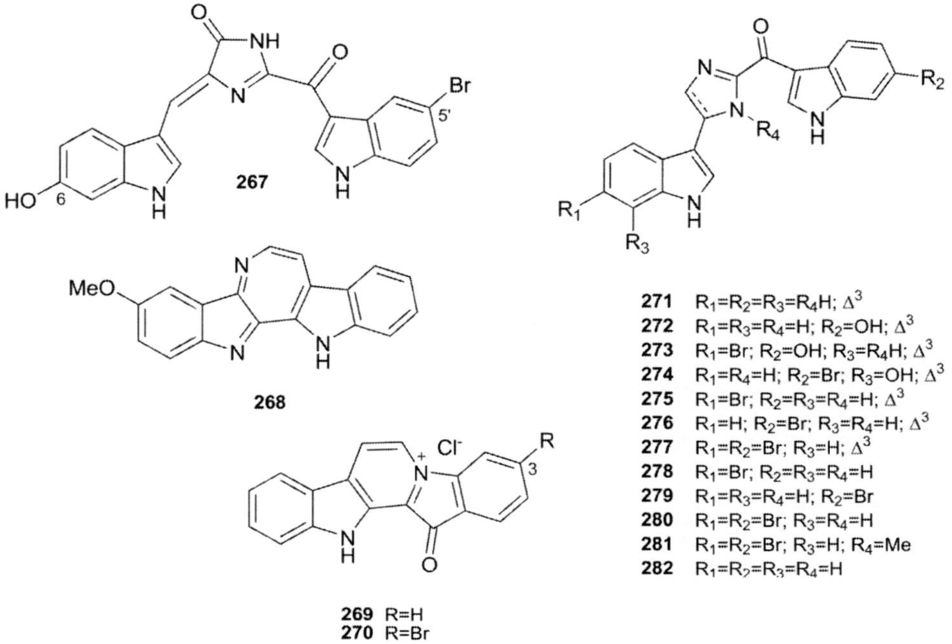

Figure 34. Structures of rhopaladins A (**267**), iheyamine A (**268**), fascaplysin (**269**) and its 3-bromoderivative (**270**), topsentins (**271**-**277**), and spongotins (**278**-**282**).

These alkaloids were shown to be strong inhibitors of several kinases, in particular protein kinase C (PKC); other activities include inhibition of platelet aggregation and smooth muscle contraction, induction of cell cycle arrest and apoptosis, and the reversal of multidrug resistance in some cancer cell line s. The potential of these derivatives as anticancer agents is supported by the example of 7-hydroxystaurosporine which is in clinical phase I trials at the NCI [238]. 11-Hydroxystaurosporine (**259**), from *Eudistoma* sp., is a PKC inhibitor about 30% more potent than staurosporine [239] whereas 3-hydroxystaurosporine (**258**), from *E. toealensis*, is one of the most active staurosporine-type inhibitors of cell proliferation described so far [239]. Derivatives **265** and **266**, from *Cystodites solitus*, displayed cytotoxic activity in the submicromolar range against three human tumour cell lines [240]. Other examples of bisindole alkaloids are rhopaladins (rhopalaadin A, **267**), from *Rhopalaea* sp., possessing an imidazolinone moiety [241], and iheyamines (iheyamine A, **268**), from *Polycitorella* sp., with a new heteroaromatic skeleton composed of an azocine unit fused onto a bisindole system (Figure 13) [242]. The 12*H*-pyrido[1,2-a:3,4-b0]diindole ring system forms the framework of the red pigment fascaplysin (**269**), which was isolated in 1988 from the sponge *Fascaplysinopsis Bergquist* sp. and successively found in other Thorectidae sponges; this compound, together with its 3-bromoderivative (**270**), has been found in some *Didemnum* collections [243, 244]. Fascaplysin and its derivatives exhibit a broad range of bioactivities including antibacterial, antifungal, antiviral, HIV -1 -RT, p56 tyrosine kinase, antimalarial, citotoxicity against numerous cancer cell lines, specific inhibition of Cdk 4 and DNA intercalation, demonstrating a huge potential for therapeutic assays [243, 244].

Topsentins (**271**-**277**) and spongotines (**278**-**282**) from a group of structurally related bisindole alkaloids in which two indole fragments are connected by a carbonylimidazole linker (Figure 34).

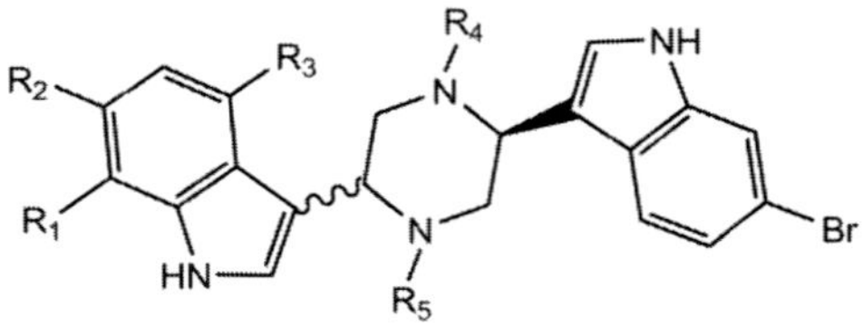

283 *trans*-, $R_1=R_2=Br$; $R_3=OH$; $R_4=R_5=H$
284 *trans*-, $R_1=R_3=R_5=H$; $R_2=Br$; $R_4=CH_3$
285 *trans*-, $R_1=R_3=H$; $R_2=Br$; $R_4=R_5=CH_3$
286 *cis*-, $R_1=R_3=R_4=R_5=H$; $R_2=Br$

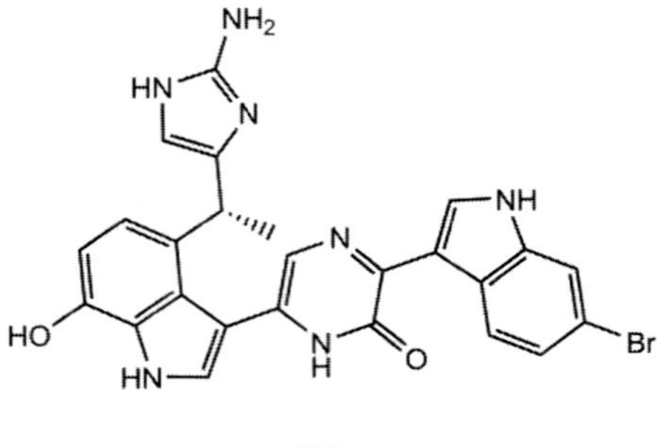

287

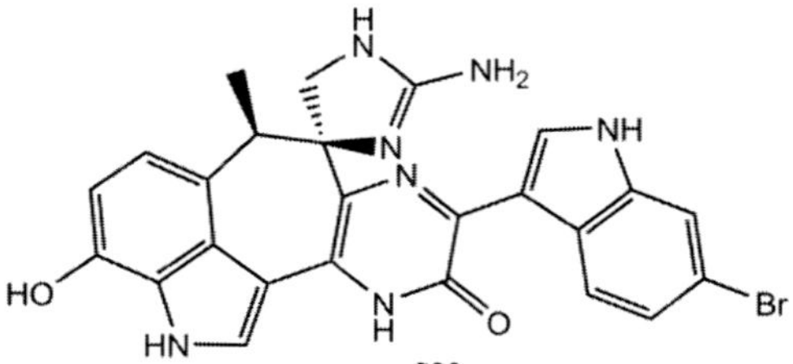

288

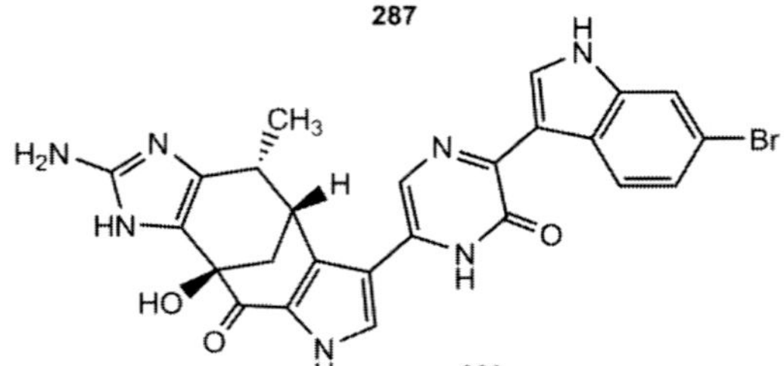

289

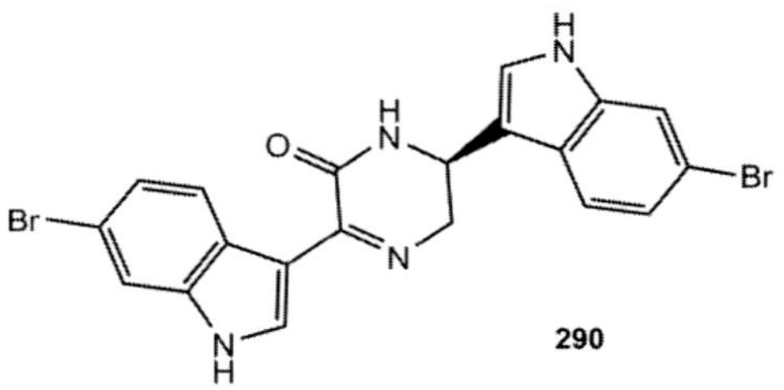

290

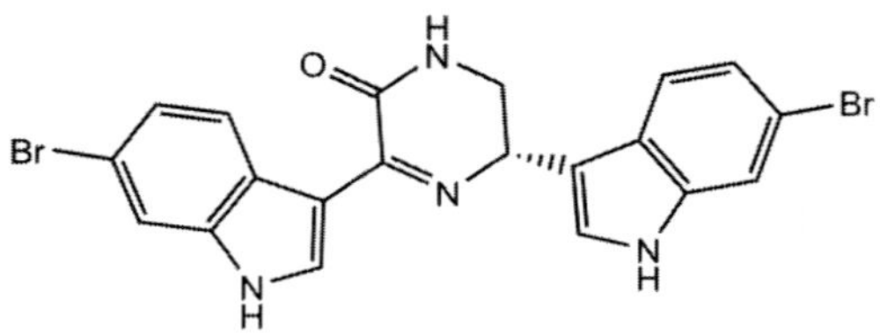

291

Figure 35. Structures of dragmacidins (**283-289**) and hamacanthins A (**290**) and B (**291**).

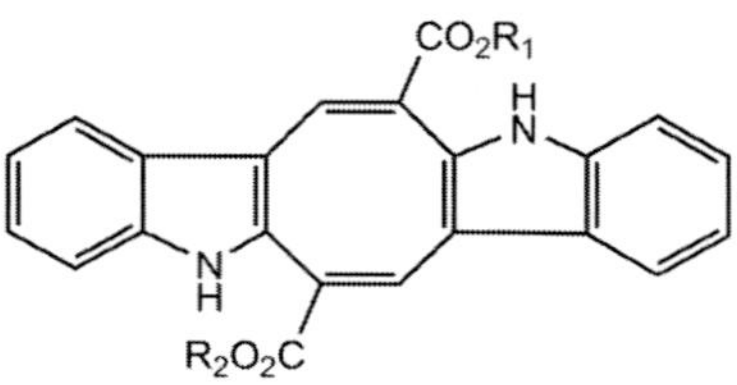

292 $R_1=R_2=Me$
293 $R_1=R_2=H$
294 $R_1=Me$; $R_2=H$

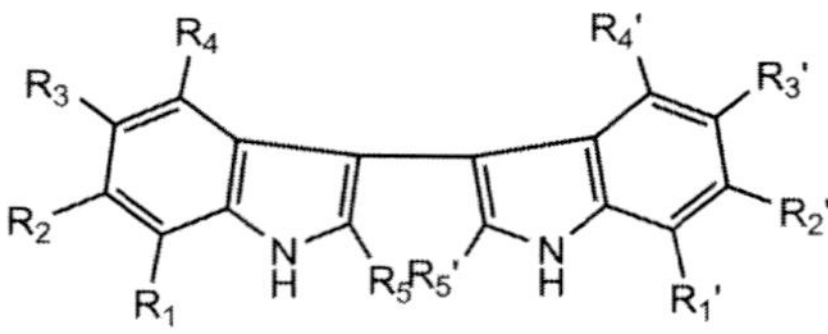

296 $R_1=R_1'=OCH_3$; $R_3=R_3'=Br$; $R_4=R_4'=Cl$; $R_2=R_2'=R_5=R_5'=H$
297 $R_2=R_2'=R_3=R_3'=R_5=R_5=Br$; $R_1=R_1'=R_4=R_4'=H$
298 $R_2=R_2'=R_4=R_4'=Br$; $R_5=R_5'=SCH_3$; $R_1=R_1'=R_3=R_3'=H$

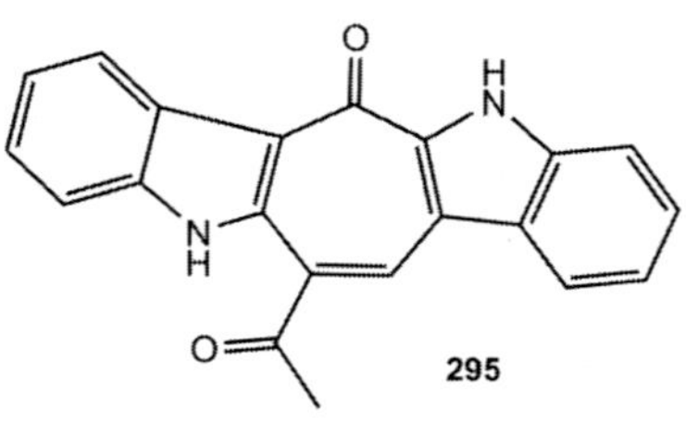

295

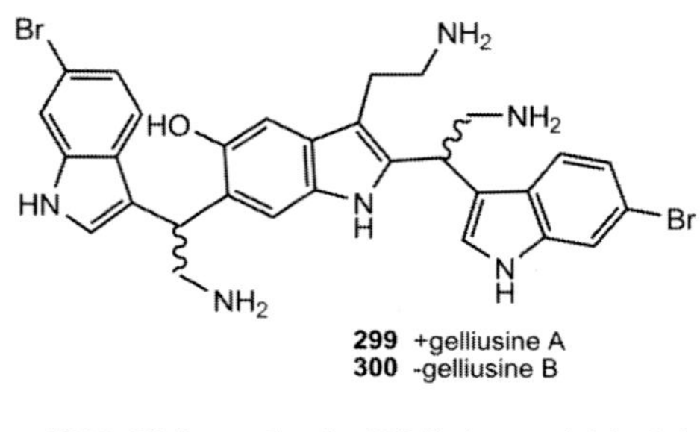

299 +gelliusine A
300 -gelliusine B

Figure 36. Structures of caulerpin and its analogues (**292-294**), caulersin (**295**), bromobisindoles **296-298**, and gelliusines A and B (**299-300**).

They have been isolated from several sponges of the genera *Topsentia Spongosorites*, *Hexadella*, *Discodermia*, and *Rhaphisia* [245-258]. Substances having cytotoxic, antiviral, anti-inflammatory, and antibacterial activity were found among the topsentin/spongotin class alkaloids [246, 250, 254, 258, 259-261]. Dragmacidins (Figure 35) can be divided into two structural types. The first is represented by four fairly simple cytotoxic bisindole alkaloids, in which the indole fragments are connected by a piperazine linker. This group includes dragmacidin itself (**283**) and dragmacidins A- C (**284-286**). They have been isolated from *Dragmacidon*, and *Hexadella* sponges [249, 262]; dragmacidin C (**286**), have been also found in the ascidian *Didemnum candidum* [203]. The second group, isolated from sponges of the *Spongosorites* and *Halicortex* genera, is represented by structurally more complex dragmacidins D (**287**) [257, 263], E (**288**) [263], and F (**289**) [264], which contain also a guanidine fragment (Figure 35). Structurally related to dragmacidins A-C are the hamacanthins (hamacanthins A and B, **290** and **291**) (Figure 35), characterized by a dihydropyrazine or piperazine ring decorated with two indole fragments, where the hamacanthins A derivatives are 3,6-di-substituted and the hamacanthins B derivatives are 3,5-disubstituted. This broad series of compounds includes also dihydro and debromo derivatives; they have been obtained from various sources [253-256, 258, 265]. Hamacanthins A (**290**) and B (**291**), isolated from a deep-water sponge of the *Hamacantha* genus, exhibited antifungal activity [269]. In addition to antifungal activity [265], some hamacanthins have shown cytotoxic and antibacterial effects [253-256]. The intriguing structure of caulerpin (**292**) contains two indole groups linked by a cyclic ring containing eight carbons with two carboxy groups. This compound has been first isolated from various *Caulerpa* sp. [266-26] along with two analogues CLP (II) (**293**) and (III) (**294**) [270] (Figure 36). It was also recovered from other algae, both green [271, 272] and red [273-273] algae. Caulersin (**295**), from *Caulerpa serrulata* is a bisindole alkaloid with a 7 members central ring and two «anti parallel » indole cores [274] (Figure 36). In compounds **296-298** (Figure 36), isolated from *Chaetomorpha basiretorsa* [275] and *Laurencia* species [276] two indoles unit are directly linked via C-3. They are brominated bisindoles, compound **297** featuring a thiomethyl group linked at C-2. *c) Trisindoles.* Gelliusines A and B (**299**, **300**) (Figure 36) are two diastereometric brominated tris-indole alkaloids occurring as enantiomeric pairs, isolated from a deep water New Caledonian sponge *Gellius* or *Orina* sp [277]. The inhibition of KB, P-388, P-388/dox, HT-29 and NSCLCN-6 cell line s for **299** and **300** was evaluated and shown to yield IC_{50} values of between 10 and 20 μg/mL. At higher concentrations (5–70 μg/mL), gelliusine A causes a serotonin -like and methysergide-sensitive contraction; while at low concentration, it is able to antagonize serotonin-induced contraction, indicating the ability to block the serotoninergic receptor.

8. β-Carboline Alkaloids

Among the β–carboline alkaloids of marine origin, two important classes are eudistomins and manzamines. The eudistomin-type alkaloids form a wide class of polysubstituted β–carbolines, dihydro-, and tetrahydro–β–carbolines; most of the members of the class have been isolated from ascidians belonging to the genus *Eudistoma*, but other sources are the genera *Ritterella*, *Pseudodistoma*, *Didemnum*, *Synoicum*, and *Lissoclinum*.

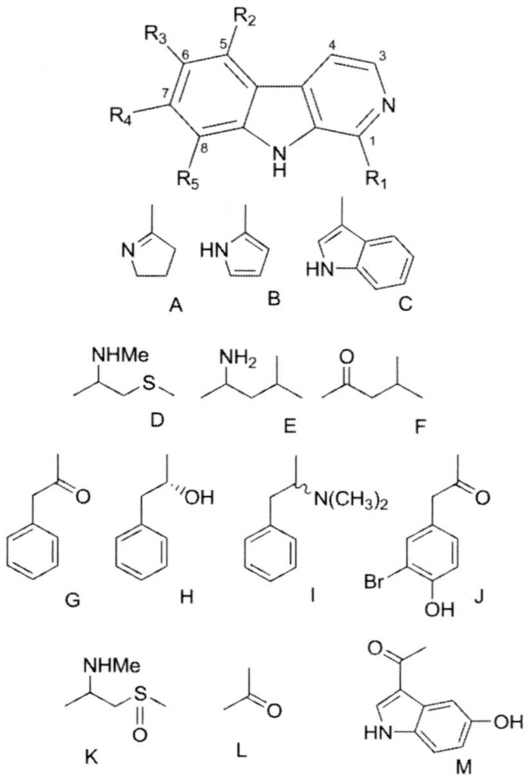

		R1	R2	R3	R4	R5
Eudistomin J	**301**	*H*	H	OH	Br	H
Eudistomin G	**302**	*A*	H	H	Br	H
Eudistomin I	**303**	*A*	H	H	H	H
Eudistomin P	**304**	*A*	H	OH	Br	H
Eudistomin V	**305**	*A*	H	Br	Br	H
Eudistomin A	**306**	*B*	H	OH	Br	H
Eudistomin M	**307**	*B*	H	OH	H	H
Eudistomin R	**308**	*G*	H	H	Br	H
Eudistomin T	**309**	*G*	H	H	H	H
Eudistomin U	**310**	*C*	H	H	H	H
Eudistomin W	**311**	*H*	H	OH	H	H
Eudistomin X	**312**	*I*	H	OH	H	H
Eudistomin D	**313**	*H*	Br	OH	H	H
Eudistomin Y4	**314**	*J*	H	Br	H	H
Eudistomidin A	**315**	*A*	H	Br	H	OH
Eudistomidin C	**316**	*D*	Br	OH	H	H
Eudistalbin A	**317**	*E*	H	H	Br	H
Eudistalbin B	**318**	*F*	H	H	Br	H
Eudistomidin J	**319**	*K*	Br	OH	H	H
Eudistomidin K	**320**	*L*	Br	OH	H	H

Figure 37. Structures of simple substituted β-carbolines of the eudistomin class.

Almost all the reported compounds are hydroxylated and/or brominated at C5, C6, C7, and C8; apart from few members of the group, which are unsubstituted at C1, they show different substituents at this position, such as pyrrole, pyrroline, or indole rings, as well as ox ygenated, aminated, or thiomethylated alkyl residues [107]. Actually, all of the β–carboline metabolites isolated thus far are related biosynthetically; the production of these metabolites is generally believed to involve the coupling of tryptophan with a second amino acid, whose nature clearly affects C1 substitution. For example, eudistomins A (**306**), M (**307**), G (**302**), and P (**304**) may be considered to be derived from tryptophan and glutamine [278], while the 2-fenylacetyl-β-carbolines, eudistomines R (**308**) and T (**309**), are, in addition to tryptophan, made up of phenylalanine or phenylpyruvic acid. Similarly, the unusual aminoacids, *p*-methylphenyl-L-alanine and S-methyl-D-cysteine have been supposed to be involved in the production of eudistomidins B (**334**) and C (**316**) [279], respectively. A proline -derived precursor is evident in eudistomins G, I, P (**302-304**), A (**306**), and M (**307**), eudistomidin A (**315**), and woodinine (**336**). *In vivo* studies with *E. olivaceum* confirmed that tryptophan and proline are the primary precursors of eudistomin I (**303**) [280]. Some of the most intriguing members of these alkaloids are those bearing an oxathiazepine ring (eudistomins C, E, F, K, and L, **344-348**) which apparently originate from condensation of tryptophan with cysteine [281]. Eudistalbins A (**317**) and B (**318**) likely derive from leucine.

Representative of simple substituted β-carbolines of the eudistomin class are eudistomins **301-314** [278, 282-285], eudistomidins A (**315**) and C (**316**) [286, 279], and eudistalbins (**317** and **318**) [287], and eudistomidins J–K (**319**, **320**) [288] (Figure 37).

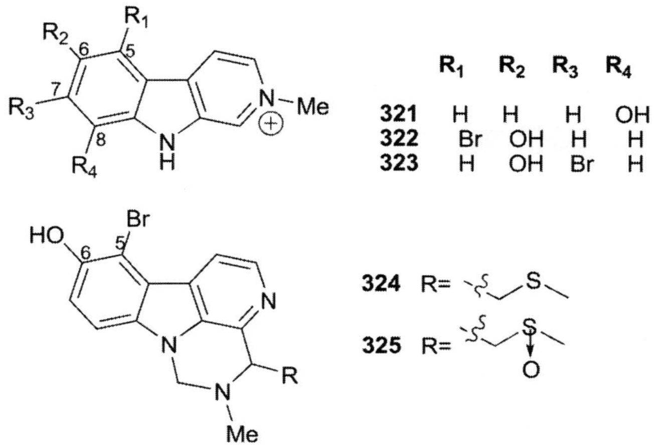

Figure 38. *N*-methyl β–carbolinium (**321-323**) and tetrahydropyrimidine ring containing (**324** and **325**) eudistomin alkaloids.

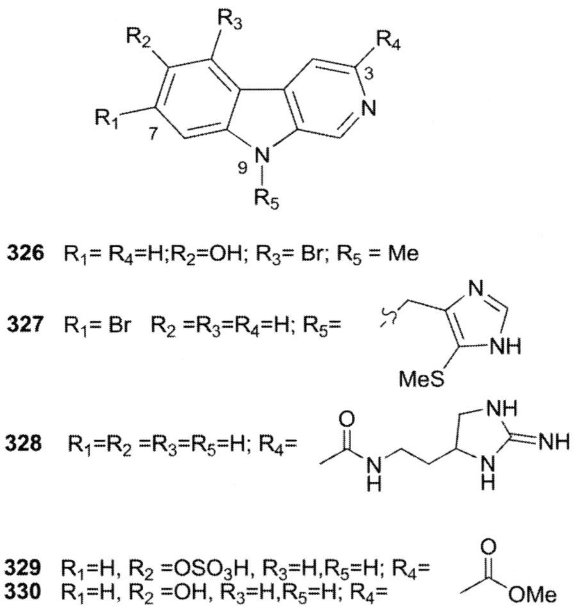

Figure 39. N9- (**326**, **327**) and C3-substituted (**328-330**) eudistomin alkaloids.

Pseudodistamine (**321**) as well as 2-methyleudistomins D (**30**) and J (**31**) are rare examples of *N*-methyl β–carbolinium derivatives described from marine sources, while eudistomidins E (**324**) and F (**325**) are structurally unique, with a tetrahydropyrimidine ring fused to the β–carboline skeleton (Figure 38) [289].

Eudistomidin D (**326**) as well as didemnolines A-D (didemnoline A, **327**) differ from the other ascidian β–carboline metabolites in that they are substituted at the N9 position of the β–carboline ring, rather than at the C1 position [279, 290].

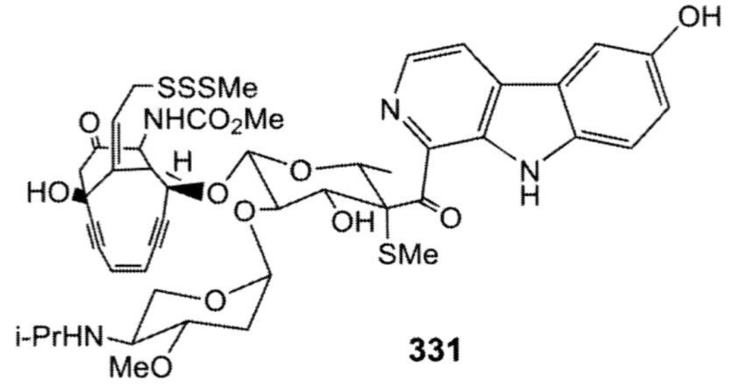

331

Figure 40. Structures of shishijimicin A (**331**).

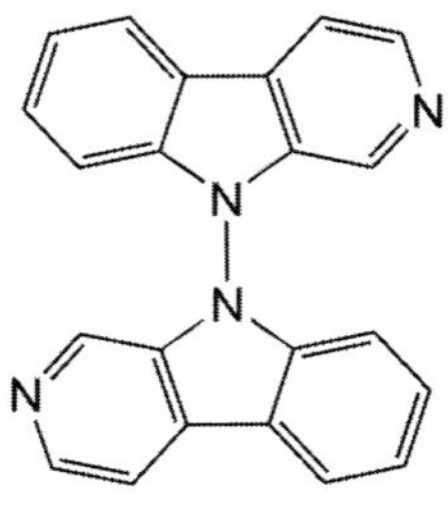

332

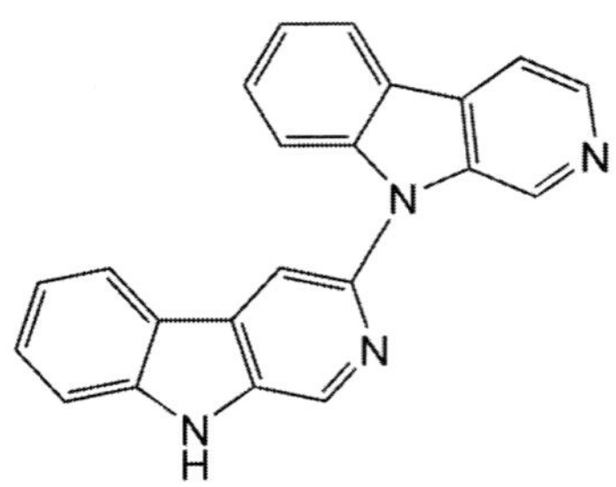

333

Figure 41. Structures of symmetrical (**332**) and non-symmetrical (**333)** β–carboline dimers.

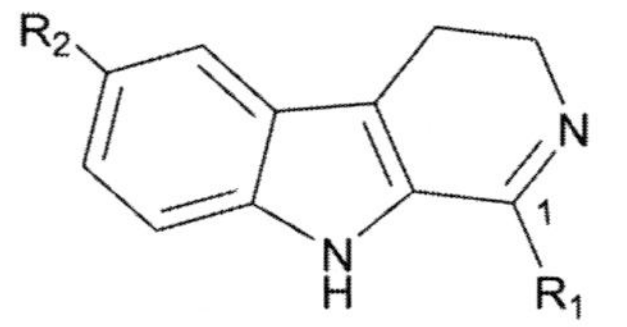

352 R_2=H R_1= 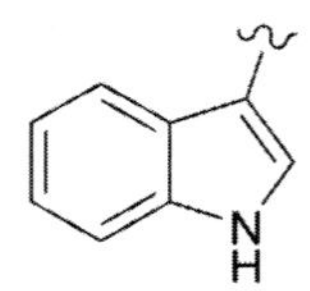

353 R_2=H R_1= 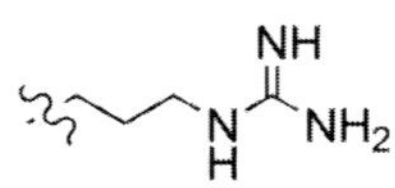

354 R_2=OH R_1= COOH

Figure 42. Structures of tetrahydro–β–carbolines.

Tiruchanduramide (**328**) is the sole 3-substituted β–carboline isolated to date from ascidians [291]. Other β–carboline alkaloids substituted at C3 have been isolated from sponges, such variabines A (**329**) and B (**330**), from *Luffariella variabilis* (Figure 39) [292]. Shishijimicins (shishijimicin A, **331**) are perhaps the most complex β–carbolines from ascidians; they belong to the enediyne class of antibiotics, which are also potent antitumor agent s (Figure 40) [293].

Both symmetrical (**332**) and non-symmetrical (**333**) β–carboline dimers have been reported from *Didemnum* sp. and their structures were confirmed by synthesis [294].

Compound **332** was known as a synthetic compound derived from photochemical dimerization of β–carboline (norharmane) but it is the first example of naturally occurring β–carboline dimer (Figure 41). Numerous tetrahydro–β–carbolines with various substituents at C1 have been reported from ascidians; they include eudistomidins B (**334**) and G (**335**) [279, 295], woodinine (**336**) [296], lissoclin C (**337**) [297], and compound **338**, the 1-carboxyl analogue of trypargine, a tetrahydro–β–carboline previously isolated from the skin of the African frog *Kassina senegalensi* [298]. Some tetrahydro–β–carbolines have been also isolated from sponges such as 1,6-dihydroxy-1,2,3,4-tetrahydro–β–carboline (**339**) and the antimalarial β–carboline, (+)-7-bromotrypargine (**340**) (Figure 42) [299]. Arborescidines B-D (**341**-**343**) [300], eudistomins C, E, F, K, L (**344**-**348**) [301], K-sulfoxide (**349**) [302], and debromoeudistomin K (**350**) [303] contain a 7-membered additional ring attached to the tetrahydro β–carboline moiety which, in the case of the eudistomins, is an oxathiazepine ring (Figure 42). Bengacarboline (**351**) is a tetrahydro–β–carboline with two indole units attached to C1 of the carboline nucleus (Figure 42); it resulted cytotoxic toward a 26 cell line human tumor panel in vitro and inhibited the catalytic activity of TOPO II [304]. There are only few examples of dihydro–β–carbolines reported marine organisms. They include isoeudistomin U (**352**) [280, 305], trypargimine (**353**), the 1,2-dehydro analogue of trypargine, isolated from a previously undescribed *Eudistoma* sp. [298], and the 1-carboxy-6-hydroxy-3,4-dihydro–β–carboline (**354**) from the tropical sponge *Hyrtios* sp. (Figure 43) [306].

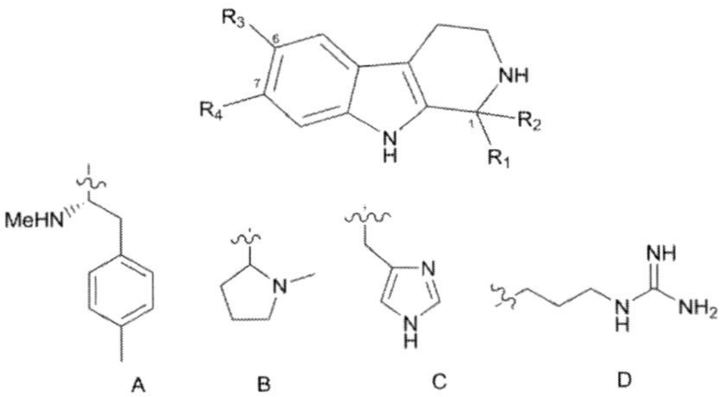

	R_1	R_2	R_3	R_4
334	*A*	H	Br	H
335	*A*	H	H	Br
336	*B*	H	Br	H
337	*C*	H	H	Br
338	*D*	CO_2H	H	H
339	OH	H	OH	H
340	D	H	H	Br

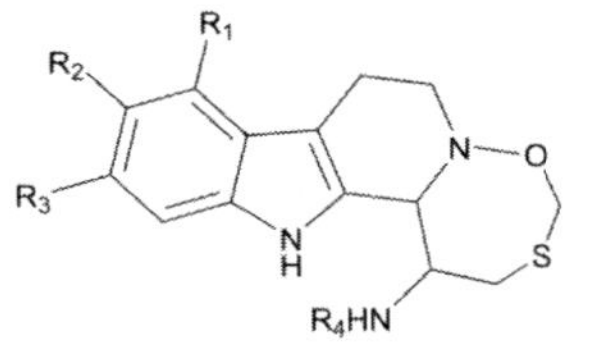

344 $R_1 = R_4 = H$, $R_2 = OH$, $R_3 = Br$
345 $R_1 = Br$, $R_2 = OH$, $R_3 = R_4 = H$
346 $R_1 = H$, $R_2 = R_3 = OH$, $R_4 = C_3H_3O_2$
347 $R_1 = R_2 = R_4 = H$, $R_3 = Br$
348 $R_1 = R_3 = R_4 = H$, $R_2 = Br$
350 $R_1 = R_2 = R_3 = R_4 = H$

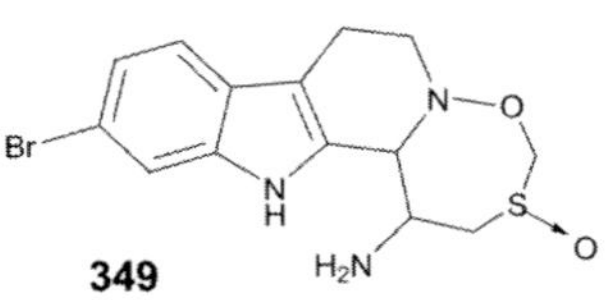

349

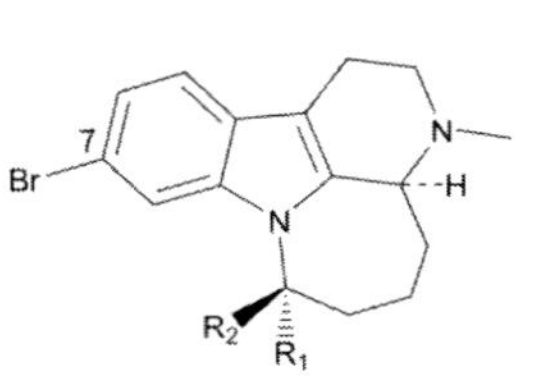

342 $R_1 = OH$, $R_2 = H$
343 $R_1 = H$, $R_2 = OH$

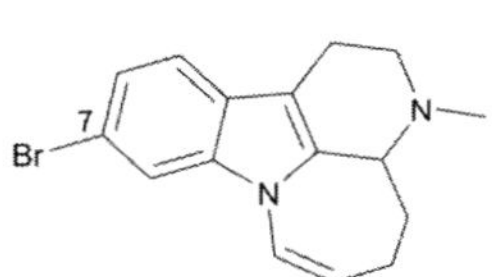

341

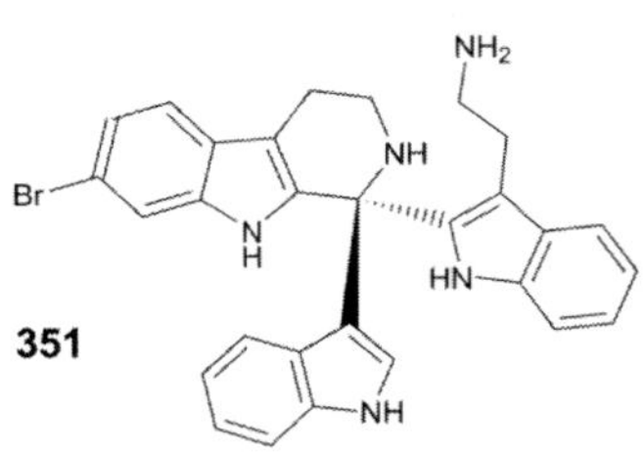

351

Figure 43. Structures of dihydro–β–carbolines (**352-354**).

The eudistomins display a variety of biological activities, including a broad spectrum of antibiotic activity and cytotoxicity, but they are notable for the strong antiviral properties displayed by some members of the family [107, 108]. The tetrahydro–β–carbolines generally exhibited higher levels of biological activity than their fully aromatic relatives; the oxathiazepino-eudistomins **341**-**347**, for example, exhibit the highest level of antiviral activity and were also endowed with antimicrobial activity. Eudistomin K (**344**) is significantly active against *Herpes simplex* Type I (HSV-1) and *Polio* virus, while the sulfoxide (**347**) and debromo (**343**) derivatives are active against both virus but less potent. Enhancement of antiviral activity correlates to bromination at C7.

Together with eudistomins, also manzamine alkaloids have received special attention due to their biological activities [2]. Manzamine alkaloids, mostly isolated from sponges, are characterized by a unique polycyclic ring system, which may biogenetically derive from ammonia, C_{10} and C_3 units, and tryptamine [307]. In the majority of the manzamine alkaloids, the multicyclic units are condensed with tryptamine to form β–carboline. The first representative of this class of alkaloids, manzamine A (**71**), was isolated as hydrochloride salt from an Okinawan sponge *Haliclona* sp. [308].

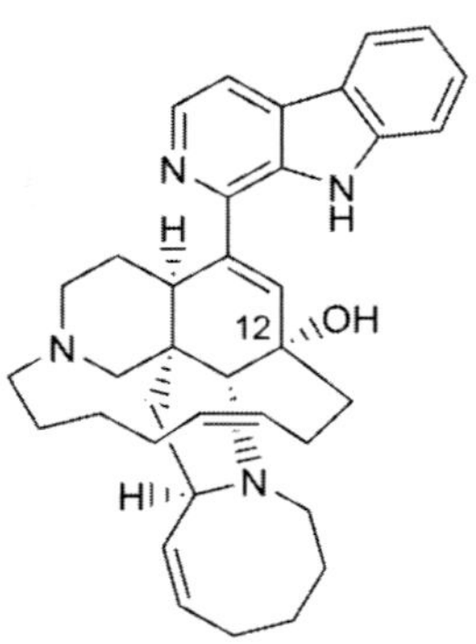

355 Manzamine A

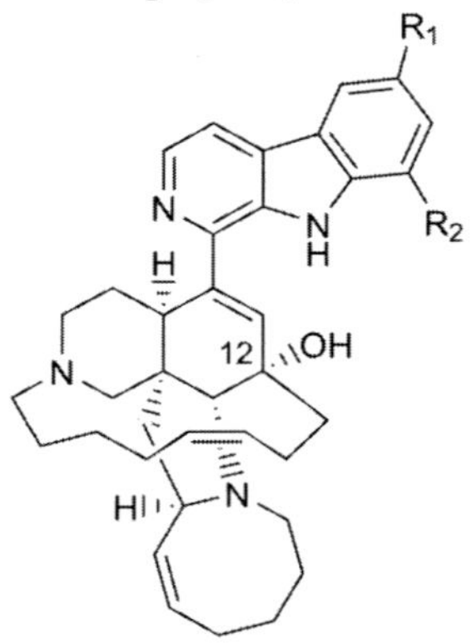

356 Manzamine E R1=R2=H
357 Manzamine F R1=OH;R2=H
358 ManzamineG R1=H; R2=OH
360 ManzamineY R1=OH; R2=H

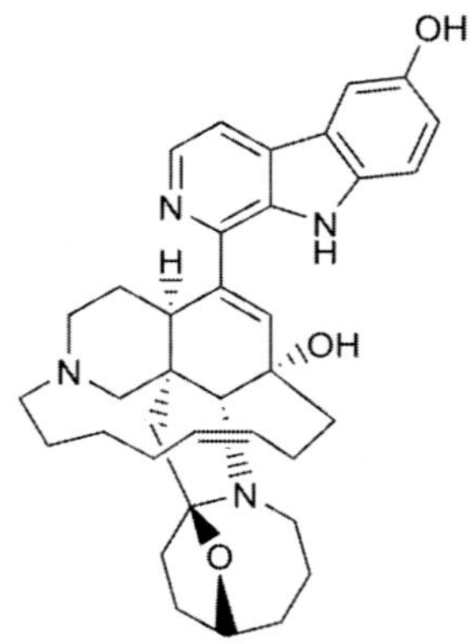

359 Manzamine X

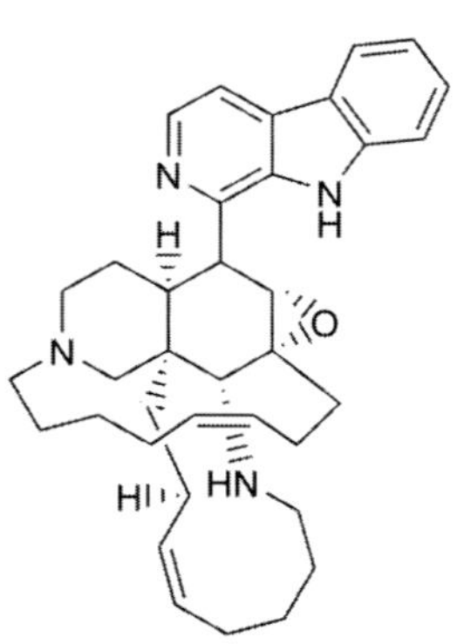

361 Manzamine B

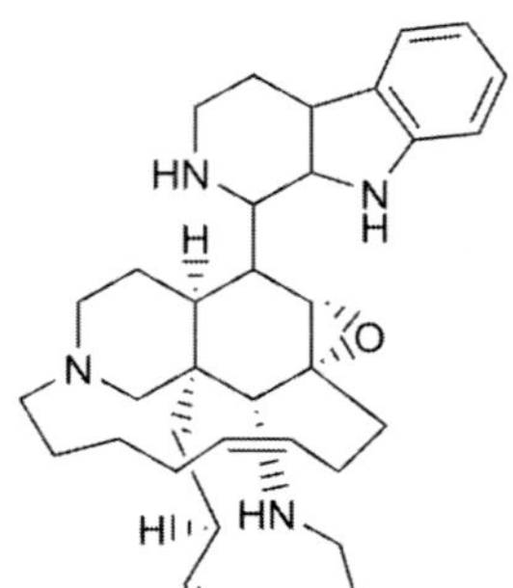

362 1,2,3,4-tetrahydromanzamine B

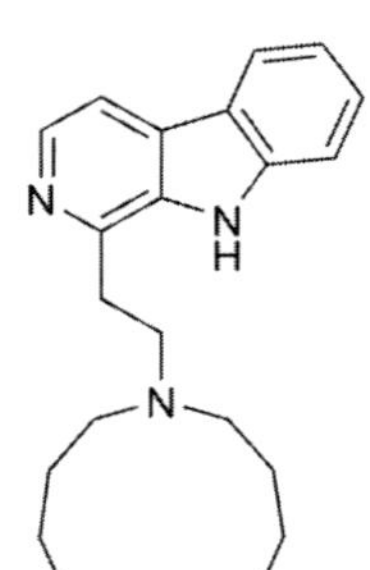

363 Manzamine C

Figure 44. Structures of manzamine alkaloids.

Manzamines have shown a variety of bioactivities including cytotoxicity, antimicrobial, pesticidal, and anti-inflammatory properties [2]; some of them demonstrated effects on HIV and AIDS opportunistic infections [2]. To date, the greatest potential for the manzamine alkaloids appears to be against malaria with manzamine A showing improved activity over the clinically used drugs chloroquine and artemisinin both *in vitro* and *in vivo* [2].

Figure 44 reports some examples of manzamine structures. Manzamines A (**355**) [308], D [309], E and F (**356**, **357**) [310], G (**358**) [311], X (**359**) [312] and Y (**360**) [313], contain a closed eight-membered bottom ring and a double bond between C-10 and C-11. Manzamines E and F differ from the others because they include a ketonic carbonyl group in the eight-membered ring portion of the molecule. Manzamine B (**361**) and its analogs 1,2,3,4-tetrahydromanzamine B (**362**) and ma'eganedin A are characterized by a bottom eight-membered ring opened [309, 314, 315]. Ma'eganedin A has a unique methylene carbon bridge between N-2 and N-27. The structure of manzamine C (**363**) features a 2-ethyl-N-azacycloundec-6-ene unit connected to C-1 of a β–carboline [316].

Conclusion

The marine environment has been explored in the search for new bioactive compounds over the last 50 years and marine life has proven to be the most promising source of alkaloids with extraordinary structures and significant biological activities. The alkaloids obtained from marine organisms frequently possess novel frameworks while in other cases terrestrially related compounds clearly exist. Marine metabolites often possess complexities such as halogen substituents, large and/or poly-substituted frameworks, high number of stereogenic centers, symmetrical or non-symmetrical dimeric and tetrameric structures. Their structure elucidation, chemical modification, stereochemistry, synthesis, and pharmacology have received a great deal of interdisciplinary attention from areas of research other than chemistry, including pharmacology, physiology, and medicine, making them promising lead compounds for various "druggable" targets.

References

[1] Magnier, E., Langlois, Y., (1998). Manzamine Alkaloids, Syntheses and Synthetic Approaches. *Tetrahedron* 54, 6201-6258.

[2] Fattorusso, E., Taglialatela-Scafati, O., Eds. (2008) *Modern Alkaloids. Structure, Isolation, Synthesis and Biology*; Wiley-VCH Verlag GmbH and Co. KGaA: Weinheim, Germany.

[3] Newman, D. J., Cragg, G. M., (2004). Marine Natural Products and Related Compounds in Clinical and Advanced Preclinical Trials. *J. Nat. Prod.* 67, 1216-1238.

[4] Forte, B., Malgesini, B., Piutti, C., Quartieri, F., Scolaro, A., Papeo, G., (2009). A Submarine Journey: The Pyrrole-Imidazole Alkaloids. *Mar. Drugs* 7, 705-753.

[5] Braekman, J. C., Daloze, D., Stoller, C., Van Soest, R. W. M., (1992). Chemotaxonomy of *Agelas* (Porifera: *Demospongiae*). *Biochem. Syst. Ecol.* 20, 417-431.

[6] Garcia, E. E., Benjamin, L. E., Fryer, R. I., (1973). Reinvestigation into the structure of oroidin, a bromopyrrole derivative from marine sponge. *J. Chem. Soc. D. Chem. Commun.*, 78-79.

[7] Assmann, M., Licthe, E., Pawlik, J. R., Kock, M., (2000). Chemical defenses of the Caribbean sponges *Agelas wiedenmayeri* and *Agelas conifera*. *Mar. Ecol Prog. Ser.* 207, 255-262.

[8] Rane, R., Sahu, N., Shah, C., Karpoormath, R., (2014). Marine Bromopyrrole Alkaloids: Synthesis and Diverse Medicinal Applications. *Curr. Top. Med. Chem.* 14, 253-273.

[9] Kobayashi, J., Ohizumi, Y., Nnakamura, H., Hirata, Y., (1986) A novel antagonist of serotonergic receptors, hymenidin, isolated from the Okinawan marine sponge *Hymeniacidon* species. *Experientia* 42, 1176-1177.

[10] Morales, J. J., Rodriguez, A. D., (1991). The structure of clathrodin, a novel alkaloid isolated from the Caribbean sea sponge *Agelas clathrodes*. *J. Nat. Prod.* 54, 629-631.

[11] Assmann, M., Zea, S., Koeck, M., (2001). Sventrin, a new bromopyrrole alkaloid from the Caribbean sponge *Agelas sventres*. *J. Nat. Prod.* 64, 1593-1595.

[12] Rosa, R., Silva, W., Escalona de Motta, G., Rodriguez, A. D., Morales, J. J., Ortiz, M., (1992). Antimuscarinic activity of a family of C11N5 compounds isolated from *Agelas* sponges. *Experientia* 48, 885-887.

[13] Rivera Rentas, A. L., Rosa, R., Rodriguez, A. D., Escalona de Motta, G., (1995). Effect of alkaloid toxins from tropical marine sponges on membrane sodium currents. *Toxicon* 33, 491-497.

[14] Cafieri, F., Fattorusso, E., Mangoni, A., Taglialatela-Scafati, O., (1996). Dispacamides, anti-histamine alkaloids from Caribbean *Agelas* sponges. *Tetrahedron Lett.* 7, 3587-3590.

[15] Cafieri, F., Carnuccio, R., Fattorusso, E., Taglialatela-Scafati, O., Vallefuoco, T., (1997). Anti-histaminic activity of bromopyrrole alkaloids isolated from Caribbean *Agelas* sponges. *Bioorg. Med. Chem. Lett.* 7, 2283-2288.

[16] Aiello, A., D'Esposito, M., Fattorusso, E., Menna, M., Mueller, W. E. G., Perovic-Ottstadt, S., Schroeder, H. C., (2006). Novel bioactive bromopyrrole alkaloids from the Mediterranean sponge *Axinella verrucosa*. *Bioorg. Med. Chem.* 14, 17-24.

[17] Uemoto, H., Tsuda, M., Kobayashi, J., (1999). Mukanadins A-C, new bromopyrrole alkaloids from marine sponge *Agelas nakamurai*. *J. Nat. Prod.* 62, 1581-1583.

[18] Hu, J.-F., Peng, J. K., Kazi, A. B., Kelly, M., Hamann, M. T., (2005). Bromopyrrole alkaloids from the Jamaican sponge *Didiscus oxeata*. *J. Chem. Res.* 7, 427-428.

[19] Chevolot, L., Padua, S., Ravi, B. N., Blyth, P. C., Scheuer, P. J., (1977). Isolation of 1-methyl-4,5-dibromopyrrole-2-carboxylic acid and its 3'-(hydantoyl)propylamide (midpacamide) from a marine sponge *Heterocycles* 7, 891-894.

[20] Jimenez, C., Crews, P., (1994). Mauritamide A and accompanying oroidin alkaloids from the sponge *Agelas mauritiana*. *Tetrahedron Lett.* 35, 1375-1378.

[21] Kobayashi, J., Inaba, K., Tsuda, M., (1997). Tauroacidins A and B, new bromopyrrole alkaloids possessing a taurine residue from *Hymeniacidon* sponge. *Tetrahedron* 53, 16679-16682.

[22] Fattorusso, E., Taglialatela-Scafati, O., (2000). Two novel pyrrole-imidazole alkaloids from the Mediterranean sponge *Agelas oroides*. *Tetrahedron Lett.* 41, 9917-9922.

[23] Tsuda, M., Uemoto, H., Kobayashi, J., (1999). Slagenins A-C, novel bromopyrrole alkaloids from marine sponge *Agelas nakamurai. Tetrahedron Lett.* 40, 5709-5712.

[24] Kobayashi, J., Ohizumi, Y., Nakamura, H., Hirata, Y., Wakamatsu, K., Miyazawa, T., (1986). Hymenin, a novel α-adrenoceptor blocking agent from the Okinawan marine sponge *Hymeniacidon* sp. *Experientia* 42, 1064-1065.

[25] Kobayashi, J., Nakamura, H., Ohizumi, Y., (1988). α-Adrenoceptor blocking action of hymenin, a novel marine alkaloid. *Experientia* 44, 86-87.

[26] Albizati, K. F., Faulkner, D. J., (1985). Stevensine, a novel alkaloid of an unidentified marine sponge. *J. Org. Chem.* 50, 4163-4164.

[27] De Nanteuil, G., Ahond, A., Guilhem, J., Poupat, C., Tran Huu Dau, E., Potier, P., Pusset, M., Pusset, J., Laboute, P., (1985). Marine invertebrates from the New Caledonian lagoon V: Isolation and identification of metabolites of a new species of sponge, *Pseudaxinyssa cantharella. Tetrahedron* 41, 6019-6033.

[28] Eder, C., Proksch, P., Wray, V., Steube, K., Bringmann, G., Van Soest, R. W. M., Sudarsono; Ferdinandus, E., Pattisina, L. A., Wiryowidagdo, S., Moka, W., (1999). New Alkaloids from the Indopacific Sponge *Stylissa carteri. J. Nat. Prod.* 62, 184-187.

[29] Wilson, D. M., Puyana, M., Fenical, W., Pawlik, J. R., (1999). Chemical defense of the Caribbean reef sponge *Axinella corrugata* against predatory fishes. *J. Chem. Ecol.* 25, 2811-2823.

[30] Sharma, G. M., Buyer, J. S., Pomerantz, M. W., (1980). Characterization of a yellow compound isolated from the marine sponge *Phakellia flabellate. J. Chem. Soc. Chem. Commun.* 435-436.

[31] Cimino, G., De Rosa, S., De Stefano, S., Mazzarella, L., Puliti, R., Sodano, G., (1982). Isolation and x-ray crystal structure of a novel bromo compound from two marine sponges. *Tetrahedron Lett.* 23, 767-768.

[32] Kitagawa, I., Kobayashi, M., Kitanaka, K., Kido, M., Kyogoku, Y., (1983). Marine natural products. XII. On the chemical constituents of the Okinawan marine sponge *Hymeniacidon aldis. Chem. Pharm. Bull.* 31, 2321-2328.

[33] Curman, D., Cinel, B., Williams, D. E., Rundle, N., Block, W. D., Goodarzi, A. A., Hutchins, J. R., Clarke, P. R., Zhou, B., Lees-Miller, S. P., Andersen, R. J., Roberge, M., (2001). Inhibition of the G2 DNA damage checkpoint and of protein kinases Chk1 and Chk2 by the marine sponge alkaloid debromohymenialdisine. *J. Biol. Chem.* 276, 17914-17919.

[34] Tasdemir, D., Mallon, R., Greenstein, M., Feldberg, L. R., Kim, S. C., Collins, K., Wojciechowicz, D., Mangalindan, G. C., Concepcion, G. P., Harper, M. K., Ireland, C. M., (2002). Aldisine Alkaloids from the Philippine Sponge *Stylissa massa* Are Potent Inhibitors of Mitogen-Activated Protein Kinase Kinase-1 (MEK -1). *J. Med. Chem.* 45, 529-532.

[35] Wan, Y., Hur, W., Cho, C. Y., Liu, Y., Adrian, F. J., Lozach, O., Bach, S., Mayer, T., Fabbro, D., Meijer, L., Gray, N., (2004). Synthesis and Target Identification of Hymenialdisine Analogs. *Chem. Biol.* 11, 247-259.

[36] Badger, A. M., Cook, M. N., Swift, B. A., Newman-Tarr, T. M., Gowen, M., Lark, M., (1999). Inhibition of interleukin-1-induced proteoglycan degradation and nitric oxide production in bovine articular cartilage /chondrocyte cultures by the natural product, hymenialdisine. *J. Pharmacol. Exp. Ther.* 290, 587-593.

[37] Sharma, V., Lansdell, T. A., Jin, G., Tepe, J. J., (2004). Inhibition of Cytokine Production by Hymenialdisine Derivatives. *J. Med. Chem.* 47, 3700-3703.

[38] Patil, A. D., Freyer, A. J., Killmer, L., Hofmann, G., Johnson, R. K., (1997). Z-axinohydantoin and debromo-Z-axinohydantoin from the sponge *Stylotella aurantium*. Inhibitors of protein kinase C. *Nat. Prod. Lett.* 9, 201-207.

[39] Inaba, K., Sato, H., Tsuda, M., Kobayashi, J., (1998). Spongiacidins A–D, New Bromopyrrole Alkaloids from *Hymeniacidon* Sponge *J. Nat. Prod.* 61, 693-695.

[40] Tsuda, M., Yasuda, T., Fukushi, E., Kawabata, J., Sekiguchi, M., Fromont, J., Kobayashi, J., (2006). Agesamides A and B, bromopyrrole alkaloids from sponge *Agelas* species: Application of DOSY for chemical screening of new metabolites. *Org. Lett.* 8, 4235-4238.

[41] Grube, A., Köck, M., (2006). Oxocyclostylidol, an Intramolecular Cyclized Oroidin Derivative from the Marine Sponge *Stylissa caribica*. *J. Nat. Prod.* 69, 1212-1214.

[42] Sharma, G. M., Burkholder, P. R., (1971). Structure of dibromophakellin, a new bromine -containing alkaloid from the marine sponge *Phakellia flabellate. J. Chem. Soc. Chem. Commun.* 151-152.

[43] Sharma, G., Magdoff-Fairchild, B., (1977). Natural products of marine sponges. 7. The constitution of weakly basic guanidine compounds, dibromophakellin and monobromophakellin. *J. Org. Chem.* 42, 4118-4124.

[44] Gautschi, J. T., Whitman, S., Holman, T. R., Crews, P., (2004). An analysis of phakellin and oroidin structures stimulated by further study of an *Agelas* sponge. *J. Nat. Prod.* 67, 1256-1261.

[45] Boyd, M. R., Pettit, G. R., McNulty, J., Herald, D. L., Doubek, D. L., Chapuis, J., Schmidt, J. M., Tackett, L. P., (1997). Antineoplastic agents. 362. isolation and X-ray crystal structure of dibromophakellstatin from the Indian Ocean sponge *Phakellia mauritiana*. *J. Nat. Prod.* 60, 180-183.

[46] Vergne, C., Boury-Esnault, N., Perez, T., Martin, M., Adeline, M., Dau, E. T. H., Al-Mourabit, A., (2006). Verpacamides A-D, a sequence of $C_{11}N_5$ diketopiperazines relating cyclo(Pro-Pro) to cyclo(Pro-Arg), from the marine sponge *Axinella vaceleti*: Possible biogenetic precursors of pyrrole-2-aminoimidazole alkaloids. *Org. Lett.* 8, 2421-2424.

[47] Kinnel, R. B., Gehrken, H. P., Scheuer, P. J., (1993). Palau'amine: a cytotoxic and immunosuppressive hexacyclic bisguanidine antibiotic from the sponge *Stylotella agminata*. *J. Am. Chem. Soc.* 115, 3376-3377.

[48] Kinnel, R. B., Gehrken, H. P., Swali, R., Skoropowski, G., Scheuer, P. J., (1998). Palau'amine and its congeners: A family of bioactive bisguanidines from the marine sponge *Stylotella aurantium*. *J. Org. Chem.* 63, 3281-3286.

[49] Fedoreev, S. A., Utkina, N. K., Il'in, S. G., Reshetnyak, M. V., Maksimov, O. B., (1986). The structure of dibromoisophakellin from the marine sponge *Acanthella carteri*. *Tetrahedron Lett.* 27, 3177-3180.

[50] Assmann, M., Kock, M., (2002). Monobromoisophakellin, a new bromopyrrole alkaloid from the Caribbean sponge *Agelas* sp. *Z. Naturforsch. C: J. Biosci.* 57, 153-156.

[51] Assmann, M., van Soest, R. W. M., Kock, M., (2001). New antifeedant bromopyrrole alkaloid from the Caribbean sponge *Stylissa caribica*. *J. Nat. Prod.* 64, 1345-1347.

[52] Tsukamoto, S., Tane, K., Ohta, T., Matsunaga, S., Fusetani, N., van Soest, R. W. M., (2001). Four new bioactive pyrrole-derived alkaloids from the marine sponge *Axinella brevistyla*. *J. Nat. Prod.* 64, 1576-1578.
[53] Goetz, G. H., Harrigan, G. G., Likos, J., (2001). Ugibohlin: A new dibromo-seco-isophakellin from *Axinella carteri*. *J. Nat. Prod.* 64, 1581-1582.
[54] Kato, T., Shizuri, Y., Izumida, H., Yokoyama, A., Endo, M., (1995). Styloguanidines, new chitinase inhibitors from the marine sponge *Stylotella aurantium*. *Tetrahedron Lett.* 36, 2133-2136.
[55] Fedoreev, S. A., Il'in, S. G., Utkina, N. K., Maksimov, O. B., Reshetnyak, M. V., Antipin, M. Y., Struchkov, Y. T., (1989). The structure of dibromoagelaspongin - a novel bromine -containing guanidine derivative from the marine sponge *Agelas* sp. *Tetrahedron* 45, 3487-3492.
[56] D'Ambrosio, M., Guerriero, A., Debitus, C., Ribes, O., Pusset, J., Leroy, S., Pietra, F., (1993). Agelastatin A, a new skeleton cytotoxic alkaloid of the oroidin family. Isolation from the axinellid sponge *Agelas dendromorpha* of the Coral Sea. *J. Chem. Soc. Chem. Comm.* 16, 1305-1306.
[57] D'Ambrosio, M., Guerriero, A., Ripamonti, M., Debitus, C., Waikedre, J., Pietra, F., (1996). The active centers of agelastatin A, a strongly cytotoxic alkaloid of the Coral sea axinellid sponge *Agelas dendromorpha*, as determined by comparative bioassays with semisynthetic derivatives. *Helv. Chim. Acta* 79, 727-735.
[58] Hong, T. W., Jimenez, D. R., Molinski, T. F., (1998). Agelastatins C and D, new pentacyclic bromopyrroles from the sponge *Cymbastela* sp., and potent arthropod toxicity of (-)-agelastatin A. *J. Nat. Prod.* 61, 158-161.
[59] Meijer, L., Thunnissen, A.-M. W. H., White A. W., Garnier, M., Nikolic, M., Tsai, L.-H., Walter, J., Cleverley, K. E., Salinas, P. C., Wu, Y.-Z., Biernat, J., Mandelkow, E.-M., Kim, S.-H., Pettit, G. R., (2000). Inhibition of cyclin-dependent kinases, GSK-3β and CK1 by hymenialdisine, a marine sponge constituent. *Chem. Biol.* 7, 51-63.
[60] Walker, Roger P., Faulkner, D. John, Van Engen, Donna, Clardy, Jon., (1981). Sceptrin, an antimicrobial agent from the sponge *Agelas sceptrum*. *J. Am. Chem. Soc.* 103, 6772-6773.
[61] Faulkner, D. John; (1983). PATENT US 3 pp. CODEN: USXXAM US 4,370,484 Appl. No.: 242728.
[62] Bernan, V. S., Roll, D. M., Ireland, Chris M., Greenstein, M., Maiese, William M., Steinberg, D. A., (1993). A study on the mechanism of action of sceptrin, an antimicrobial agent isolated from the South Pacific sponge *Agelas mauritiana*. *J. Antimicrob. Chemother.* 32, 539-550.
[63] Peng, J., Shen, X., El Sayed, K. A., Dunbar, D. C., Perry, T. L., Wilkins, S. P., Hamann, M. T., Bobzin, S., Huesing, J., Camp, R., Prinsen, M., Krupa, D., Wideman, M. A., (2003). Marine natural products as prototype agrochemical agents. *J. Agric. Food Chem.* 51, 2246-2252.
[64] Keifer, P. A., Schwartz, R. E., Koker, M. E. S., Hughes, R. G., Jr., Rittschof, D., Rinehart, K. L., (1991). Bioactive bromopyrrole metabolites from the Caribbean sponge *Agelas conifera*. *J. Org. Chem.* 56, 2965-2975.
[65] Vassas, A., Bourdy, G., Paillard, J. J., Lavayre, J., Pais, M., Quirion, J. C., Debitus, C., (1996). Naturally occurring somatostatin and vasoactive intestinal peptide inhibitors. Isolation of alkaloids from two marine sponges. *Planta Medica*, 62, 28-30.

[66] Rinehart, Kenneth L., Jr, (1988). PATENT US15 pp. CODEN: USXXAM US 4,737, 510 Appl. No.: 913819.
[67] Kobayashi, J., Tsuda, M., Ohizumi, Y., (1991). A potent actomyosin ATPase activator for the Okinawan marine sponge *Agelas cf. nemoechinata*. *Experientia* 47, 301-304.
[68] Shen, X., Perry, T. L., Dunbar, C. D., Kelly-Borges, M., Hamann, M. T., (1998). Debromosceptrin, an alkaloid from the Caribbean sponge *Agelas conifera*. *J. Nat. Prod.* 61, 1302-1303.
[69] Assmann, M., Kock, M., (2002). Bromosceptrin, an alkaloid from the marine sponge *Agelas conifera*. *Z. Naturforsch. C: J. Biosci.* 57, 157-160.
[70] Bickmeyer, U., Drechsler, C., Kock, M., Assmann, M., (2004). Brominated pyrrole alkaloids from marine *Agelas* sponges reduce depolarization -induced cellular calcium elevation. *Toxicon* 44, 45-51.
[71] Kobayashi, J., Suzuki, M., Tsuda, M., (1997). Konbu'acidin A, a new bromopyrrole alkaloid with cdk4 inhibitory activity from *Hymeniacidon* sponge. *Tetrahedron* 53, 15681-15684.
[72] Urban, S., de Almeida Leone, P., Carroll, A. R., Fechner, G. A., Smith, J., Hooper, J. N. A., Quinn, R. J., (1999). Axinellamines A-D, Novel Imidazo-Azolo-Imidazole Alkaloids from the Australian Marine Sponge *Axinella* sp. *J. Org. Chem.* 64, 731-735.
[73] Nishimura, S., Matsunaga, S., Shibazaki, M., Suzuki, K., Furihata, K., Van Soest, R. W. M., Fusetani, N., (2003). Massadine, a Novel Geranylgeranyltransferase Type I Inhibitor from the Marine Sponge *Stylissa* aff. *massa*. *Org. Lett.* 5, 2255-2257.
[74] Grube, A., Köck, M., (2006). Stylissadines A and B: The first tetrameric pyrrole-imidazole alkaloids. *Org. Lett.* 8, 4675-4678.
[75] Cafieri, F., Fattorusso, E., Mangoni, A., Taglialatela-Scafati, O., (1996). Clathramides, unique bromopyrrole alkaloids from the Caribbean sponge *Agelas clathrodes*. *Tetrahedron* 52, 13713-13720.
[76] Jahn, T., Koning, G. M., Wright, A. D., (1997). Manzacidin D: an unprecedented secondary metabolite from the "living fossil" sponge *Astrosclera willeyana*. *Tetrahedron Lett.* 38, 3883-3884.
[77] Assmann, M., Lichte, E., Van Soest, R. W. M., Koeck, M., (1999). New bromopyrrole alkaloid from the marine sponge *Agelas wiedenmeyeri*. *Org. Lett.* 1, 455-457.
[78] Fujita, M., Nakao, Y., Matsunaga, S., Seiki, M., Itoh, Y., Yamashita, J., Van Soest, R. W. M., Fusetani, N., (2003). Ageladine A: An antiangiogenic matrix metalloproteinase inhibitor from the marine sponge *Agelas nakamurai*. *J. Am. Chem. Soc.* 125, 15700-15701.
[79] Linington, R. G., Williams, D. E., Tahir, A., Van Soest, R. W., Andersen, R. J., (2003). Latonduines A and B, new alkaloids isolated from the marine sponge *Stylissa carteri*: Structure elucidation, synthesis, and biogenetic implications. *Org. Lett.* 5, 2735-2738.
[80] Bailly, C., (2004). Lamellarins, from A to Z: A family of anticancer marine pyrrole alkaloids. *Curr. Med. Chem. Anti-Cancer Agents* 4, 363-378.
[81] Pla, D., Albericio, F., Alvarez, M., (2011). Progress on lamellarins. *Med. Chem. Commun.* 2, 689-697.
[82] Menna, M., (2014). Important classes of bioactive alkaloids from marine ascidians: structures, isolation and bioactivity. *Curr. Top. Med. Chem.* 14, 207-223.

[83] Andersen, R. J., Faulkner, D. J., He, C. H., Van Duyne, G. D., Clardy, J., (1985). Metabolites of the marine prosobranch mollusk *Lamellaria* sp. *J. Am. Chem. Soc.* 107, 5492-5495.

[84] Lindquist, N., Fenical, W., Van Duyne, G. D., Clardy, J., (1988). New alkaloids of the Lamellarin class from the marine ascidian *Didemnum chartaceum. J. Org. Chem.* 53, 4570-4574.

[85] Carroll, A. R., Bowden, B. F., Coll, J. C., (1993). Studies of Australian ascidians. I. Six new lamellarin class alkaloids from a colonial ascidian, *Didemnum* sp. *Aust. J. Chem.* 46, 489-501.

[86] Davis, R. A., Carroll, A. R., Pierens, G. K., Quinn, R. J., (1999). New lamellarin alkaloids from the Australian ascidian *Didemnum chartaceum J. Nat. Prod.* 62, 419-424.

[87] Reddy, M. V. R., Rao, M. R., Rhodes, D., Hansen, M. S. T., Rubins, K., Bushman, F. D., Venkateswarlu, Y., Faulkner, D. J., (1999). Lamellarin α-20 sulfate, an inhibitor of HIV -1 integrase active against HIV-1 virus in cell culture *J. Med. Chem.* 42, 1901-1907.

[88] Krishnaiah, P., Reddy, V. L. N., Venkataramana, G., Ravinder, K., Srinivasulu, M., Raju, T. V., Ravikumar, K., Chandrasekar, D., Ramakrishna, S., Venkateswarlu, Y., (2004). New lamellarin alkaloids from the Indian ascidian *Didemnum obscurum* and their antioxidant properties. *J. Nat. Prod.* 67, 1168-1171.

[89] Reddy, S. M., Srinivasulu, M., Satyanarayana, N., Kondapi, A. K., Venkateswarlu, Y., (2005). New potent cytotoxic lamellarin alkaloids from Indian ascidian *Didemnum obscurum. Tetrahedron* 61, 9242-9247.

[90] Urban, S., Butler, M. S., Capon, R. J., (1994). Lamellarins O and P: New aromatic metabolites from the Australian marine sponge *Dendrilla cactos. Aust. J. Chem.* 47, 1919-1924.

[91] Urban, S., Hobbs, L., Hooper, J. N. A., Capon, R. J., (1995). Lamellarins Q and R: New aromatic metabolites from an Australian marine Sponge *Dendrilla cactos. Aust. J. Chem.* 48, 1491-1494.

[92] Facompre, M., Tardy, C., Bal-Mahieu, C., Colson, P., Perez, C., Manzanares, I., Cuevas, C., Bailly, C., (2003). Lamellarin D: a novel potent inhibitor of topoisomerase I, *Cancer Res.* 63, 7392-7399.

[93] Marco, E., Laine, W., Tardy, C., Lansiaux, A., Iwao, M., Ishibashi, F., Bailly, C., Gago, F., (2005). Molecular determinants of topoisomerase I poisoning by lamellarins: comparison with camptothecin and structure -activity relationships, *J. Med. Chem.* 48, 3796-3807.

[94] Kluza, J., Gallego, M. A., Loyens, A., Beauvillain, J. C., Sousa-Faro, J. M., Cuevas, C., Marchetti, P., Bailly, C., (2006). Cancer cell mitochondria are direct proapoptotic targets for the marine antitumor drug lamellarin D. *Cancer Res.* 66, 3177-3187.

[95] Gallego, M. A., Ballot, C., Kluza, J., Hajji, N., Martoriati, A., Castéra, L., Cuevas, C., Formstecher, P., Joseph, B., Kroemer, G., Bailly, C., Marchetti, P., (2008). Overcoming chemoresistance of non-small cell lung carcinoma through restoration of an AIF-dependent apoptotic pathway. *Oncogene* 27, 1981-1992.

[96] Ballot, C., Kluza, J., Lancel, S., Martoriati, A., Hassoun, S., Mortier, L., Vienne, J. C., Briand, G., Formstecher, P., Bailly, C., Neviere, R., Marchetti, P., (2010). Inhibition of

mitochondrial respiration mediates apoptosis induced by the anti-tumoral alkaloid lamellarin D. *Apoptosis* 15, 769-781.

[97] Quesada, A. R., Garcia Gravalos, M. D., Fernandez Puentes, J. L., (1996). Polyaromatic alkaloids from marine invertebrates as cytotoxic compounds and inhibitors of multidrug resistance caused by P-glycoprotein. *Br. J. Cancer.* 74, 677-682.

[98] Vanhuyse, M., Kluza, J., Tardy, C., Otero, G., Cuevas, C., Bailly, C., Lansiaux, A., (2005). Lamellarin D: a novel pro-apoptotic agent from marine origin insensitive to P-glycoprotein-mediated drug efflux. *Cancer Lett.* 22, 165-175.

[99] Ridley, C. P., Reddy, M. V. R., Rocha, G., Bushman, F. D., Faulkner, D. J., (2002). Total synthesis and evaluation of lamellarin alpha 20-Sulfate analogues, *Bioorg. Med. Chem.* 10, 3285-3290.

[100] Baunbæk, D., Trinkler, N., Ferandin, Y., Lozach, O., Ploypradith, P., Rucirawat, S., Ishibashi, F., Iwao, M., Meijer, L., (2008). Anticancer Alkaloid Lamellarins Inhibit Protein Kinases, *Mar. Drugs* 6, 514-527.

[101] Chittchang, M., Gleeson, M. P., Ploypradith, P., Ruchirawat, S., (2010). Assessing the drug-likeness of lamellarins, a marine-derived natural product class with diverse oncological activities. *Eur. J. Med. Chem.* 45, 2165-2172.

[102] Yoshida, W. Y., Lee, K. K., Carroll, A. R., Scheuer, P. J., (1992). A complex pyrrolo-oxazinone and its iodo derivative isolated from a tunicate. *Helv. Chim. Acta.* 75, 1721-1725.

[103] Rudi, A., Goldberg, I., Stein, Z., Frolow, F., Benayahu, Y., Schleyer, M., Kashman, Y., (1994). Polycitone A and Polycitrins A and B: New alkaloids from the marine Ascidian *Polycitor* sp. *J. Org. Chem.* 59, 999-1003.

[104] Rudi, A., Evan, T., Aknin, M., Kashman, Y., (2000). Polycitone B and Prepolycitrin A: Two novel alkaloids from the marine ascidian *Polycitor africanus. J. Nat. Prod.* 63, 832-833.

[105] Kang, H., Fenical, W., (1997). Ningalins A–D: Novel aromatic alkaloids from a western Australians ascidian of the genus *Didemnum. J. Org. Chem.* 62, 3254-3262.

[106] Fan, H., Peng, J., Hamann, M. T., Hu, J. F., (2008). Lamellarins and related pyrrole-derived alkaloids from marine organisms. *Chem. Rev.* 108, 264-287.

[107] Menna, M., Fattorusso, E., Imperatore, C., (2011). Alkaloids from Marine Ascidians. *Molecules* 16, 8694-8732.

[108] Menna, M., Aiello, A., (2012). The Chemistry of Marine Tunicates. In: *Handbook of Marine Natural Products,* vol. 2, Fattorusso, E., Gerwick, W. H., Taglialatela-Scafati, O. (Eds.), Springer Science, 295-385.

[109] Moser, B. R., 2008. Review of cytotoxic cephalostatins and ritterazines: Isolation and synthesis. *J. Nat. Prod.* 71, 487-491.

[110] Lee, S., LaCour, T. G., Fuchs, P. L., 2009. Chemistry of trisdecacyclic pyrazine antineoplastics: The cephalostatins and ritterazines. *Chem. Rev.* 109, 2275-2314.

[111] Komiya, T., Fusetani, N., Matsunaga, S., Kubo, A., Kaye, F. J., Kelley, M. J., Tamura, K., Yoshida, M., Fukuoka, M., Nakagawa, K., 2003. Ritterazine B, a new cytotoxic natural compound, induces apoptosis in cancer cells. *Cancer Chemother. Pharmacol.* 51, 202-208.

[112] Ganesan, A., 1996. The dimeric steroid-pyrazine marine alkaloids: challenges for isolation, synthesis, and biological studies. *Angew. Chem. Int. Ed. Engl.* 35, 611-615.

[113] López-Antón, N., Rudy, A., Barth, N., Schmitz, L. M., Pettit, G. R., Schulze-Osthoff, K., Dirsch, V. M., Vollmar, A. M., 2006. The marine product cephalostatin 1 activates and ER stress -specific and apoptosome- independent apoptotic signaling pathway. *J. Biol. Chem.* 281, 33078-33086.

[114] Rudy, A., López-Antón, N., Dirsch, V. M., Vollmar, A. M., 2008. The Cephalostatin Way of Apoptosis. *J. Nat. Prod.* 71, 482-486.

[115] Rudy, A., López-Antón, N., Barth, N., Pettit, G. R., Dirsch, V. M., Schulze-Osthoff, K., Rehm, M., Prehn, J. H. M., Vogler, M., Fulda, S., Vollmar, A. M., 2008. Role of Smac in cephalostatin-induced cell death. *Cell Death Differ.* 15, 1930-1940.

[116] Cuevas, C., Francesch, A., 2009. Development of Yondelis® (trabectedin, ET-743). A semisynthetic process solves the supply problem. *Nat. Prod. Rep.* 26, 322-337.

[117] Leal, J. F. M., Martínez-Díez, M., García-Hernández, V., Moneo, V., Domingo, A., Bueren-Calabuig, J. A., Negri, A., Gago, F., Guillén-Navarro, M. J., Avilés, P., Cuevas, C., García-Fernández, L. F., Galmarini, G. M., 2010. PM01183, a new DNA minor groove covalent binder with potent *in vitro* and *in vivo* anti-tumour activity. *British J. Pharmacol.* 161, 1099-1110.

[118] Rinehart, K. L., Holt, T. G., Fregeau, N. L., Stroh, J. G., Keifer, P. A., Sun, F., Li, L. H., Martin, D. G., 1990. Ecteinascidins 729, 743, 745, 759A. 759B, and 770: potent antitumor agent s from the Caribbean tunicate *Ecteinascidia turbinata*. *J. Org. Chem.* 55, 4512-4515.

[119] Wright, A. E., Forleo, D. A., Gunawardana, G. P., Gunasekera, S. P., Koehn, F. E., McConnell, O. J., 1990. Antitumor tetrahydroisoquinoline alkaloids from the colonial ascidian *Ecteinascidia turbinata*. *J. Org. Chem.* 55, 4508-4512.

[120] Guan, Y., Sakai, R., Rinehart, K. L., Wang, A. H. J., 1993. Molecular and crystal structure s of ecteinascidins: potent antitumor compounds from the Caribbean tunicate *Ecteinascidia turbinata*. *J. Biomol. Struct. Dyn.* 10, 793-818.

[121] Sakai, R., Rinehart, K. L., Guan, Y., Wang, A. H. J., 1992. Additional antitumor ecteinascidins from a Caribbean tunicate: Crystal structures and activities *in vivo*. *Proc. Natl. Acad. Sci., US* 89, 11456-11460.

[122] Rinehart, K. L., Sakai, R., 2004. Isolation, structure elucidation, and bioactivities of novel ecteinascidins from *Ecteinascidia turbinata*. *US Pat. Appl. Publ.* 0059112.

[123] Suwanborirux, K., Charupant, K., Amnuoypol, S., Pummangura, S., Kubo, A., Saito, N., 2002. Ecteinascidins 770 and 786 from the Thai Tunicate *Ecteinascidia thurstoni*. *J. Nat. Prod.* 65, 935-937.

[124] Sakai, R., Jares-Erijman, E. A., Manzanares, I., Elipe, M. V. S., Rinehart, K. L., 1996. Ecteinascidins: putative biosynthetic precursors and absolute stereochemistry *J. Am. Chem. Soc.* 118, 9017-9023.

[125] Scott, J. D., Williams, R. M., 2002. Chemistry and Biology of the Tetrahydroisoquinoline Antitumor Antibiotics. *Chem. Rev.* 102, 1669-1730.

[126] Soares, D. G., Larsen, A. K., Escargueil, A. E., 2012. The DNA damage response to monofunctional anticancer DNA binders. *Drug Discovery Today: Disease Models* 9(2), e59-e67.

[127] Aune, G. J., Furuta, T., Pommier, Y., 2002. Ecteinascidin 743: a novel anticancer drug with a unique mechanism of action. *Anticancer Drugs* 13, 545-555.

[128] Gajdos, C., Elias, A., 2011. Trabectedin: Safety and efficacy in the treatment of advanced sarcoma. *Clin. Med. Ins. Oncol.* 5, 35-43.

[129] http://www.yondelis.com.

[130] Carballo, J. L., Naranho, S., Kukurtzu, B., De La Calle, F., Hernandez-Zanuy, A., 2000. Production of *Ecteinascidia turbinata*(Ascidiacea: Perophoridae) for obtaining anticancer compounds. *J. World Aquacult. Soc.* 31, 481-490.

[131] Cuevas, C., Pérez, M., Martin, M. J., Chicharro, J. L., Fernández-Rivas, C., Flores, M., Francesch, A., Gallego, P., Zarzuelo, M., de la Calle, F., Garcia, J., Polanco, C., Rodrigue,z I., Manzanares, I., (2000). Synthesis of ecteinascidin ET-743 and phthalascidin Pt-650 from cyanosafracin B. *Org Lett.* 2, 2545-2548.

[132] Menchaca, R., Martínez, V., Rodríguez, A., Rodríguez, N., Flores, M., Gallego, P., Manzanares, I., Cuevas, C., (2003). Synthesis of Natural Ecteinascidins (ET-729, ET-745, ET-759B, ET-736, ET-637, ET- 594) from Cyanosafracin B. *J. Org. Chem.* 68, 8859-8866.

[133] Martinez, E. J., Corey, E. J., Owa, T., (2001). Antitumor activity- and gene expression - base d profiling of ecteinascidin Et743 and phtalascidin Pt 650. *Chem. Biol.* 8, 1151-1160.

[134] Soares, D. G., Machado, M. S., Rocca, C. J., Poindessous, V., Ouaret, D., Sarasin, A., Galmarini, C. M., Henriques, G. A. P., Escargueil, A. E., Larsen, A. K., (2011). Trabectedin and Its C Subunit ModifiedAnalogue PM01183 Attenuate Nucleotide Excision Repair and Show Activity toward Platinum -Resistant Cells. *Mol. Cancer. Ther.* 10, 1481-1489.

[135] http://www.pharmamar.com.

[136] Kazlauskas, R., Murphy, P. T., Quinn, R. J., Wells, R. J., (1977). Aplysinopsin, a new tryptophan derivative from a sponge. *Tetrahedron Lett.* 61-64.

[137] Hollenbeak, K. H., Schmitz, F. J., (1977). Aplysinopsin: antineoplastic tryptophan derivative from the marine sponge *Verongia spengelii*. *J. Nat. Products* 40, 479-481.

[138] Tymiak, A. A., Rinehart, K. L. Jr., Bakus, G. J., (1985). Constituents of morphologically similar sponges. *Aplysina* and *Smenospongia* species. *Tetrahedron* 41, 1039-1047.

[139] Seagraves, N. L., Crews, P., (2005). Investigation of brominated tryptophan alkaloids from two *Thorectidae* sponges: *Thorectandra* and *Smenospongia*. *J. Nat. Prod.* 68, 1484-1488.

[140] Fattorusso, E., Lanzotti, V., Magno, S., Novellino, E., (1985). Tryptophan derivatives from a Mediterranean anthozoan, *Astroides calycularis*. *J. Nat. Prod.* 48, 924-927.

[141] Fusetani, N., Asano, Matsunaga, S., Hashimoto, K., (1986). Bioactive marine metabolites. XV. Isolation of aplysinopsin from the scleractinian coral *Tubastrea aurea* as an inhibitor of development of fertilized sea urchin eggs. *Comp. Biochem. Phys. B* 85B, 845-846.

[142] Okuda, R. K., Klein, D., Kinnel, R. B., Li, M., Scheuer, P. J., (1982). Marine natural products: the past twenty years and beyond. *Pure Appl. Chem.* 54, 1907-1914.

[143] Guella, G., Mancini, I., Zibrowius, H., Pietra, F., (1989). Aplysinopsin-type alkaloids from *Dendrophyllia* sp., a scleractinian coral of the family *Dendrophylliidae* of the Philippines. Facile photochemical (Z/E) photoisomerization and thermal reversal. *Helv. Chim. Acta* 72, 1444-1450.

[144] Kondo, K., Nishi, J., Ishibashi, M., Kobayashi, J., (1994). Two new tryptophan -derived alkaloids from the Okinawan marine sponge *Aplysina* sp. *J. Nat. Prod.* 57, 1008-1011.

[145] Aoki, S., Ye, Y., Higuchi, K., Takashima, A., Tanaka, Y., Kitagawa, I., Kobayashi, M., (2001). Novel neuronal nitric oxide synthase (nNOS) selective inhibitors, aplysinopsin-type indole alkaloids, from marine sponge Hyrtios erecta. *Chem. Pharm. Bull.* 49, 1372-1374.

[146] Baker, J. T., Wells, R. J., (1982). Biologically active substances from Australian marine organisms, In: Beal, J. L. and Reinhard, E. (Eds.) *Natural Products as Medicinal Agents*, Hippokrates, Stuttgard, 281-318.

[147] Baker, J. T., (1984). Modern drug research: the potential and the problems of marine natural products, In: Krogsgaard-Larsen, P. Brøgger Christersen, S. Kofod H. (Eds.) *Natural Products and Drug Development*, Munksgaard, Copenhagen, 145-163.

[148] Debitus, C., Cesario, M., Gilhem, J., Pascard, C., Pais, M. (1989). Corallistine, a new polynitrogen compound from the sponge *Corallistes fulvodesmus* L. and L. *Tetrahedron Lett.* 30, 1535-1538.

[149] Edrada, R. A., Stessman, C. C., Crews, P., (2003). Uniquely modified imidazole alkaloids from a calcareous *Leucetta* sponge. *J. Nat. Prod.* 66, 939-942.

[150] Ralifo, P., Crews, P., (2004). A new structural theme in the imidazole-containing alkaloids from a calcareous *Leucetta* sponge. *J. Org. Chem.* 69, 9025-9029.

[151] Crews, P., Clark, D. P., Tenney, K., (2003). Variation in the alkaloids among Indo-Pacific *Leucetta* sponges. *J. Nat. Prod.* 66, 177-182.

[152] Perry, N. B., Ettouati, L., Litaudon, M., Blunt, J. W., Munro, M. H. G., Parkin, S., Hope, H., (1994). Alkaloids from the antarctic sponge *Kirkpatrickia varialosa*. Part 1: Variolin B, a new antitumor and antiviral compound. *Tetrahedron* 50, 3987-3992.

[153] Trimurtulu, G., Faulkner, D. J., Perry, N. B., Ettouati, L., Litaudon, M., Blunt, J. W., Munro, M. H. G., Jameson, G. B., (1994). Alkaloids from the antarctic sponge Kirkpatrickia varialosa. Part 2: Variolin A and N(3')-methyl tetrahydrovariolin B. *Tetrahedron* 50, 3993-4000.

[154] Loukaci, A., Guyot, M., Chiaroni, A., Riche, C., (1998). A new indole alkaloid from the marine tunicate *Dendrodoa grossularia. J. Nat. Prod.* 61, 519-522.

[155] Van Wagoner, R. M., Jompa, J., Tahir, R., Ireland, C. M., (1999). Trypargine alkaloids from a previously undescribed *Eudistoma* sp. Ascidian. *J. Nat. Prod.* 62, 794-797.

[156] Pearce, A. N., Appleton, D. R., Babcock, R. C., Copp, B. R., (2003). Distomadines A and B, novel 6-hydroxyquinoline alkaloids from the New Zealand ascidian, *Pseudodistoma aureum. Tetrahedron Lett.* 44, 3897-3899.

[157] García, A., Vásquez, M. J., Quiñoá, E., Riguera, R., Débitus, C., (1996). New amino acid derivatives from the marine ascidian *Leptoclinides dubius. J. Nat. Prod.* 59, 782-785.

[158] Sölter, S., Dieckmann, R., Blumenberg, M., Francke, W., (2002). Barettin, revisited? *Tetrahedron Lett.*, 43, 3385-3386.

[159] Carroll, A. R., Pierens, G. K., Fechner, G., Leone, P. A., Ngo, A., Simpson, M., Hyde, E., Hooper, J. N. A., Boström, S.-L., Musil, D., Quinn, R. J., (2002). Dysinosin A: A novel inhibitor of factor VIIa and thrombin from a new genus and species of Australian sponge of the family *Dysideidae. J. Am. Chem. Soc.* 124, 13340-13341.

[160] Fusetani, N., Kakao, Y., Matsunaga, S., (1991). Bioactive marine metabolites. 39. Nazumamide A, a thrombin -inhibitory tetrapeptide, from a marine sponge, *Theonella* sp. *Tetrahedron Lett.* 32, 7073-7074.

[161] Nienaber, V. L., Amparo, E. C., (1996). A Noncleavable Retro-Binding Peptide That Spans the Substrate Binding Cleft of Serine Proteases. Atomic Structure of Nazumamide A: Human Thrombin. *J. Am. Chem. Soc.* 118, 6807-6810.
[162] Coleman, J. E., de Silva, E. D., Kong, F., Andersen, R. J., Allen, T. M., (1995). Cytotoxic peptides from the marine sponge *Cymbastela* sp. *Tetrahedron* 51, 10653-10662.
[163] Swersey, J. C., Ireland, C. M., Cornell, L. M., Peterson, R. W., (1994). Eusynstyelamide, a highly modified dimer peptide from the ascidian *Eusynstyela misakiensis*. *J. Nat. Prod.* 57, 842-845.
[164] Expósito, M. A., López, B., Fernández, R., Vásquez, M. J., Débitus, C., Iglesias, T., Jiménez, C., Quiñoá, E., Riguera, R., (1998). Minalemines A-F: sulfamic acid peptide guanidine derivatives isolated from the marine tunicate *Didemnun rodriguesi*. *Tetrahedron* 54, 7539-7550.
[165] Casapullo, A., Finamore, E., Minale, L., Zollo, F., (1993). A dimeric peptide alkaloid of a completely new type, anchinopeptolide A, from the marine sponge *Anchinoe tenacior*. *Tetrahedron Lett.* 34, 6297-6300.
[166] Casapullo, A., Minale, L., Zollo, F., Lavayre, J., (1994). Four new dimeric peptide alkaloids, anchinopeptolides B-D, and cycloanchinopeptolide C, congeners of anchinopeptolide A, from the Mediterranean marine sponge *Anchinoe tenacior*. *J. Nat. Prod.* 57, 1227-1233.
[167] Fusetani, N., Matsunaga, S., Matsumoto, H., Takebayashi, Y., (1990). Bioactive marine metabolites. 33. Cyclotheonamides, potent thrombin inhibitors, from a marine sponge *Theonella* sp. *J. Am. Chem. Soc.* 112, 7053-7054.
[168] Nakao, Y., Matsunaga, S., Fusetani, N., (1995). Bioactive marine metabolites. Part 65. Three more cyclotheonamides, C, D, and E, potent thrombin inhibitors from the marine sponge *Theonella swinhoei*. *Bioorgan. Med. Chem.* 3, 1115-1122.
[169] Nakao, Y., Masuda, A., Matsunaga, S., Fusetani, N., (1999). Pseudotheonamides, serine protease inhibitors from the marine sponge *Theonella swinhoei*. *J. Am. Chem. Soc.* 121, 2425-2431.
[170] Murakami, Y., Takei, M., Shindo, K., Kitazume, C., Tanaka, J., Higa, T., Fukamachi, H., (2002). Cyclotheonamide E4 and E5, new potent tryptase inhibitors from an *Ircinia* species of sponge. *J. Nat. Prod.* 65, 259-261.
[171] Zampella, A., D'Auria, M. V., Paloma, L. G., Casapullo, A., Minale, L., Débitus, C., Henin, Y., (1996). Callipeltin A, an anti-HIV cyclic depsipeptide from the New Caledonian Lithistida sponge *Callipelta* sp. *J. Am. Chem. Soc.* 118, 6202-6209.
[172] D'Auria, M. V., Zampella, A., Paloma, L. G., Minale, L., Débitus, C., Roussakis, C., Le Bert, V., (1996). Callipeltins B and C; bioactive peptides from a marine Lithistida sponge *Callipelta* sp. *Tetrahedron*, 52, 9589-9596.
[173] Zampella, A., D'Auria, M. V., (2002). Stereoselective synthesis of (2*R*,3*R*,4*R*)-3-hydroxy-2,4,6-trimethylheptanoic acid and determination of the absolute stereochemistry of the natural product from callipeltin A. *Tetrahedron-Asymmetr.* 13, 1237-1239.
[174] Zampella, A., Randazzo, A., Borbone, N., Luciani, S., Trevisi, L., Debitus, U., D'Auria, M. V., (2002). Isolation of callipeltins A-C and of two new open-chain derivatives of callipeltin A from the marine sponge *Latrunculia* sp. A revision of the stereostructure of callipeltins. *Tetrahedron Lett.* 43, 6163-6166.

[175] Bassarello, C., Zampella, A., Monti, M. C., Gomez-Paloma, L., D'Auria, M. V., Riccio, R., Bifulco, G., (2006). Quantum mechanical calculation of coupling constants in the configurational analysis of flexible systems: determination of the configuration of callipeltin A. *Eur. J. Org. Chem.* 604-609.

[176] Trevisi, L., Bova, S., Cargnelli, G., Danieli-Betto, D., Floreani, M., Germinario, E., D'Auria, M. V., Luciani, S., (2000). Callipeltin A, a Cyclic Depsipeptide Inhibitor of the Cardiac Sodium-Calcium Exchanger and Positive Inotropic Agent. *Biochem. Bioph. Res. Co.* 279, 219-222.

[177] Gulavita, N. K., Gunasekera, S. P., Pomponi, S. A., Robinson, E. V., (1992). Polydiscamide A: a new bioactive depsipeptide from the marine sponge *Discodermia* sp. *J. Org. Chem.* 57, 1767-1772.

[178] Matsunaga, S., Fusetani, N., Konosu, S., (1985). Bioactive marine metabolites, IV. Isolation and the amino acid composition of discodermin A, an antimicrobial peptide, from the marine sponge *Discodermia kiiensis*. *J. Nat. Prod.* 48, 236-241.

[179] Matsunaga, S., Fusetani, N., Konosu, S., (1984). Structure elucidation of discodermin A, an antimicrobial peptide from the marine sponge *Discodermia kiiensis*. *Tetrahedron Lett.* 25, 5165-5168.

[180] Matsunaga, S., Fusetani, N., Konosu, S., (1985). Structures of discodermins B, C, and D, antimicrobial peptides from the marine sponge *Discodermia kiiensis*. *Tetrahedron Lett.* 26, 855-856.

[181] Ryu, G., Matsunaga, S., Fusetani, N., (1994). Discodermin E, a cytotoxic and antimicrobial tetradecapeptide, from the marine sponge *Discodermia kiiensis*. *Tetrahedron Lett.* 35, 8251-8254.

[182] Ryu, G., Matsunaga, S., Fusetani, N., (1994). Discodermins F-H, cytotoxic and antimicrobial tetradecapeptides from the marine sponge *Discodermia kiiensis*: Structure revision of discodermins A-D. *Tetrahedron* 50, 13409-13416.

[183] Sato, K., Horibe, K., Amano, K., Mitusi-Saito, M., Hori, M., Matsunaga, S., Fusetani, N., Ozaki, H., Karaki, H., (2001). Membrane permeabilization induced by discodermin A, a novel marine bioactive peptide. *Toxicon* 39, 259-264.

[184] Williams, D. E., Austin, P., Diaz-Marrero, A. R., Van Soest, R., Matainaho, T., Roskelley, C. D., Roberge, M., Andersen, R. J., (2005). Neopetrosiamides, Peptides from the Marine Sponge *Neopetrosia* sp. That Inhibit Amoeboid Invasion by Human Tumor Cells. *Org. Lett.* 7, 4173-4176.

[185] Carter, G. T., Rinehart, K. L., (1978). Acarnidines, novel antiviral and antimicrobial compounds from the sponge *Acarnus erithacus* (de Laubenfels). *J. Am. Chem. Soc.* 100, 4302-4304.

[186] Kourany-Lefoll, E., Laprevote, O., Sevenet, T., Montagnac, A., Pais, M., Debitus, C., (1994). Phloeodictines A1-A7 and C1-C2, antibiotic and cytotoxic guanidine alkaloids from the New Caledonian sponge, *Phloeodictyon* sp. *Tetrahedron* 50, 3415-3426.

[187] Sakemi, S., Ichiba, T., Kohmoto, S., Saucy, G., Higa, T., (1988). Isolation and structure elucidation of onnamide A, a new bioactive metabolite of a marine sponge, *Theonella* sp. *J. Am. Chem. Soc.*110, 4851-4853.

[188] Matsunaga, S., Fusetani, N., Nakao, Y., (1992). Eight new cytotoxic metabolites closely related to onnamide A from two marine sponges of the genus Theonella. *Tetrahedron* 48, 8369-8376.

[189] Kobayashi, J., Itagaki, F., Shigemori, H., Sasaki, T., (1993). Three new onnamide congeners from the Okinawan marine sponge *Theonella* sp. *J. Nat. Prod.* 56, 976-981.

[190] Lee, K. H., Nishimura, S., Matsunaga, S., Fusetani, N., Horinouchi, S., Yoshida, M., (2005). Inhibition of protein synthesis and activation of stress -activated protein kinases by onnamide A and theopederin B, antitumor marine natural products. *Cancer Sci.* 96, 357-364.

[191] Harbour, Gary C., Tymiak, Adrienne A., Rinehart, Kenneth L., Shaw, Paul D., Robert, Hughes Mizsak Stephen A., Coats, John H., Zurenko, Gary E., Li, Li H., Kuentzel, Sandra L., (1981). Ptilocaulin and isoptilocaulin, antimicrobial and cytotoxic cyclic guanidines from the Caribbean sponge *Ptilocaulis* aff. *P. spiculifer* (Lamarck, 1814). *J. Am. Chem. Soc.* 103, 5604-5606.

[192] Kashman, Y., Hirsh, S., McConnell, O. J., Ohtani, I., Kusumi, T., Kakisawa, H., (1989). Ptilomycalin A: a novel polycyclic guanidine alkaloid of marine origin. *J. Am. Chem. Soc.* 111, 8925-8926.

[193] Ohtani, I., Kusumi, T., Kakisawa, H., (1992). An insight into the conformation of ptilomycalin A. The NMR properties of trifluoroacetylated spermidine analogs. *Tetrahedron Lett.* 33, 2525-2528.

[194] Ohtani, I., Kusumi, T., Kakisawa, H., Kashman, Y., Hirsh, S., (1992). Structure and chemical properties of ptilomycalin A. *J. Am. Chem. Soc.* 114, 8472-8479.

[195] Black, G. P., Coles, S. J., Hizi, A., Howard-Jones, A. G., Hursthouse, M. B., McGown, A. T., Loya, S., Moore, C. G., Murphy, P. J., Smith, N. K., Walshe, N. D. A., (2001). Synthesis and biological activity of analogues of ptilomycalin A. *Tetrahedron Lett.* 42, 3377-3381.

[196] Ohizumi, Y., Sasaki, S., Kusumi, T., Ohtani, I., (1996). Ptilomycalin A, a novel Na+, K+ - or Ca2+ -ATPase inhibitor, competitively interacts with ATP at its binding site. *Eur. J. Pharmacol.* 310, 95-98.

[197] Jares-Erijman, E. A., Sakai, R., Rinehart, K. L., (1991). Crambescidins: new antiviral and cytotoxic compounds from the sponge *Crambe crambe*. *J. Org. Chem.* 56, 5712-5715.

[198] Berlinck, R. G. S., Braekman, J. C., Daloze, D., Bruno, I., Riccio, R., Ferri, S., Spampinato, S., Speroni, E., (1993). Polycyclic guanidine alkaloids from the marine sponge *Crambe crambe* and calcium channel blocker activity of crambescidin 816. *J. Nat. Prod.* 56, 1007-1015.

[199] Jares-Erijman, E. A., Ingrum, A. L., Carney, J. R., Rinehart, K. L., Sakai, R., (1993). Polycyclic guanidine-containing compounds from the Mediterranean sponge *Crambe crambe*: the structure of 13,14,15-isocrambescidin 800 and the absolute stereochemistry of the pentacyclic guanidine moieties of the crambescidins. *J. Org. Chem.*58, 4805-4808.

[200] Patil, A. D., Kumar, N. V., Kokke, W. C., Bean, M. F., Freyer, A. J., Debrosse, C., Mai, S., Truneh, A., Faulkner, D. J., Carte´, B., Breen, A. L., Hertzberg, R. P., Johnson, R. K., Westley, J. W., Potts, B. C. M., (1995). Novel Alkaloids from the Sponge *Batzella* sp.: Inhibitors of HIV gp120-Human CD4 Binding. *J. Org. Chem.* 60, 1182-1188.

[201] Patil, A. D., Freyer, A. J., Taylor, P. B., Carte, B., Zuber, G., Johnson, R. K., Faulkner, D. J., (1997). Batzelladines F-I, novel alkaloids from the sponge *Batzella* sp.: inducers of p56lck-CD4 dissociation. *J. Org. Chem.* 62, 1814-1819.

[202] Kobayashi, J., Murayama, T., Ishibashi, M., Kosuge, S., Takamatsu, M., Ohizumi, Y., Kobayashi, H., Ohta, T., Nozoe, S., Sasaki, T., (1990). Hyrtiosins A and B, new indole alkaloids from the Okinawan marine sponge *Hyrtios erecta*. *Tetrahedron* 46, 7699-7702.

[203] Fahy, E., Potts, B. C. M., Faulkner, D. J., Smith, K., (1991). 6-Bromotryptamine derivatives from the Gulf of California tunicate *Didemnum candidum*. *J. Nat. Prod.* 54, 564-569.

[204] Aiello, A., Borrelli, F., Capasso, R., Fattorusso, E., Luciano, P., Menna, M., (2003). Conicamin, a novel histamine antagonist from the Mediterranean tunicate *Aplidium conicum*. *Bioorg. Med. Chem. Lett.* 13, 4481-4483.

[205] Roll, D. M., Ireland, C. M., (1985). Citorellamine, a new bromoindole derivative from *Polycitorella mariae*. *Tetrahedron Lett.* 26, 4303-4306.

[206] Moriarty, R. M., Roll, D. M., Ku, Y. Y., Nelson, C., Ireland, C. M., (1987). A revised structure for the marine bromoindole derivative citorellamine. *Tetrahedron Lett.* 28, 749-752.

[207] Lindquist, N., Fenical, W., (1990). Polyandrocarpamides A-D, novel metabolites from the marine ascidian *Polyandrocarpa* sp. *Tetrahedron Lett.* 31, 2521-2524.

[208] Heitz, S., Durgeat, M., Guyot, M., Brassy, C., Bachet, B., (1980). New indolic derivative of 1,2,4-thiadiazole, isolated from a tunicate (*Dendrodoa grossularia*). *Tetrahedron Lett.* 21, 1457-1458.

[209] Guyot, M., Meyer, M., (1986). An 3-indolyl-4H-imidazol-4-one from the tunicate *Dendrodoa grossularia*. *Tetrahedron Lett.* 27, 2621-2622.

[210] Bergmann, T., Schories, D., Steffan, B., (1997). Alboinon, an oxadiazinone alkaloid from the ascidian *Dendrodoa grossularia*. *Tetrahedron* 53, 2055-2060.

[211] Franco, L. H., Joffe, E. B., Puricelli, L., Tatian, M., Seldes, A. M., Palermo, J. A., (1998). Indole alkaloids from the tunicate *Aplidium meridianum*. *J. Nat. Prod.* 61, 1130-1132.

[212] Seldes, A. M., Brasco, M. F. R., Franco, L. H., Palermo, J. A., (2007). Identification of two meridianins from the crude extract of the tunicate *Aplidium meridianum* by tandem mass spectrometry. *Nat. Prod. Res.* 2, 555-563.

[213] Gompel, M., Leost, M., De Kier, J. E. B., Puricelli, L., Franco, L. H., Palermo, J., Meijer, L., (2004) Meridianins, a new family of protein kinase inhibitors isolated from the ascidian *Aplidium meridianum*. *Bioorg. Med. Chem. Lett.* 14, 1703-1707.

[214] Reyes, F., Fernandez, R., Rodriguez, A., Francesch, A., Taboada, S., Avila, C., Cuevas, C., (2008). Aplicyanins A-F, new cytotoxic bromoindole derivatives from the marine tunicate *Aplidium cyaneum*. *Tetrahedron* 64, 5119-5123.

[215] Appleton, D. R., Page, M. J., Lambert, G., Berridge, M. V., Copp, B. R., (2002). Kottamides A-D: novel bioactive imidazolone-containing alkaloids from the New Zealand ascidian *Pycnoclavella kottae*. *J. Org. Chem.* 67, 5402-5404.

[216] Copp, B. R., Ireland, C. M., Barrows, L. R., (1991). Wakayin: a novel cytotoxic pyrroloiminoquinone alkaloid from the ascidian *Clavelina* species. *J. Org. Chem.* 56, 4596-4597.

[217] Appleton, D. R., Copp, B. R., (2003). Kottamide E, the first example of a natural product bearing the amino acid 4-amino-1,2-dithiolane-4-carboxylic acid (Adt). *Tetrahedron Lett.* 44, 8963-8965.

[218] Vervoort, H. C., Richards-Gross, S. E., Fenical, W., Lee, A. Y., Clardy, J., (1997). Didemnimides A-D: novel predator-deterrent alkaloids from the Caribbean mangrove ascidian *Didemnum conchyliatum*. *J. Org. Chem.* 62, 1486-1490.
[219] Berlinck, R. G. S., Britton, R., Piers, E., Lim, L., Roberge, M., Moreira da Rocha, R., Andersen, R. J., (1998). Granulatimide and Isogranulatimide, Aromatic Alkaloids with G2 Checkpoint Inhibition Activity Isolated from the Brazilian Ascidian *Didemnum granulatum*: Structure Elucidation and Synthesis. *J. Org. Chem.* 63, 9850-9856.
[220] Vervoort, H. C., Fenical, W., Keifer, P. A., (1999). A Cyclized Didemnimide Alkaloid from the Caribbean Ascidian *Didemnum conchyliatum*. *J. Nat. Prod.* 62, 389-391.
[221] Britton, R., de Oliveira, J. H. H. L., Andersen, R. J., Berlinck, R. G. S., (2001). Granulatimide and 6-bromogranulatimide, minor alkaloids of the Brazilian ascidian *Didemnum granulatum*. *J. Nat. Prod.* 64, 254-255.
[222] Henon, H., Messaoudi, S., Anizon, F., Aboab, B., Kucharczyk, N., Leonce, S., Golsteyn, R. M., Pfeiffer, B., Prudhomme, M., (2007). Bis-imide granulatimide analogues as potent checkpoint 1 kinase inhibitors. *Eur. J. Pharmacol.* 554, 106-112.
[223] Kirkup, M. P., Moore, R. E., (1983). Indole alkaloids from the marine red alga *Martensia fragilis*. *Tetrahedron Lett.* 24, 2087-2090.
[224] Takamatsu, S., Hodges, T. W., Rajbhandari, I., Gerwick, W. H., Hamann, M. T., Nagle, D.G., (2003). Marine natural products as novel antioxidant prototypes. *J. Nat. Prod.* 66, 605-608.
[225] Takahashi, S., Matsunaga, T., Hasegawa, C., Saito, H., Fujita, D., Kiuchi, F., Tsuda, Y., (1997). Studies on bioactive substances in natural resources. VI. Isolation and structure elucidation of martefragin A. *Toyama-ken Yakuji Kenkyusho Nenpo* 24, 53-57.
[226] Nishida, A., Fuwa, M., Fujikaw, Y., (1998). First total synthesis of Martefragin A, a potent inhibitor of lipid peroxidation. *Tetrahedron Lett.* 39, 5983-5986.
[227] Murakami, H., Kato, T., Mimura, A., Takahara, Y., (1994). New indole derivatives from *Martensia denticulata* seaweed. *Biosci. Biotechnol. Biochem.* 58, 535-538.
[228] Carter, G. T., Rinehart, K. L., Jr., Li, L. H., Kuentzel, S. L., Connor, J. L., (1978). Brominated indoles from *Laurencia brongniartii*. *Tetrahedron Lett.* 46, 4479-4482.
[229] Tanaka, J., Higa, T., Bernardinelli, G., Jefford, C. W., (1989). Sulfur-containing polybromoindoles from the red alga *Laurencia brongniartii*. *Tetrahedron* 45, 7301-7310.
[230] Masuda, M., Kawaguchi, S., Takahashi, Y., Okamoto, K., Suzuki, M., (1999). Halogenated secondary metabolites of *Laurencia similis* (*Rhodomelaceae*, *Rhodophyta*). *Bot. Mar.* 42, 199-202.
[231] Ji, N.-Y., Li, X.-M., Cui, C.-M., Wang, B.-G., (2007). Terpenes and polybromoindoles from the marine red alga *Laurencia decumbens* (*Rhodomelacea*). *Helv. Chim. Acta* 90, 1731-1736.
[232] Sun, H. H., Sakemi, S., (1991). A brominated (aminoimidazolinyl)indole from the sponge *Discodermia polydiscus*. *J. Org. Chem.* 56, 4307-4308.
[233] Cohen, J., Paul, G. K., Gunasekera, S. P., Longley, R. E., Pomponi, S. A., (2004). 6-Hydroxydiscodermindole, a new discodermindole from the marine sponge *Discodermia polydiscus*. *Pharm. Biol.* 42, 59-61.
[234] Capon, R. J., Peng, C., Dooms, C., (2008). Trachycladindoles A-G: cytotoxic heterocycles from an Australian marine sponge, *Trachycladus laevispirulifer*. *Org. Biomol. Chem.* 6, 2765-2771.

[235] Blunt, J. W., Copp, B. R., Munro, M. H. G., Northcote, P. T., Prinsep, M. R., (2006). Marine natural products. *Nat. Prod. Rep.* 23, 26-78.

[236] Cragg, G. M., Grothaus, P. G., Newman, D. J., (2009). Impact of natural products on developing new anti-cancer agents. *Chem. Rev.* 109, 3012-3043.

[237] Omura, S., Sasaki, Y., Iwai, Y., Takeshima, H., (1995). Staurosporine, a potentially important gift from a microorganism. *J. Antibiot.* 48, 535-548.

[238] Schupp, P., Steube, K., Meyer, C., Proksch, P., (2001). Anti-proliferative effects of new staurosporine derivatives isolated from a marine ascidian and its predatory flatworm. *Cancer Lett.* 174, 165-172.

[239] Kinnel, R. B., Scheuer, P. J., (1992). 11-Hydroxystaurosporine: a highly cytotoxic, powerful protein kinase C inhibitor from a tunicate. *J. Org. Chem.* 57, 6327-6329.

[240] Reyes, F., Fernandez, R., Rodriguez, A., Bueno, S., de Eguilior C., Francesch, A., Cuevas, C., (2008). Cytotoxic staurosporines from the marine ascidian *Cystodytes solitus*. *J. Nat. Prod.* 71, 1046-1048.

[241] Sato, H., Tsuda, M., Watanabe, K., Kobayashi, J., (1998). Rhopaladins A-D, new indole alkaloids, from marine tunicate *Rhopalaea* sp. *Tetrahedron* 54, 8687-8690.

[242] Sasaki, T., Ohtani, I. I., Tanaka, J., Higa, T., (1999). Iheyamines, new cytotoxic bisindole pigments from a colonial ascidian, *Polycitorella* sp. *Tetrahedron Lett.* 40, 303-306.

[243] Segraves, N. L., Lopez, S., Johnson, T. A., Said, S. A., Fu, X., Schmitz, F. J., Pietraszkiewicz, H., Valeriote, F. A., Crews, P., (2003). Structures and cytotoxicities of fascaplysin and related alkaloids from two marine phyla-*Fascaplysinopsis* sponges and *Didemnum* tunicates. *Tetrahedron Lett.* 4, 3471-3475.

[244] Segraves, N. L., Robinson, S. J., Garcia, D., Said, S. A., Fu, X., Schmitz, F. J., Pietraszkiewicz, H., Valeriote, F. A., Crews, P., (2004). Comparison of fascaplysin and related alkaloids: A study of structures, cytotoxicities, and sources. *J. Nat. Prod.* 67, 783-792.

[245] Bartik, K., Braekman, J.-C., Daloze, D., Stoller, C., Huysecom, J., Vandevyver, G., Ottinger, R., (1987). Topsentins, new toxic bis-indole alkaloids from the marine sponge *Topsentia genitrix*. *Can. J. Chem.* 65, 2118-2121.

[246] Tsujii, S., Rinehart, K. L., Gunasekera, S. P., Kashman, Y., Cross, S. S., Lui, M. S., Pomponi, S. A., Diaz, M. C., (1988). Topsentin, bromotopsentin, and dihydrodeoxybromotopsentin: antiviral and antitumor bis(indolyl)imidazoles from Caribbean deep-sea sponges of the family *Halichondriidae*. Structural and synthetic studies. *J. Org. Chem.* 53, 5446-5453.

[247] Morris, S. A., Andersen, R. J., (1989). Nitrogenous metabolites from the deep water sponge *Hexadella* sp. *Can. J. Chem.* 67, 677-681.

[248] Braekman, J. C., Daloze, D., Moussiaux, B., Stoller, C., Deneubourg, F., (1989). Sponge secondary metabolites: new results. *Pure Appl. Chem.* 61, 509-512.

[249] Morris, S. A., Andersen, R. J., (1990). Brominated bis(indole) alkaloids from the marine sponge *Hexadella sp. Tetrahedron* 46, 715-720.

[250] Sakemi, S., Sun, H. H., (1991). Nortopsentins A, B, and C. Cytotoxic and antifungal imidazolediylbis[indoles] from the sponge *Spongosorites ruetzleri*. *J. Org. Chem.* 56, 4304-4307.

[251] Murray, L. M., Lim, T. K., Hooper, J. N. A., Capon, R. J., (1995). Isobromotopsentin: a new bis(indole) alkaloid from a deep-water marine sponge *Spongosorites* sp. *Aust. J. Chem.* 48, 2053-2058.

[252] Shin, J., Seo, Y., Cho, K. W., Rho, J.-R., Sim, C. J., (1999). New Bis(Indole) Alkaloids of the Topsentin class from the sponge *Spongosorites genitrix*. *J. Nat. Prod.* 62, 647-649.

[253] Casapullo, A., Bifulco, G., Bruno, I., Riccio, R., (2000). New bisindole alkaloids of the Topsentin and Hamacanthin classes from the Mediterranean marine sponge *Rhaphisia lacazei*. *J. Nat. Prod.* 63, 447-451.

[254] Bao, B., Sun, Q., Yao, X., Hong, J., Lee, C.-O., Sim, C. J., Im, K. S., Jung, J. H., (2005). Cytotoxic bisindole alkaloids from a marine sponge *Spongosorites* sp. *J. Nat. Prod.* 68, 711-715.

[255] Jung, J. H., Shinde, P. B., Hong, J., Liu, Y., Sim, C. J., (2006). Secondary metabolites from a marine sponge *Discodermia calyx*. *Biochem. Syst. Ecol.* 35, 48-51.

[256] Bao, B., Sun, Q., Yao, X., Hong, J., Lee, C.-O., Cho, H. Y., Jung, J. H., (2007). Bisindole alkaloids of the Topsentin and hamacanthin classes from a marine sponge *Spongosorites* sp. *J. Nat. Prod.* 70, 2-8.

[257] Wright, A. E., Pomponi, S. A., Cross, S. S., McCarthy, P., (1992). A new bis-(indole) alkaloid from a deep-water marine sponge of the genus *Spongosorites*. *J. Org. Chem.* 57, 4772-4775.

[258] Oh, K.-B., Mar, W., Kim, S., Kim, J.-Y., Oh, M.-N., Kim, J.-G., Shin, D., Sim, C. J., Shin, J., (2005). Bis(indole) alkaloids as sortase A inhibitors from the sponge *Spongosorites* sp. *Bioorg. Med. Chem. Lett.* 15, 4927-4931.

[259] Gunasekera, S. P., Cross, S. S., Kashman, Y., Lui, M. S., K. Rinehart, L., Tsujii, S., (1989). Topsentin compounds effective against viruses and certain tumors. *US Pat. Appl.* 4866084 A 19890912.

[260] Oh, K.-B., Mar, W., Kim, S., Kim, J.-Y., Lee, T.-H., Kim, J.-G., Shin, D., Sim, C. J., Shin, J., (2006). Antimicrobial activity and cytotoxicity of bis(indole) alkaloids from the sponge *Spongosorites* sp. *Biol. Pharm. Bull.* 29, 570-573.

[261] Zoraghi, R., Worrall, L., See, R. H., Strangman, W., Popplewell, W. L., Gong, H., Samaai, T., Swayze, R. D., Kaur, S., Vuckovic, M., Finlay, B. B., Brunham, R. C., McMaster, W. R., Davies-Coleman, M. T., Strynadka, N. C., Andersen, R. J., Reiner, N. E., (2011). Methicillin-resistant *Staphylococcus aureus* (MRSA) pyruvate kinase as a target for bis-indole alkaloids with antibacterial activities. *J. Biol. Chem.* 286, 44716-44725.

[262] Kohmoto, S., Kashman, Y., McConnell, O. J., Rinehart, K. L. Jr., Wright, A., Koehn, F., (1988). Dragmacidin, a new cytotoxic bis(indole) alkaloid from a deep water marine sponge, *Dragmacidon* sp. *J. Org. Chem.* 53, 3116-3118.

[263] Capon, R. J., Rooney, F., Murray, L. M., Collins, E., Sim, A. T. R., Rostas, J. A. P., Butler, M. S., Carroll, A. R., (1998). Dragmacidins: new protein phosphatase inhibitors from a Southern Australian deep-water marine sponge, *Spongosorites* sp. *J. Nat. Prod.* 61, 660-662.

[264] Cutignano, A., Bifulco, G., Bruno, I., Casapullo, A., Gomez-Paloma, L., Riccio, R., (2000). Dragmacidin F: a new antiviral bromoindole alkaloid from the Mediterranean sponge *Halicortex* sp. *Tetrahedron* 56, 3743-3748.

[265] Gunasekera, S. P., McCarthy, P. J., Kelly-Borges, M., (1994). Hamacanthins A and B, new antifungal bis indole alkaloids from the deep-water marine sponge, *Hamacantha* sp. *J. Nat. Prod.* 57, 1437-1441.

[266] Aguilar-Santos, G., Doty, M. S., (1968). Chemical studies on three species of the marine algal genus *Caulerpa*. In: *Drugs from the Sea*; Freudenthal, H. D., Ed.; Marine Technology Society: Washington, DC, US, 173-176.

[267] Aguilar-Santos, G. (1970). Caulerpin, a new red pigment from green alga e of the genus *Caulerpa*. *J. Chem. Soc. Perkin C: Org*. 6, 842-843.

[268] Vest, S. E., Dawes, C. J., Romeo, J. T., (1983). Distribution of caulerpin and caulerpicin in eight species of the green alga *Caulerpa* (*Caulerpales*). *Bot. Mar.* 26, 313-316.

[269] Capon, R. J., Ghisalberti, E. L., Jefferies, P. R., (1983). Metabolites of the green alga e, *Caulerpa* species. *Phytochemistry* 22, 1465-1467.

[270] Anjaneyulu, A. S. R., Prakash, C. V. S., Mallavadhani, U. V., (1991). Two caulerpin analogues and a sesquiterpene from *Caulerpa racemosa*. *Phytochemistry* 30, 3041-3042.

[271] Yan, S., Su, J., Wang, Y., Zeng, L. (1999). Studies on chemical constituents of *Halimeda incrassata*. *Redai Haiyang* 18, 91-94.

[272] Xu, X. H., Su, J. Y., Zeng, L. M., Wang, M. Y., (1998). Chemical constituents of the alga *Caloglossa leprieurii*. *Gaodeng Xuexiao Huaxue Xuebao* 19, 249-251.

[273] Govenkar, M. B., Wahidulla, S., (2000). Constituents of *Chondria armata*. *Phytochemistry* 54, 979-981.

[274] Su, J. Y., Zhu, Y., Zeng, L. M., Xu, X. H., (1997). A new bisindole from alga *Caulerpa serrulata*. *J. Nat. Prod.* 60, 1043-1044.

[275] Shi, D. Y., Han, L. J., Sun, J., Li, S., Wang, S. J., Yang, Y. C., Fan, X., Shi, J. G., (2005). A new halogenated biindole and a new apo-carotenone from green alga *Chaetomorpha basiretorsa* Setchell. *Chin. Chem. Lett.* 16, 777-780.

[276] Su, H., Yuan, Z. H., Li, J., Guo, S. J., Deng, L. P., Han, L. J., Zhu, X. B., Shi, D. Y., (2009). Two new bromoindoles from red alga *Laurencia similis*. *Chin. Chem. Lett.* 20, 456-458.

[277] Bifulco, G., Bruno, I., Minale, L., Riccio, R., Calignano, A., Debitus, C., (1994). (±)-Gelliusines A and B, two diastereomeric brominated tris-indole alkaloids from a deep water New Caledonian marine sponge (*Gellius* or *Orina* sp.). *J. Nat. Prod.* 57, 1294-1299.

[278] Kobayashi, J., Harbour, G. C., Gilmore, J., Rinehart, K. L. Jr., (1984). Eudistomins A, D, G, H, I, J, M, N, O, P, and Q, bromo, hydroxy, pyrrolyl and iminoazepino β - carbolines from the antiviral Caribbean tunicate *Eudistoma olivaceum*. *J. Am. Chem. Soc.* 106, 1526-1528.

[279] Kobayashi, J., Cheng, J. F., Oht, T., Nozoe, S., Ohizumi, Y., Sasaki, T., (1990). Eudistomidins B, C, and D: Novel antileukemic alkaloids from the Okinawan marine tunicate *Eudistoma glaucus*. *J. Org. Chem.* 55, 3666-3670.

[280] Shen, G. Q., Baker, B. J., (1994). Biosynthetic studies of the eudistomins in the tunicate *Eudistoma olivaceum*. *Tetrahedron Lett.* 35, 1141-1144.

[281] Rinehart, K. L., Kobayashi, J., Harbour, G. C., Hughes, R. G., Jr., Mizsak, S. A., Scahill, T. A., (1984). Eudistomins C, E, K, and L, potent antiviral compounds

containing a novel oxathiazepine ring from the Caribbean tunicate *Eudistoma olivaceum*. *J. Am. Chem. Soc.* 106, 1524-1526.
[282] Rinehart, K. L. Jr., Kobayashi, J., Harbour, G. C., Gilmore, J., Mascal, M., Holt, T. G., Shield, L. S., Lafargue, F., (1987). Eudistomins A-Q, β -carbolines from the antiviral Caribbean tunicate *Eudistoma olivaceum*. *J. Am. Chem. Soc.* 109, 3378-3387.
[283] Kinzer, K. F., Cardellina, J. H. II, (1987). Three new β -carbolines from the Bermudian tunicate *Eudistoma olivaceum*. *Tetrahedron Lett.* 28, 925-926.
[284] Schupp, P., Poehner, T., Edrada, R., Ebel, R., Berg, A., Wray, V., Proksch, P., (2003). Eudistomins W and X, two new β -carbolines from the Micronesian tunicate *Eudistoma* sp. *J. Nat. Prod.* 66, 272-275.
[285] Wang, W., Nam, S., Lee, B., Kang, H., (2008). β–Carboline alkaloids from a Korean tunicate *Eudistoma* sp. *J. Nat. Prod.* 71, 163-166.
[286] Kobayashi, J., Nakamura, H., Ohizumi, Y., Hirata, Y., (1986). Eudistomidin-A, a novel calmodulin antagonist from the Okinawan tunicate *Eudistoma glaucus*. *Tetrahedron Lett.* 27, 1191-1194.
[287] Adesanya, S. A., Chbani, M., Pais, M., Debitus, C., (1992). Brominated β–carbolines from the marine tunicate *Eudistoma album*. *J. Nat. Prod.* 55, 525-527.
[288] Suzuki, T., Kubota, T., Kobayashi, J., (2011). Eudistomidins H-K, new β–carboline alkaloids from the Okinawan marine tunicate *Eudistoma glaucus*. *Bioorg. Med. Chem. Lett.* 21, 4220-4223.
[289] Murata, O., Shigemori, H., Ishibashi, M., Sugama, K., Hayashi, K., Kobayashi, J., (1991). Eudistomidins E and F, new β–carboline alkaloids from the Okinawan marine tunicate *Eudistoma glaucus*. *Tetrahedron Lett.* 32, 3539-3542.
[290] Schumacher, R. W., Davidson, B. S., (1995). Didemnolines A-D, new N9-substituted β -carbolines from the marine ascidian *Didemnum* sp. *Tetrahedron* 51, 10125-10130.
[291] Ravinder, K., Reddy, A. V., Krishnaiah, P., Ramesh, P., Ramakrishna, S., Laatsch, H., Venkateswarlu, Y., (2005). Isolation and synthesis of a novel β-carboline guanidine derivative tiruchanduramine from the Indian ascidian *Synoicum macroglossum*. *Tetrahedron Lett.* 46, 5475-5478.
[292] Sakai, E., Kato, H., Rotinsulu, H., Losung, F., Mangindaan, R. E. P., de Voogd, N. J., Yokosawa, H., Tsukamoto, S., (2014). Variabines A and B: new β–carboline alkaloids from the marine sponge *Luffariella variabilis*. *J. Nat. Med.* 68, 215-219.
[293] Oku, N., Matsunaga, S., Fusetani, N., (2003). Shishijimicins A-C, novel enediyne antitumor antibiotics from the ascidian *Didemnum proliferum*. *J. Am. Chem. Soc.* 125, 2044-2045.
[294] Kearns, P. S., Coll, J. C., Rideout, J. A., (1995). A β–carboline dimer from an ascidian, *Didemnum* sp. *J. Nat. Prod.* 58, 1075-1076.
[295] Takahashi, Y., Ishiyama, H., Kubota, T., Kobayashi, J., (2010). Eudistomidin G, a new β–carboline alkaloid from the Okinawan marine tunicate *Eudistoma glaucus* and structure revision of eudistomidin B. *Bioorg. Med. Chem. Lett.* 20, 4100-4103.
[296] Debitus, C., Laurent, D., Pais, M., (1988). Alkaloids from an ascidian of New Caledonia, *Eudistoma fragum*. *J. Nat. Prod.* 51, 799-801.
[297] Searle, P. A., Molinski, T. F., (1994). Five new alkaloids from the tropical ascidian, *Lissoclinum* sp. lissoclinotoxin A is chiral. *J. Org. Chem.* 59, 6600-6605.

[298] Van Wagoner, R. M., Jompa, J., Tahir, A., Ireland, C. M., (1999). Trypargine alkaloids from a previously undescribed *Eudistoma* sp. ascidian. *J. Nat. Prod.* 62, 794-797.

[299] Davis, R. A., Duffy, S., Avery, V. M., Camp, D., Hooper, J. N. A., Quinn, R. J, (2010). (+)-7-Bromotrypargine: an antimalarial β–carboline from the Australian marine sponge *Ancorina* sp. *Tetrahedron Lett.* 51, 583-585.

[300] Chbani, M., Pais, M., Delauneux, J. M., Debitus, C., (1993). Brominated indole alkaloids from the marine tunicate *Pseudodistoma arborescens*. *J. Nat. Prod.* 56, 99-104.

[301] Rinehart, K. L., Kobayashi, J., Harbour, G. C., Hughes, R. G. Jr., Mizsak, S. A., Scahill, T. A., (1984). Eudistomins C, E, K, and L, potent antiviral compounds containing a novel oxathiazepine ring from the Caribbean tunicate *Eudistoma olivaceum*. *J. Am. Chem. Soc.* 106, 1524-1526.

[302] Lake, R. J., Brennan, M. M., Blunt, J. W., Munro, M. H. G., Pannell, L. K., (1988). Eudistomin K sulfoxide. An antiviral sulfoxide from the New Zealand ascidian *Ritterella sigillinoides*. *Tetrahedron Lett.* 29, 2255-2256.

[303] Lake, R. J., Blunt, J. W., Munro, M. H. G., (1989) Eudistomins from the New Zealand ascidian *Ritterella sigillinoides*. *Aust. J. Chem.* 42, 1201-1206.

[304] Foderaro, T. A., Barrows, L. R., Lassota, P., Ireland, C. M., (1997). Bengacarboline, a new β–carboline from a marine ascidian *Didemnum* sp. *J. Org. Chem.* 62, 6064-6065.

[305] Massiot, G., Nazabadioko, S., Bliard, C., (1995). Structural Revision of Isoeudistomin U by Total Synthesis. *J. Nat. Prod.* 58, 1636-1639.

[306] Lee, H.-S., Yoon, K.-M., Han, Y.-R., Lee, K. J., Chung, S.-C., Kim, T.-I., Lee, S.-H., Shin, J., Oh, K.-B., (2009). 5-Hydroxyindole-type alkaloids, as *Candida albicans* isocitrate lyase inhibitors, from the tropical sponge *Hyrtios* sp. *Bioorg. Med. Chem. Lett.* 19, 1051-1053.

[307] Baldwin, J. E., Whitehead, R. C., (1992). On the Biosynthesis of Manzamines. *Tetrahedron Lett.* 33, 2059-2062.

[308] Sakai, R., Higa, T., Jefford, C. W., Bernardinelli, G., (1986). Manzamine A, a novel antitumor alkaloid from a sponge. *J. Am. Chem. Soc.* 108, 6404-6405.

[309] Higa, T., Sakai, R., Kohmoto, S., Lui, M. S., (1988) European Patent Application. EP 272056 (Cl. C07D471/04), *US Appl.* 943, 609, 18 Dec. 1986; 14 pp.

[310] Ichiba, T., Sakai, R., Kohmoto, S., Saucy, G., Higa, T., (1988). New manzarnine alkaloids from a sponge of the genus *Xestospongia*. *Tetrahedron Lett.* 29, 3083-3086.

[311] Ichiba, T., Corgiat, J. M., Scheuer, P. J., Kelly-Borges, M., (1994). 8-Hydroxymanzamine A, a beta-carboline alkaloid from a sponge *Pachypellina* sp. *J. Nat. Prod.* 57, 168-170.

[312] Kobayashi, M., Chen, Y. J., Aoki, S., In, Y., Ishida, T., Kitagawa, I., (1995). Four new β-carboline alkaloids isolated from two Okinawan marine sponges of *Xestospongia* sp. and *Haliclona* sp. *Tetrahedron* 51, 3727-3736.

[313] Kobayashi, J., Tsuda, M., Kawasaki, N., Sasaki, T., Mikami, Y., (1994). 6-Hydroxymanzamine A and 3,4-Dihydromanzamine A, New Alkaloids from the Okinawan Marine Sponge *Amphimedon* sp. *J. Nat. Prod.* 57, 1737-1740.

[314] Tsuda, M., Watanabe, D., Kobayashi, J., (1999). A New Manzamine Congener from Marine Sponge *Amphimedon* sp. *Heterocycles* 50, 485-488.

[315] Tsuda, M., Watanabe, D., Kobayashi, J., (1998). Ma'eganedin A, a new manzamine alkaloid from *Amphimedon* sponge. *Tetrahedron Lett.* 39, 1207-1210.

[316] Sakai, R., Kohmoto, S., Higa, T., Jefford, C. W., Bernardinelli, G., (1987). Manzamine B and C, two novel alkaloids from the sponge *haliclona* sp. *Tetrahedron Lett.* 28, 5493-5496.

In: Alkaloids
Editor: Eduardo Sobarzo-Sánchez
ISBN: 978-1-63482-074-5

Chapter 8

TRABECTEDIN: A MARINE ALKALOID WITH ANTICANCER EFFECTS AGAINST OVARIAN CANCER

***Seyed Fazel Nabavi*[1], *Maria Daglia*[2], *Eduardo Sobarzo-Sánchez*[3] *and Seyed Mohammad Nabavi*[1]**
[1]Applied Biotechnology Research Center,
Baqiyatallah University of Medical Sciences, Tehran, Iran
[2]Department of Drug Sciences, Medicinal Chemistry and Pharmaceutical Technology Section, University of Pavia, Italy
[3]Department of Pharmaceutical Technology, Faculty of Pharmacy, University of Santiago de Compostela, Santiago de Compostela, Spain

ABSTRACT

Trabectedin is one of the most important marine alkaloids derived from *Ecteinascidia turbinate*. It is used as anticancer agent, especially for ovarian cancer treatment.

Nowadays, it is recommended as a combination therapy in patients with platinum - sensitive and/or platinum-resistant ovarian cancer. Some clinical studies showed that trabectedin has beneficial effect s against the progression of platinum-sensitive advanced or recurrent ovarian cancer in women.

In view of its manageable tolerability profile, it can be recommended as a standard anticancer drug for patients who suffer from platinum -sensitive and/or -resistant ovarian cancer. In this chapter we critically reviewed the anticancer effects of trabectedin against platinum-sensitive and/or -resistant ovarian cancer so far described in literature. Furthermore, we will provide information about chemistry, molecular mechanisms and the recommended doses of trabectedin required to achieve the anticancer effects.

Keywords: anticancer, chemotherapy, ovarian cancer, trabectedin

Introduction

The word cancer refers to a group of diseases caused by abnormal cell growth that can invade or spread to other body organs [1]. According to statistical reports, cancer affects 14.1 million people annually worldwide [2]. Among them, ovarian cancer is known as one of the leading causes of cancer-related deaths in women, causing 125.000 deaths every year [3]. Current therapeutic strategies against ovarian cancer include surgical procedures and platinum -based chemotherapy [4, 5]. Although the combination of surgery and chemotherapy reduces the disease progression and improves life quality and clinical outcomes, there is a high risk of resistance to such therapies and recurrence [6]. Therefore, new therapeutic strategies are needed to develop more improved treatments. For example, the addition or combination of cytotoxic agents to platinum-based chemotherapy increases its efficacy in the platinum-sensitive and/or -resistant ovarian cancer [7]. One of the most important and common cytotoxic agents largely used in treating platinum-sensitive and/or -resistant ovarian cancer is represented by the marine anticancer trabectedin [8].

Since ancient times, we are dependent on nature to continue our lives and prepare our basic requirements such as pharmaceutical compounds [9]. The use of natural compound s such as therapeutic strategies to fight against diseases is identified through folklore and is developed and approved by a plethora of recent studies. So, many natural products are currently in use [10]. During the last decades a revolution has occurred in the field of search for natural products with anticancer effect [11]. This phenomenon leads to an upsurge in finding of thousands of natural products with alkaloid, terpenoid, flavonoid, etc. skeletons which show anticancer properties in cellular and animal model system s and in humans [11].

In order to find effective natural products with anticancer activity, the national cancer institute of USA evaluated numerous plant extracts from thousands of plant species as well as other natural source [12]. These efforts led to find several anticancer drug s which are commercially available and approved by the FDA. According to statistics more than 60 % of anticancer drugs are of natural origin [13].

The oceans cover 71 % of the Earth's surface and the marine ecosystems represent 95 % of the biosphere that comprises gigantic biodiversity; with the exception of two phyla, all 28 major animal phyla are represented [14]. The tremendous marine flora may be useful in finding novel drugs with greater efficacy to fight against human diseases, especially certain types of cancers. First efforts to extract drugs from marine organisms started in the late 1960s [15]. Up to now, over 10000 new compounds from marine organisms have been discovered and some of them have been candidates to enter clinical trials [16]. Most of these organisms are invertebrates including nudibranchs, sponges, sea hares, tunicates, sea fans, bryazoans, and soft corals which produce the natural product as secondary metabolites and most these toxic compounds is produce as a form of defense system [17]. Many of these products have potent pharmacological activity [17]. In spite of the development of marine drugs is still in its infancy, this tremendous biodiversity has attracted the attention of many scientists to marine natural products all over the world and several governments design plans and funds to support researchers direct towards the finding new marine based drugs [18].

Trabectedin is one of the most important marine anticancer agents isolated from *Ecteinascidia turbinate,* a tunicate from Caribbean Sea that grows on mangrove roots [19]. There are several reports about its anticancer effects against different types of solid tumors

[20]. In this chapter we critically reviewed the literature on the anticancer effects of trabectedin against platinum -sensitive and/or -resistant ovarian cancer so far described in literature.

Chemistry of Trabectedin

Chemical structure of trabectedin (Figure 1) was reported for the first time in 1990 by Rhinehart. Its chemical structure contains three fused tetrahydroisoquinoline rings, two of which (A and B) are covalently bonded to the DNA [21]; C ring instead, interacts with some of the transcription factors. $C_{39}H_{43}N_3O_{11}S$ is trabectedin molecular formula, and its IUPAC name is (1'R,6R,6aR,7R,13S,14S,16R)-6',8,14-trihydroxy-7',9-dimethoxy-4,10,23-trimethyl-19-oxo-3',4',6,7,12,13,14,16-octahydrospiro[6,16-(epithiopropano-oxymethano)-7,13-imino-6aH-1,3-dioxolo[7,8]isoquino[3,2-b][3]benzazocine-20,1'(2'H)-isoquinolin]-5-yl acetate [22].

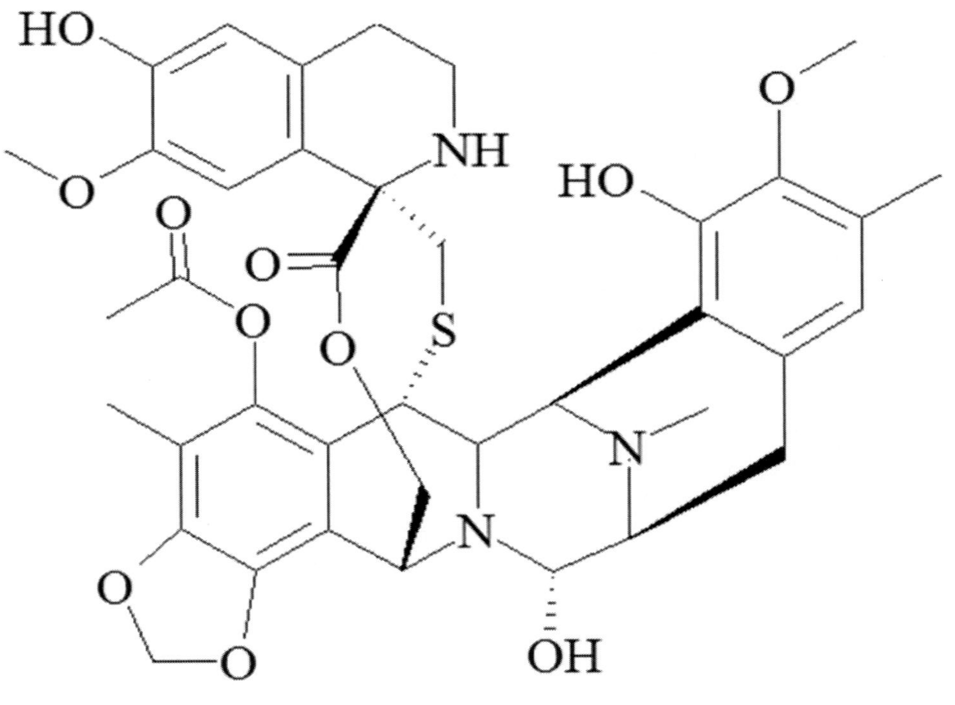

Figure 1. Chemical structure of trabectedin.

Toxicity of Trabectedin

Despite its powerful anticancer effects, trabectedin presents serious adverse event s [23]. In phase I and II clinical trials, the administration of trabectedin caused dose-dependent reversible myelosuppression [24]. Neutropenia is another dose-dependent side effect [25]. It has also been reported that trabectedin induces hepatotoxicity in patients by increasing serum levels of transaminases, bilirubin and alkaline phosphatase [26]; hepatotoxicity has been reported in different animal models [27, 28]. Furthermore, trabectedin major metabolite, *N*-demethyl-trabectedin, causes hepatotoxicity in different animal models as well as in phase I and II clinical trials [27]. Moreover, a close correlation between the levels of trabectedin and *N*-demethyl-trabectedin in the plasma, liver and bile with liver injuries in experimental animals has been reported [27].

Beneficial Effects of Trabectedin on Ovarian Cancer

Trabectedin enters phase I clinical trial (table 1) due to its efficacy and therapeutic outcome in preclinical studies against several types of cancers. According to the obtained results, hepatotoxicity (elevation of alkaline phosphatase in half of patient receiving trabectedin treatment) and non-cumulative hematological toxicity were the most commonly reported adverse effects [29]. However, vomiting, fatigue, asthenia, and reversible transaminitis were other dose-limiting toxicity of treatment in phase I (0.75 mg/m i.v. on day 1 every 3 weeks) [30, 31].

A phase II clinical study in 147 patients was carried out by Krasner et al. [32] to evaluate 3-h infusion of trabectedin weekly for 3 weeks and 1 week rest in recurrent ovarian cancer, previously treated with platinum -containing regimens. The results showed that a 3-week treatment with 0.58 mg/m^2 of trabectedin significantly mitigates recurrent ovarian cancer. The overall response rate and median progression-free survival were 29.0 % and 5.1 months for platinum sensitive group, respectively (62 patients). For platinum-resistant/refractory groups (79 patients) response rate and median progression-free survival were 6.3 % and 2.0 months, respectively. Krasner et al. concluded that trabectedin can be used as an effective treatment for platinum-resistant recurrent ovarian cancer [32]. Del Campo et al. [33] performed a randomized, open-label, phase II randomized study to determine the optimal dose of trabectedin administered every 3 weeks in platinum sensitive recurrent ovarian cancer. They found that 1.5 mg/m^2 (for 24 h) and 1.3 mg/m^2 (for 3 h) trabectedin administration regime were effective with objective response rate equal to 38.9 % and 35.8 %, respectively. Sessa et al. [34] performed a study to determine effectiveness and toxicity of trabectedin in 59 women suffered from ovarian cancer (resistant n=30 or sensitive n=29 to prior platinum and taxanes) after treatment with platinum and taxanes fails. They found that a 3-week treatment with 1.3 mg/m^2 trabectedin has beneficial effect s in patients with advanced ovarian cancer. Monk et al. [35] compared the anticancer effects of pegylated liposomal doxorubicin versus trabectedin plus pegylated liposomal doxorubicin in patients with platinum-resistant recurrent ovarian cancer. They found that the intravenous administration of pegylated liposomal doxorubicin (30 mg/m^2 plus 3-hour infusion of 1.1 mg/m^2 of trabectedin every 3 weeks) has more beneficial effects compared to the administration of 50 mg/m^2 of pegylated liposomal doxorubicin every 4 weeks in women who suffered from platinum-resistant recurrent ovarian cancer. Kaye et al. [36] showed that the combination of pegylated liposomal doxorubicin with trabectedin in patients with platinum-sensitive ovarian cancer significantly delays third-line chemotherapy in comparison with single-agent pegylated liposomal doxorubicin.

A phase III randomized trial by Poveda et al. [37] showed that trabectedin combined with pegylated liposomal doxorubicin reduces the progression of relapsed ovarian cancer and/or death (35 %) and significantly reduces the risk of death (41 %).

Table 1. Completed clinical trials on Trabectedin (http://clinicaltrial.gov/)

Identifier	Title
NCT00113607	An Efficacy and Safety Study for Yondelis (Trabectedin) in Patients With Advanced Relapsed Ovarian Cancer
NCT00786838	A Study to Assess the Potential Effects of a Single-Dose Administration of Trabectedin on the QT Intervals of the Electrocardiogram
NCT00147212	ET 743 (Yondelis) in Men With Advanced Prostate Cancer
NCT00072670	A Phase 2 Study of Trabectedin (Yondelis) in Adult Male Participants With Advanced Prostate Cancer
NCT00070109	Trabectedin in Treating Young Patients With Recurrent or Refractory Soft Tissue Sarcoma or Ewing's Family of Tumors
NCT00050427	A Study of ET-743 (Trabectedin) in Patients With Advanced Breast Cancer
NCT00102609	A Safety Study Utilizing Yondelis and Doxorubicin in Patients With a Type of Cancer Called Soft Tissue Sarcoma
NCT00580112	An Efficacy Study of Trabectedin in the Treatment of Participants With Specific Subtypes of Metastatic Breast Cancer
NCT01267084	A Study to Assess the Potential Effects of Ketoconazole on the Pharmacokinetics of Trabectedin in Patients With Advanced Malignancies
NCT00579501	Safety and Efficacy Study of Trabectedin for the Treatment of Localized Myxoid / Round Cell Liposarcoma
NCT00050440	Phase 2 Study of ET-743 (Trabectedin) in Patients With Persistent or Recurrent Endometrial Carcinoma
NCT00050414	A Study of Trabectedin in Patients With Advanced Ovarian Cancer
NCT00060944	A Study to Assess Treatment With 2 Different Dosing Schedules of Trabectidin Administered to Patients With Advanced Cancer
NCT01273480	A Study to Assess the Potential Effects of Rifampin on the Pharmacokinetics of Trabectedin in Patients With Advanced Malignancies
NCT02131480	Study of Doxorubicin and Trabectedin in First Line Treatment on Patients With Metastatic Leiomyosarcoma
NCT00379145	Trabectedin in Treating Patients With Advanced, Persistent, or Recurrent Leiomyosarcoma of the Uterus
NCT01453283	Tabectedin to Treat Children and Adolescents With Cancer
NCT00569673	Docetaxel, Trabectedin, and G-CSF or Pegfilgrastim in Treating Patients With Recurrent or Persistent Ovarian Epithelial Cancer, Primary Peritoneal Cavity Cancer, or Fallopian Tube Cancer
NCT00006463	Ecteinascidin 743 in Treating Children With Refractory Solid Tumors
NCT00002904	Ecteinascidin 743 in Treating Adults With Advanced Solid Tumors
NCT00005625	Ecteinascidin 743 in Treating Patients With Previously Treated Metastatic Osteosarcoma
NCT00027508	Ecteinascidin 743 in Treating Patients With Malignant Mesothelioma
NCT00003939	Ecteinascidin 743 in Treating Patients With Advanced Soft Tissue Sarcoma
NCT00017030	Ecteinascidin 743 in Treating Patients With Unresectable Advanced or Metastatic Soft Tissue Sarcoma
NCT00796120	A Study of the Safety and Effectiveness of Trabetedin Versus Doxorubicin-based Chemotherapy in Patients With Translocation-Related Sarcomas (TRS)

In phase II of treatment hepatic failure, renal failure, pancytopenia, and rhabdomyolysis were common adverse effects. Pharmacokinetic and pharmacodynamic studies show that biliary function can be served as reference parameter to indicate whether patient receive full dose of trabectedin or not. It is also reported that alkaline phosphatase and bilirubin can be served as a high risk indicator in full dose treatment. Phase II clinical trials show that trabectedin has greater efficacy in patient who suffered from breast and ovarian cancer and refractory soft tissue sarcoma. A search in Pubmed shows that there are several studies reporting the results of clinical trials (phase I, II and III) on the efficacy of trabectedin against ovarian cancer. Ferrandina et al. [38] performed retrospective, multicenter study to determine the potential efficacy of trabectedin as single agent in 98 patients (67 platinum sensitive, and 31 platinum refractory/resistant) who suffered from heavily relapsed ovarian cancer. They found that trabectedin treatment can prolong the survival of patients (18 month for treatment vs. 14 month for SD). They also found that hepatotoxicity and rare cardiotoxicity were the most important adverse effects of trabectedin treatment [38].

Monk et al. [39] evaluated the effects of trabectedin plus pegylated liposomal doxorubicin (PLD) versus PLD in survival of platinum sensitive patient ovarian cancer. They found that median overall survival for PLD + trabectedin group (22.2 months) was higher than PLD alone (18.9 months) [39]. Kaye et al. [36] also reported that PLD + trabectedin group delayed 2.5 months versus PLD alone and this enhanced benefit might be due to a prolongs platinum-free interval. Krasner et al. [40] also designed phase III clinical trial to evaluate outcome of PLD + trabectedin versus PLD in 672 patients who suffered from relapsed ovarian cancer. They found that PLD + trabectedin resulted in an improvement of progression-free survival and manageable toxicity in comparison with PLD alone group. Fisher and Gore [41] estimated cost -effectiveness of PLD + trabectedin versus PLD regime for the treatment of relapsed platinum-sensitive ovarian cancer patient in UK (October 2012). They found that trabectedin + PLD increased overall survival by 9.7 months compared with PLD alone. Trabectedin + PLD resulted in enhancing of incremental cost-effectiveness per quality-adjusted life-year.

It has been also reported that 24h infusion of trabectedin (1.5 mg/m^2) in advanced breast cancer patients pretreated with taxanes or anthracyclines leads to prolong survival and tumor control with manageable toxicity [42]. Monk et al. [43] designed phase II clinical trial to study the effectiveness of one hour intravenous administration of docetaxel (60mg/m^2) accompanied with three hours administration of trabectedin (1.1mg/m^2) with pegfilgrastim, filgrastim, or sargramostim every 3weeks in 71 patients who suffered from recurrent ovarian cancer. They found that the combination therapy is more active than single taxane therapy with overall survival equal to 16.9 months [43].

Mechanisms of Actions

There is growing evidence that trabectedin is a DNA minor groove binding agent thanks to a triplet rich in GC and alkylated guanine at N2 position resulting in the DNA bending towards the major groove [21]. Extensive reports show the important role of the transcription -coupled nucleotide excision repair responsible for trabectedin cytotoxicity [44]. Transcription-coupled nucleotide excision repair mechanism seems to detect trabectedin induced DNA binding and, then, to irreversibly induce single strand breaks [45]. It is known that deficiency in transcription-coupled nucleotide excision repair mechanism leads to trabectedin resistance. Herrero et al. [46] used fission yeast model *Schizosaccharomyces pombe* to examine the effect of nucleotide excision and the homologous recombination that occurs during DNA repair on trabectedin anticancer effect. They found that deficiency in Rad13 (a homolog of the human xeroderma pigmentosum complementation group G, nucleotide excision repair endonuclease) causes resistance to trabectedin. However, an inactive Rad51 and Rad54 (the proteins which play important roles in the homologous recombination that occurs during DNA and double strand break repair) enhanced trabectedin. They concluded that Rad13 protein binding to DNA formed more toxic complexes [46].

Soares et al. [47] found that human cell line s with homologous recombination repair protein deficiency were up to 23 times more sensitive to trabectedin. RNA polymerase II-mediated gene transcription is another target of trabectedin. This drug can selectively inhibit the activation of the drug resistance genes [48]. This drug also modulates tumor -induced chemokines and cytokines production and selective depletion of monocyte/macrophage leading to the inhibition of tumor angiogenesis [49]. It is believed that trabectedin causes selective cytotoxicity and activates caspase-8 dependent apoptosis in mononuclear phagocytes through decoying the TNF -related apoptosis-inducing ligand receptors [50].

Resistance to Trabectedin

There are few reports about cell line resistance to trabectedin [51, 52]. Colorectal carcinoma ER5 cell line is shown to be resistant to trabectedin. This is associated with loss of heterozygosity of chromosome 13q33 (which is known as location of xeroderma pigmentosum group G gene encoding) [21]. According to sequential analysis, xeroderma pigmentosum group G gene inserts adenine at codon 240 and leads to a stop codon 243 [21]. In addition, human surgically resected chondrosarcoma cells are another cell line resistant to trabectedin [53]. In these cells an abnormality in the cytoskeleton, which is correlated with modified types I and IV collagen expression, is the molecular mechanism of trabectedin resistance [54].

Furthermore, among ovarian cancer cells, the exposure to high doses of trabectedin causes high resistance, as well as P-glycoprotein expression [38].

Completed and Ongoing Clinical Trials

A research on http://clinicaltrial.gov/ writing “trabectedin” as keyword (June 20, 2014) showed that there are 46 clinical trials on this substance. Our search confirmed 10 recruited studies and 5 not yet recruited studies. Our research also confirmed the existence of 25 completed studies and 4 unknown ones. Furthermore, 1 study was terminated and 1 study suspended. Details of the studies completed are summarized in Table 1.

Recommended Doses

With respect to many completed clinical trials on the efficacy of trabectedin, a decision can be easily taken and conclusions drawn on the most effective and safe dose for the beneficial role of trabectedin against ovarian cancer.

According to previous studies, recommended dose for intravenous infusion is 1.5 mg/m^2 body surface over 24 hours with 3 week intervals in the cycle [54]. Central venous catheter is also recommended for the administration. Prior to trabectedin administration (30 min), dexamethasone should be given intravenously at 20 mg, both as anti-emetic and hepatoprotective agent [55]. Moreover, dexamethasone, given as pre-medication 24 hours before trabectedin treatment, has been used for decreasing myelosuppression as well as hepatotoxicity [47]. Although dexamethasone causes nausea and vomiting, some of the debilitating adverse effects such as cardiotoxicity and neurotoxicity associated with the administration of doxorubicin and ifosfamide were avoided [56, 57]. Its unopened package can be used 36 months after production. However, its chemical and physical characteristics are stable for only 30 hours at room temperature. For packaging of trabectedin, polyvinylchloride and polyethylene and titanium packages are recommended [24].

Conclusion

In the present chapter, we have explained the beneficial role of trabectedin treatment against ovarian cancer s and we have shown its manageable adverse effects. We have concluded that compared to the significant risks of recurrence and resistance of ovarian cancer to the available therapeutic strategies, trabectedin can be used as an alternative therapeutic method. Moreover, there are numerous clinical studies on the efficacy and side effects of trabectedin and, therefore, we can easily take a decision about its most effective doses needed for obtaining anticancer effects. To conclude, it seems that transcription -coupled nucleotide excision repair mechanism plays a pivotal role in the anticancer effects of trabectedin.

References

[1] Ruddon, R.W., (2007). *Cancer biology*. Oxford University Press.

[2] International Agency for Research on Cancer. [Feb 18, 2014]. http://globocan.iarc.fr/Default.aspx.

[3] Ferlay, J., Shin, H.R., Bray, F., Forman, D., Mathers, C., Parkin, D.M. (2010). Estimates of worldwide burden of cancer in 2008: GLOBOCAN 2008. *Int. J. cancer*, 127(12), 2893-2917.

[4] Bristow, R.E., Chi, D.S., (2006). Platinum -based neoadjuvant chemotherapy and interval surgical cytoreduction for advanced ovarian cancer: a meta-analysis. *Gynecol. oncol.*, 103(3), 1070-1076.

[5] Vergote, I., Tropé, C.G., Amant, F., Kristensen, G.B., Ehlen, T., Johnson, N., Verheijen, R.H., van der Burg, M.E., Lacave, A.J., Panici, P.B., Kenter, G.G., Casado,

A., Mendiola, C., Coens, C., Verleye, L., Stuart, G.C., Pecorelli, S., Reed, N.S. (2010). Neoadjuvant chemotherapy or primary surgery in stage IIIC or IV ovarian cancer. *New Engl. J. Med.*, 363(10), 943-953.

[6] Deraco, M., Rossi, C. R., Pennacchioli, E., Guadagni, S., Somers, D.C., Santoro, N., Raspagliesi, F., Kusamura, S., Vaglini, M., (2001). Cytoreductive surgery followed by intraperitoneal hyperthermic perfusion in the treatment of recurrent epithelial ovarian cancer: a phase II clinical study. *Tumori*, 87(3), 120-126.

[7] Covens, A., Blessing, J., Bender, D., Mannel, R., Morgan, M., (2006). A phase II evaluation of tirapazamine plus cisplatin in the treatment of recurrent platinum - sensitive ovarian or primary peritoneal cancer: a Gynecologic Oncology Group study. *Gynecol. oncol.*, 100(3), 586-590.

[8] Cvetkovic, R. S., Figgitt, D. P., Plosker, G. L., (2002). ET-743. *Drugs*, 62(8), 1185-1192.

[9] Schwartsmann, G., da Rocha, A. B., Berlinck, R. G., Jimeno, J., (2001). Marine organisms as a source of new anticancer agents. *Lancet Oncol.*, 2(4), 221-225.

[10] Qin, J. J., Nag, S., Voruganti, S., Wang, W., Zhang, R., (2012). Natural product MDM2 inhibitors: anticancer activity and mechanisms of action. *Curr Med. Chem.*, 19(33), 5705-5725.

[11] Newman, D. J., Cragg, G. M., (2007). Natural Products as Sources of New Drugs over the Last 25 Years⊥. *J. Nat. Prod.*, 70(3), 461-477.

[12] Cragg, G. M., Boyd, M., (1996). Drug discovery and development at the National Cancer Institute: the role of natural products of plant origin. *Medicinal Plant Resources of the Tropical Forest*, 101-136.

[13] Cragg, G. M., Newman, D. J., Snader, K. M., (1997). Natural products in drug discovery and development. *J. nat. prod.*, 60(1), 52-60.

[14] Jimeno, J., Faircloth, G., Sousa-Faro, J. M., Scheuer, P., Rinehart, K., (2004). New Marine Derived Anticancer Therapeutics— A Journey from the Sea to Clinical Trials. *Mar. Drugs*, 2(1), 14-29.

[15] Sithranga Boopathy, N., Kathiresan, K., (2011). Anticancer drugs from marine flora: an overview. *J. Oncol.*, 2010.

[16] Gerwick, W. H., Moore, B. S., (2012). Lessons from the past and charting the future of marine natural products drug discovery and chemical biology. *Chem. Biol.*, 19(1), 85-98.

[17] Blunt, J. W., Copp, B. R., Keyzers, R. A., Munro, M. H., Prinsep, M. R., (2012). Marine natural products. *Nat. Prod. Rep.*, 29(2), 144-222.

[18] Kingston, D. G., (2010). Modern natural products drug discovery and its relevance to biodiversity conservation. *J. Nat. Prod.*, 74(3), 496-511.

[19] Van Kesteren, C., De Vooght, M.M.M., Lopez-Lazaro, L., Mathôt, R.A.A., Schellens, J.H.M., Jimeno, J. M., Beijnen, J.H., (2003). Yondelis®(trabectedin, ET-743): the development of an anticancer agent of marine origin. *Anti-cancer drugs,* 14(7), 487-502.

[20] Cioffi, A., LeCesne, A., Blay, J. Y., Delaloge, S., Yovine, A., Maki, R., Misset, J.L., Frontelo, P., Nieto, A., Jiao, J.J., Demetri, G. D. (2009). Trabectedin phase II clinical trials: pooled analysis of safety in patients with solid tumors. *J. Clin. Oncol.*, 27, e13510.

[21] D'Incalci, M., Galmarini, C.M., (2010). A review of trabectedin (ET-743): a unique mechanism of action. *Mol. Cancer Therap.*, 9(8), 2157-2163.

[22] Gajdos, C., Elias, A. (2011). Trabectedin: safety and efficacy in the treatment of advanced sarcoma. *Clin. Med. Insights. Oncol.*, 5, 35-43.

[23] Carter, N. J., Keam, S. J. (2007). Trabectedin. *Drugs*, 67(15), 2257-2276.

[24] Thornton, K.A. (2009). Trabectedin: the evidence for its place in therapy in the treatment of soft tissue sarcoma. *Core evid.*, 4, 191-198.

[25] Hing, J., Perez-Ruixo, J.J., Stuyckens, K., Soto-Matos, A., Lopez-Lazaro, L., Zannikos, P., (2007). Mechanism-based pharmacokinetic/pharmacodynamic meta-analysis of trabectedin (ET-743, Yondelis) induced neutropenia. *Clin. Pharmacol. Therap.*, 83(1), 130-143.

[26] Lee, J.K., Leslie, E.M., Zamek-Gliszczynski, M.J., Brouwer, K.L., (2008). Modulation of trabectedin (ET-743) hepatobiliary disposition by multidrug resistance -associated proteins (Mrps) may prevent hepatotoxicity. *Toxicol. Aappl. Pharmacol.*, 228(1), 17-23.

[27] Beumer, J.H., Schellens, J.H.M., Beijnen, J.H., (2005). Hepatotoxicity and metabolism of trabectedin: a literature review. *Pharmacol. Res.*, 51(5), 391-398.

[28] Beumer, J.H., Franke, N. E., Tolboom, R., Buckle, T., Rosing, H., Lopez-Lazaro, L., Schellens, J.H., Beijnen, J.H., van Tellingen, O., (2010). Disposition and toxicity of trabectedin (ET-743) in wild-type and mdr1 gene (P-gp) knock-out mice. *Invest. New drugs,* 28(2), 145-155.

[29] Vidal, L., Magem, M., Barlow, C., Pardo, B., Florez, A., Montes, A., Garcia, M., Judson, I., Lebedinsky, C., Kaye, S. B., Salazar, R., (2012). Phase I clinical and pharmacokinetic study of trabectedin and carboplatin in patients with advanced solid tumors. *Invest. New Drugs*, 30(2), 616-628.

[30] Sessa, C., Del Conte, G., Christinat, A., Cresta, S., Perotti, A., Gallerani, E., Lardelli, P., Kahatt, C., Alfaro, V., Iglesias, J. L., Fernández-Teruel, C., Gianni, L. (2013). Phase I clinical and pharmacokinetic study of trabectedin and cisplatin given every three weeks in patients with advanced solid tumors. *Invest. New Drugs*, 31(5), 1236-1243

[31] Pardo, B., Salazar, R., Ciruelos, E., Cortés-Funes, H., García, M., Majem, M., Montes, A., Cuadra, C., Soto-Matos, A., Lebedinsky, C., Alfaro, V., Paz-Ares, L. (2012). Phase I and pharmacokinetic study of trabectedin 3-hour infusion every three weeks in patients with advanced cancer and alteration of hepatic function. *Med Oncol.*, 29(3), 2240-2250.

[32] Krasner, C.N., McMeekin, D.S., Chan, S., Braly, P.S., Renshaw, F.G., Kaye, S., Provencher, D.M., Campos, S., Gore, M.E., (2007). A Phase II study of trabectedin single agent in patients with recurrent ovarian cancer previously treated with platinum - based regimens. *Brit. J. Cancer*, 97(12), 1618-1624.

[33] Del Campo, J.M., Roszak, A., Bidzinski, M., Ciuleanu, T.E., Hogberg, T., Wojtukiewicz, M.Z., Poveda, A., Boman, K., Westermann, A.M., Lebedinsky, C., (2009). Phase II randomized study of trabectedin given as two different every 3 weeks dose schedules (1.5 mg/m2 24 h or 1.3 mg/m2 3 h) to patients with relapsed, platinum - sensitive, advanced ovarian cancer. *Ann. Oncol*, 20(11), 1794-1802.

[34] Sessa, C., De Braud, F., Perotti, A., Bauer, J., Curigliano, G., Noberasco, C., Zanaboni, F., Gianni, L., Marsoni, S., Jimeno, J., D'Incalci, M., Dall'ó, E., Colombo, N., (2005). Trabectedin for women with ovarian carcinoma after treatment with platinum and taxanes fails. *J. Clin. Oncol.*, 23(9), 1867-1874.

[35] Monk, B. J., Herzog, T. J., Kaye, S. B., Krasner, C. N., Vermorken, J. B., Muggia, F. M., Pujade-Lauraine, E., Lisyanskaya, A.S., Makhson, A.N., Rolski, J., Gorbounova, V.A., Ghatage, P., Bidzinski, M., Shen, K., Ngan, H.Y., Vergote, I.B., Nam, J.H., Park, Y.C., Lebedinsky, C.A., Poveda, A.M., (2010). Trabectedin plus pegylated liposomal doxorubicin in recurrent ovarian cancer. *J. Clin. Oncol.*, 28(19), 3107-3114.

[36] Kaye, S.B., Colombo, N., Monk, B.J., Tjulandin, S., Kong, B., Roy, M., Chan, S., Filipczyk-Cisarz, E., Hagberg, H., Vergote, I., Lebedinsky, C., Parekh, T., Santabárbara, P., Park, Y.C., Nieto, A., Poveda, A., (2011). Trabectedin plus pegylated liposomal doxorubicin in relapsed ovarian cancer delays third-line chemotherapy and prolongs the platinum -free interval. *Ann. Oncol.*, 22(1), 49-58.

[37] Poveda, A., Vergote, I., Tjulandin, S., Kong, B., Roy, M., Chan, S., Filipczyk-Cisarz, E., Hagberg, H., Kaye, S.B., Colombo, N., Lebedinsky, C., Parekh, T., Gómez, J., Park, Y.C., Alfaro, V., Monk, B.J., (2011). Trabectedin plus pegylated liposomal doxorubicin in relapsed ovarian cancer: outcomes in the partially platinum -sensitive (platinum-free interval 6–12 months) subpopulation of OVA-301 phase III randomized trial. *Ann Oncol.*, 22(1), 39-48.

[38] Ferrandina, G., Salutari, V., Vincenzi, B., Marinaccio, M., Naglieri, E., Loizzi, V., Carpano, S., Amadio, G., Tonini, G., Scambia, G., Lorusso, D., (2013). Trabectedin as single agent in the salvage treatment of heavily treated ovarian cancer patients: a retrospective, multicenter study. *Gynecol Oncol.*, 130(3), 505-510

[39] Monk, B. J., Herzog, T. J., Kaye, S. B., Krasner, C. N., Vermorken, J. B., Muggia, F. M., Pujade-Lauraine, E., Park, Y. C., Parekh, T. V., Poveda, A. M., (2012). Trabectedin plus pegylated liposomal doxorubicin (PLD) versus PLD in recurrent ovarian cancer: overall survival analysis. *Eur. J. Cancer.*, 48(15), 2361-2368.

[40] Krasner, C. N., Poveda, A., Herzog, T. J., Vermorken, J. B., Kaye, S. B., Nieto, A., Claret, P. L., Park, Y. C., Parekh, T., Monk, B. J., (2012). Patient-reported outcomes in relapsed ovarian cancer: results from a randomized Phase III study of trabectedin with pegylated liposomal doxorubicin (PLD) versus PLD alone.*Gynecol Oncol.*, 127(1), 161-167

[41] Fisher, M., Gore, M., (2013). Cost-effectiveness of trabectedin plus pegylated liposomal doxorubicin for the treatment of women with relapsed platinum -sensitive ovarian cancer in the UK: analysis based on the final survival data of the OVA-301 trial. *Value Health* 16(4), 507-516.

[42] Zelek, L., Yovine, A., Brain, E., Turpin, F., Taamma, A., Riofrio, M., Spielmann, M., Jimeno, J., Misset, J. L., (2006). A phase II study of Yondelis (trabectedin, ET-743) as a 24-h continuous intravenous infusion in pretreated advanced breast cancer. *Brit. J. Cancer*, 94(11), 1610-1614.

[43] Monk, B. J., Sill, M. W., Hanjani, P., Edwards, R., Rotmensch, J., De Geest, K., Bonebrake, A. J., Walker, J. L., (2011) Docetaxel plus trabectedin appears active in recurrent or persistent ovarian and primary peritoneal cancer after up to three prior regimens: a phase II study of the Gynecologic Oncology Group.*Gynecol Oncol.*, 120(3), 459-463

[44] Herrero, A.B., Martín-Castellanos, C., Marco, E., Gago, F., Moreno, S., (2006). Cross-talk between nucleotide excision and homologous recombination DNA repair pathways in the mechanism of action of antitumor trabectedin. *Cancer Res.*, 66(16), 8155-8162.

[45] Gullotta, F., De Marinis, E., Ascenzi, P., di Masi, A., (2010). Targeting the DNA double strand breaks repair for cancer therapy. *Curr. Med. Chem.*, 17(19), 2017-2048.

[46] Herrero, A. B., Martín-Castellanos, C., Marco, E., Gago, F., Moreno, S., (2006). Cross-talk between nucleotide excision and homologous recombination DNA repair pathways in the mechanism of action of antitumor trabectedin. *Cancer Res.*, 66(16), 8155-8162.

[47] Soares, D. G., Escargueil, A. E., Poindessous, V., Sarasin, A., de Gramont, A., Bonatto, D., Henriques, J.A., Larsen, A.K., (2007). Replication and homologous recombination repair regulate DNA double-strand break formation by the antitumor alkylator ecteinascidin 743. *Proc. Natl. Acad. Sci.*, 104(32), 13062-13067.

[48] Beumer, J. H., Buckle, T., Ouwehand, M., Franke, N.E., Lopez-Lazaro, L., Schellens, J.H., Beijnen, J.H., van Tellingen, O., (2007). Trabectedin (ET-743, Yondelis™) is a substrate for P-glycoprotein, but only high expression of P-glycoprotein confers the multidrug resistance phenotype. *Invest. New Drugs*, 25(1), 1-7.

[49] Germano, G., Frapolli, R., Belgiovine, C., Anselmo, A., Pesce, S., Liguori, M., Erba, E., Uboldi, S., Zucchetti, M., Pasqualini, F., Nebuloni, M., van Rooijen, N., Mortarini, R., Beltrame, L., Marchini, S., Fuso Nerini, I., Sanfilippo, R., Casali, P.G., Pilotti, S., Galmarini, C.M., Anichini, A., Mantovani, A., D'Incalci, M., Allavena, P., (2013). Role of macrophage targeting in the antitumor activity of trabectedin. *Cancer cell*, 23(2), 249-262.

[50] Harati, K., Chromik, A. M., Bulut, D., Goertz, O., Hahn, S., Hirsch, T., Klein-Hitpass, L., Lehnhardt, M., Uhl, W., Daigeler, A., (2012). TRAIL and taurolidine enhance the anticancer activity of doxorubicin, trabectedin and mafosfamide in HT1080 human fibrosarcoma cells. *Anticancer Res.*, 32(7), 2967-2984.

[51] Manara, M.C., Perdichizzi, S., Serra, M., Pierini, R., Benini, S., Hattinger, C.M., Astolfi, A., Bagnati, R., D'Incalci, M., Picci, P., Scotlandi, K., (2005). The molecular mechanisms responsible for resistance to ET-743 (Trabectidin; Yondelis) in the Ewing's sarcoma cell line, TC-71. *Int. J. Oncol.*, 27(6), 1605-1616.

[52] Duan, Z., Choy, E., Harmon, D., Yang, C., Ryu, K., Schwab, J., Mankin, H., Hornicek, F.J., (2009). ZNF93 Increases Resistance to ET-743 (Trabectedin; Yondelis®) and PM00104 (Zalypsis®) in Human Cancer Cell Lines. *PloS one*, 4(9), e6967.

[53] Marchini, S., Marrazzo, E., Bonomi, R., Chiorino, G., Zaffaroni, M., Weissbach, L., Hornicek, F.J., Broggini, M., Faircloth, G.T., D'Incalci, M., (2005). Molecular characterisation of two human cancer cell line s selected *in vitro* for their chemotherapeutic drug resistance to ET-743. *Eur. J. Cancer*, 41(2), 323-333.

[54] Beumer, J.H., Rademaker-Lakhai, J.M., Rosing, H., Lopez-Lazaro, L., Beijnen, J.H., Schellens, J.H.M., (2005). Trabectedin (YondelisTM, formerly ET-743), a mass balance study in patients with advanced cancer. *Invest. New Drugs*, 23(5), 429-436.

[55] Demetri, G. D., Chawla, S. P., von Mehren, M., Ritch, P., Baker, L.H., Blay, J.Y., Hande, K.R., Keohan, M.L., Samuels, B.L., Schuetze, S., Lebedinsky, C., Elsayed, Y.A., Izquierdo, M.A., Gómez, J., Park, Y.C., Le Cesne, A. (2009). Efficacy and safety of trabectedin in patients with advanced or metastatic liposarcoma or leiomyosarcoma after failure of prior anthracyclines and ifosfamide: results of a randomized phase II study of two different schedules. *J. Clin. Oncol.*, 27(25), 4188-4196.

[56] Paz-Ares, L., López-Pousa, A., Poveda, A., Balañá, C., Ciruelos, E., Bellmunt, J., del Muro, J.G., Provencio, M., Casado, A., Rivera-Herrero, F., Izquierdo, M.A., Nieto, A., Tanović, A., Cortes-Funes, H., Buesa, J.M., (2012). Trabectedin in pre-treated patients with advanced or metastatic soft tissue sarcoma: a phase II study evaluating co-treatment with dexamethasone. *Invest. New Drugs*, 30(2), 729-740.

[57] Wang, H., Li, M., Rinehart, J.J., Zhang, R. (2004). Dexamethasone as a chemoprotectant in cancer chemotherapy: hematoprotective effects and altered pharmacokinetics and tissue distribution of carboplatin and gemcitabine. *Cancer Chemother. Pharmacol.*, 53(6), 459-467.

In: Alkaloids
Editor: Eduardo Sobarzo-Sánchez
ISBN: 978-1-63482-074-5

Chapter 9

OXOISOAPORPHINES AS MEDICAL TOOL: IN TREATMENT OF ANTINEOPLASTIC DISEASES AND MENTAL DISORDERS

***Eduardo Sobarzo-Sánchez*[1,*]*, Seyed Mohammad Nabavi*[2] *and Eugenio Uriarte Villares*[1]**

[1]Laboratory of Pharmaceutical Chemistry-Department of Organic Chemistry, Faculty of Pharmacy, University of Santiago de Compostela, Spain
[2]Applied Biotechnology Research Center, Baqiyatallah University of Medical Sciences, Tehran, Iran

ABSTRACT

Oxoisoaporphine derivatives have been studied through the pharmacological activity of certain cancer cell line s and in the treatment either of mood behavior or motor symptoms. Our data and other references confirm that oxoisoaporphine and its several analogues can be used as drug delivery in the administration of controlled manner of this drug against depression and, maintaining their biological qualities, these analogues have an important role in the treatment of both Parkinson's and Alzheimer's disease. Also, the medical application of organ ometallic complexes by using oxoisoaporphine derivatives against several cancer cell lines, and the formation of nanocapsules with natural or synthetic polymers that include in the core these alkaloids for treatment of mental disorder s are shown, which to make necessary to highlight the importance of studying this small group of compounds as a novel therapeutic alternative. Some examples of antineoplastic activity and MAO-A / Cholinesterase inbihitor by oxoisoaporphines are highlighted with several examples and therapeutic applications.

Keywords: oxoisoaporphine, 7*H*-dibenzo[*de,h*]quinolin-7-one, antineoplastic, depression, Parkinson's disease, Alzheimer's disease

* Corresponding author: Eduardo Sobarzo-Sánchez, Laboratory of Pharmaceutical Chemistry-Department of Organic Chemistry, Faculty of Pharmacy, University of Santiago de Compostela. Spain. E-mail: e.sobarzo@usc.es; esobarzo@gmail.com.

INTRODUCTION

The nature has afforded a diversity of organic material that has been used for diverse ends, food, for the housing construction and clearly in the isolation of compounds of natural sources that, along our history, have been used as therapeutic tools for treatment of diverse ailments and diseases. In this context, a wide variety of compounds from bark, sheets, roots of trees or shrubs have improved our quality life though the traditional medicine. Among these compounds we have chosen perhaps the group that more studies and industrial applications have in our daily life. Thus, the alkaloids are the group of more abundant compounds with an extensive variety of chemical structure, simple and complex, existing in nature several heterocyclic systems, which include aporphines and their oxidized analogues, "**oxoaporphines**" (7*H*-dibenzo[*de,g*]quinolin-7-one). Nevertheless, a small group of isoquinoline alkaloids has been few investigated and whose main natural source to date from the creepers of *Menispermum dauricum* DC. (Menispermaceae) collected in Kyoto, Japan and later in China [1].

Thus, there were reported several reviews on the isolation of isoquinoline alkaloids and that, according to convincing NMR spectroscopic studies and synthesis, they possess a 7*H*-dibenzo[*de,h*]quinolin-7-one skeleton unprecedented in nature. Due to its structural resemblance with oxoaporphines, the above mentioned compounds were named "**oxoisoaporphines**". These novel nitrogen -based heterocycles that are present in the rhizome of these plants, extensively used in the chinese herbal medicine [2,3] are: Menisporphine (**1**) [4-8], Dauriporphine (**2**) [9-12], Bianfugecine (**3**) [11,13] Bianfugedine (**4**) [11], Dauriporphinoline (**5**) [14], 2,3-Dihydro-menisporphine (**6**) [11], and 6-O-Demethylmenisporphine (**7**) [7]. However, later a research reported the isolation of 2,3-dihydro-dauriporphine (**8**) and four oxoisoaporphines linked at C-6 to amino group (**9-12**), being these compounds called "**daurioxoisoaporphines**" [12,14], which lakshminina (**13**) is added from a new natural source, *Sciadotenia toxifera* [15], belonging also to the family of Menispermaceae (Figure 1).

In this sense, oxoisoaporphines, structurally similar to oxoaporphines, which naturally have been reported from vegetable sources, also exist in previous references in the production of dyes of diverse coloration depending on the extension of the carbon system, and that were called "**azabenzanthrone**" [16-20]. The above mentioned compounds were in several decades used in the manufacture of materials of painting and textiles without thinking that at the end of the XXth century, these compounds would be isolated from Asian natural sources. Nevertheless, some applications in the scope of the photochemistry and energy transfer were discovered recently for this type of heterocycle. Thus, some examples of the energy transfer by using oxoisoaporphines are derived to the formation of radical cations from either generated by a thermal reaction or generated radiolytically in argon -saturated, 1,2-dichloroethane, oxygen -saturated, acetone, and acetonitrile. The radical cations of oxoisoaporphines are insensitive to oxygen and very stable either when are formed via the electron -transfer reaction with TCNE or when generated in isolation in pulse radiolysis of Ar-saturated 1,2-dichloroethane [21].

Hence, these compounds are photoreduced efficiently in the presence of aliphatic tertiary amines through a stepwise mechanism of electron -proton-electron transfer. This photoreaction generates an AH^- anion hydrogenated, whose long-lived metastable photoproducts revert thermally to the initial oxoisoaporphines nearly quantitative. This characteristic allowed adding deuterium in specific positions of the oxoisoaporphine chemical structure. In the presence of D_2O, regioselective exchange of the aromatic protons at positions 4 and 6 takes place to an extent greater than 90% under very mild conditions. This facile isotopic exchange reaction might be advantageously used to introduce deuterium, and likely tritium at these positions of oxoisoaporphines (Figure 2) [22]. Therefore, the possibility of being able to locate these compounds marked like a probe for treatment of cancer and its application in photodinamic therapy is now possible.

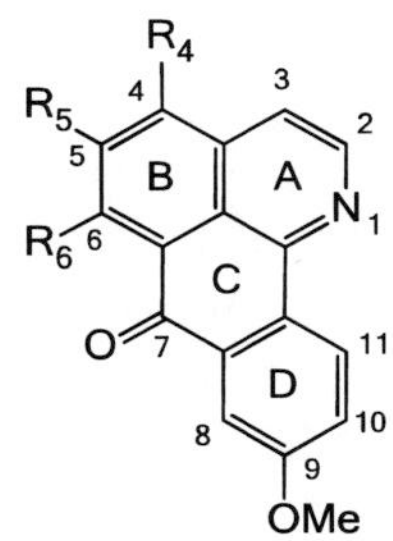

(1); $R_4 = H$, $R_5 = R_6 = OMe$
(2); $R_4 = R_5 = R_6 = OMe$
(3); $R_4 = H$, $R_5 = OMe$, $R_6 = H$
(4); $R_4 = H$, $R_5 = R_6 = -OCH_2O-$
(5); $R_4 = R_5 = OMe$, $R_6 = OH$
(6); $R_4 = H$, $R_5 = OMe$, $R_6 = OH$

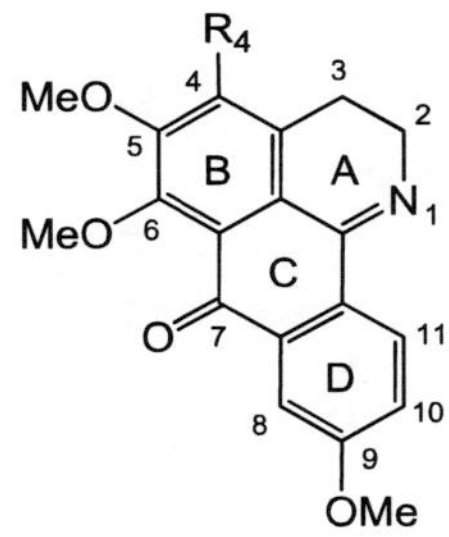

(7); $R_4 = H$
(8); $R_4 = OMe$

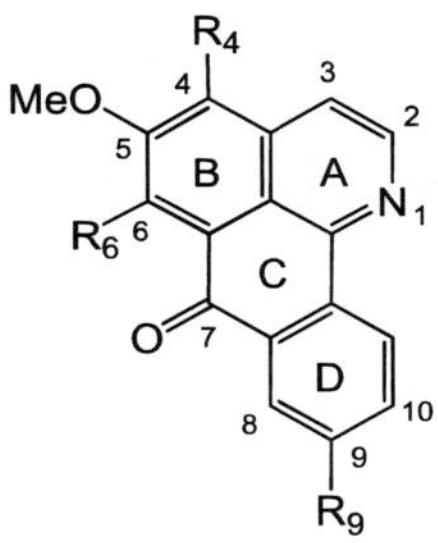

(9); $R_4 = OMe$, $R_6 = NHCH_2CH_2Ph(4\text{-}OH)$, $R_9 = OMe$
(10); $R_4 = OMe$, $R_6 = NH_2$, $R_9 = OMe$
(11); $R_4 = H$, $R_6 = NHCH_2CH_2Ph(4\text{-}OH)$, $R_9 = OMe$
(12); $R_4 = H$, $R_6 = NHMe$, $R_9 = OMe$
(13); $R_4 = H$, $R_6 = NH_2$, $R_9 = H$

Figure 1.

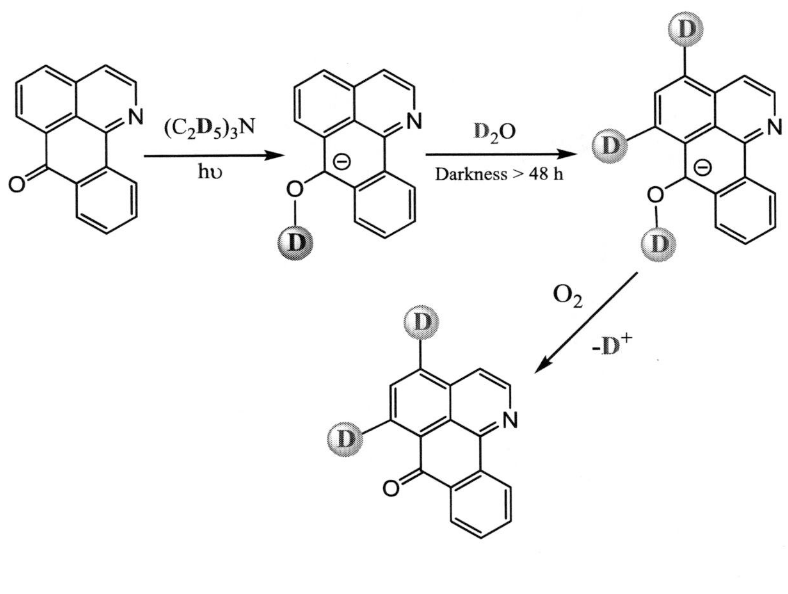

Figure 2.

Due to this fact, oxoisoaporphine derivatives present singlet oxygen quantum yields near to unity, and greater photostability than phenalenone as study pattern, these compounds may be considered as photosensitizers for singlet oxygen production and used as new universal pattern for medical applications related in diseases such as cancer or as antibacterial/antimycotic agents, for instance. In this sense, the study of oxoisoaporphines that have low cytotoxicity and high reactivity in antineoplastic chemotherapy can be interesting for this type of alkaloid and its obvious medical application (Figure 3) [23,24].

With these precedents, it is noteworthy to suppose the huge potentiality of these compounds that were isolated of natural sources, though in low concentrations, and that in the past represented a speedy route for the dyes formation in the textile and paintings industry. At present, the synthesis of diverse derivatives has led to the search of new alternatives of medical use, considering its low cytotoxicity. Among the therapeutic applications most extended given its wide field of research, they are the study of varied carcinogenic cellular lines and, among least investigated given its ignorance and possible medical properties, they are the treatment of behavioral diseases and neuro-degenerative disorders. Nevertheless, to find alternatives that diminish the side effects that some drugs produce as addictions or few interaction of the drug with the organism, for instance, at present this represents a high interest in the research of the medicinal chemistry.

Considering that oxoisoaporphine (7*H*-dibenzo[*de,h*]quinolin-7-one) have obtained the interest for its versatility in the synthesis of new derivatives in diverse medical action fields as its use in such parasitic diseases like Malaria [25] or Leishmania [26], the study of behavioral diseases as depression through compounds that inhibit the human monoamino oxidase A (hMAO-A) or to reverse the depressed state in reserpinized animal testing, has been the first impulse to compile references of this type of alkaloid in the therapeutic scope under the observation of the neoplastic and psychiatric diseases.

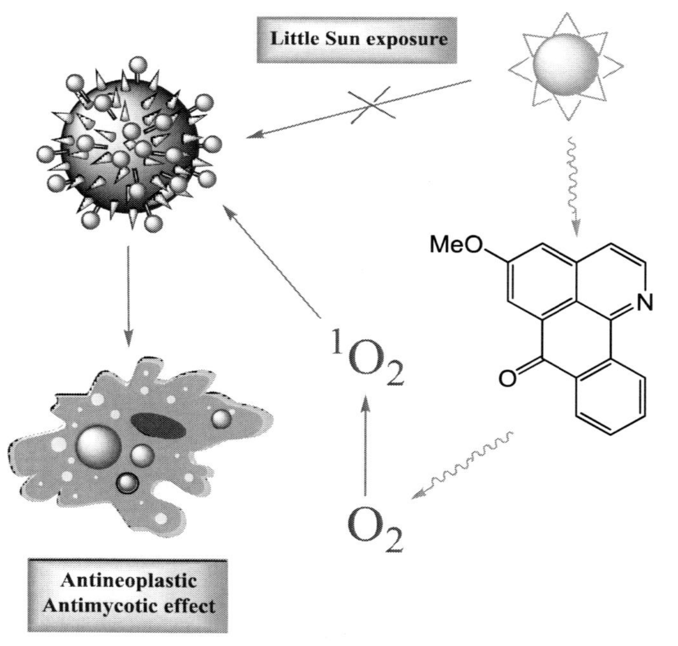

Figure 3.

A. Medical Application of Oxoisoaporphine Derivatives as Antineoplastic Agents

The first approach of the use of oxoisoaporphines in the field of antineoplastic drugs is some derivatives with chemical structure of anthraquinones that are well-known as antiproliferative agents, even in the chemotherapy. Thus, one of the most recent reports are some families of annulated anthraquinone analogs that have been examminated for antiproliferative activity. In this sense, a series of 1-azabenzanthrone derivatives, their 2,3-dihydro analogs, and congruently substituted 9,10-anthracenediones were tested against normal human fibroblasts and four human cancer cell line s. In this research, the most of the heterocyclic compounds proved to be weakly to moderately antiproliferative with IC_{50} values extending down to 0.86 μM, and exhibited up to 30-fold selectivity between cancer and normal cells. Both 1-azabenzanthrones and 1-aza-2,3-dihydrobenzanthrones were more potent than their anthraquinone counterparts, and almost without exception, 2,3-dihydro compounds were more potent than the fully aromatic 1-azabenzanthrones (Figure 4) [27].

In this sense, other authors propose a general hypothesis that any *in vitro* or *in vivo* biological activity can be explained only in terms of local reactivity indexes and tested for the inhibition cell growth by several 1-azabenzanthrone derivatives. In all of them, good structure -activity relationships have been obtained and some requirements for potent biological activity. This methodology to understand the whole process and, in the case of multi-step processes, is unable to relate the different reactivity indexes appearing in the SAR equations to any particular step [28].

As is known, many aromatic compounds can form organ ometallic complexes with some metals by interactions of hydrogen bonding either with carbonyl groups or heteroatoms. Hence, the reactivity *per se* of the compound can become in outstanding when the pharmacological properties are tested. Thus, a new 1-azabenzanthrone-platinum (II) complex, that is chloro(DMSO)(1-aza benzanthrone)platinum(II) is shown with an important biological activity. The complex is preparated by taking an equal amount of 1-azabenzanthrone and dichlorobis(DMSO)platinum(II), dissolving in a polar solvent, heating or refluxing for the reaction to obtain the desired product, specifically can be synthesized by solution method, and can also be synthesized by solvothermal method. The results from examining the inhibitory effect of the complex on the proliferation of HepG2, SK-OV-3, SK-OV-3/DDP, BEL-7,404, T-24, HL-7,702, and other human tumor cell line s show that the complex has significant anti-tumor activity *in vitro,* good potential medicinal value, and is expected for the preparation of a variety of anticancer drug s [29].

Other examples of this interesting characteristic can be in a series of new 6-hydroxyoxoisoaporphine rare earth complexes. The 6-hydroxyoxoisoaporphine rare earth complex has structural formula $Ln(HL)_2(NO_3)_3$, in which Ln is La, Ce, Pr, Nd, Sm, Eu, Gd, Tb or Dy. The complex uses 6-hydroxyoxoisoaporphine and rare earth nitrate of Ln(NO)•xHO as raw material, and synthesizes under heating or refluxing (Figure 5). Hence, 6-hydroxyoxoisoaporphine rare earth complexes was used in preparation of antitumor agent. The studies of inhibition activity of the complexes on SK-OV-3, SK-OV-3/DDP, T-24 and other human tumor cell line s proliferation were evaluated, wherein the complexes have more significant antitumor activity *in vitro,* especially 6-6-hydroxyoxoisoaporphine lanthanum complex that has very significant inhibition activity on cervical cancer cells SK-OV-3;

furthermore except 6-hydroxyoxoisoaporphine rare europium complex, other complexes have better inhibition effect on human ovarian cancer /cisplatin-resistant strains SK-OV-3/DDP [30]. Another example is the use of 6-aminooxoisoaporphine and the formation of 6-aminooxoisoaporphine platinum (II) complex. Through observing the inhibition effect of the complex on the proliferation of human tumor cell lines HepG2, BEL-7404 and NCI-H460, the complex shows significant antitumor activity *in vitro* against 3 kinds of tumor cell lines. These examples show an interesting tool the use of this alkaloid as metallic complex in the treatment of antineoplastic diseases [31].

Dauriporphines (**9-12**) are some oxoisoaporphine derivatives that have been tested against cancer cell line with moderate success [32]. Recently, some oxoisoaporphine derivatives with two methylene groups (n = 2), and piperidine or ethanolamine functional group in the side chain as Figure 6 have been synthesized and its DNA binding affinity and cytotoxicity tested. Thus, novel oxoisoaporphine alkaloid derivatives, 9-aminoalkanamido-1-azabenzanthrone (general formula Ar-NHCO$(CH_2)_n$$NR_2$, Ar = 1-azabenzanthrone, n = 1, 2 or 3 had significantly higher DNA binding affinity with calf thymus DNA, and higher potent cytotoxicity against different tumor cell lines. These compounds exhibited highest DNA binding affinity and cytotoxicity, which indicates us the pharmacological potentiality to add some substituted carbon chains with piperidine or ethanolamine groups in the interaction with certain cancerous strains [33].

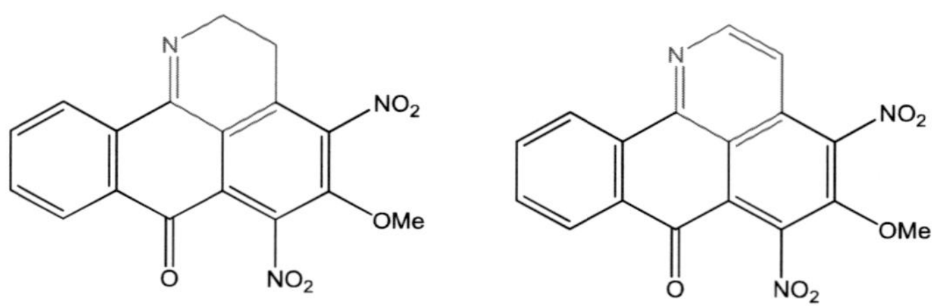

Human fibroblasts (MRC-5), IC_{50} (μM)
36.4

Human fibroblasts (MRC-5), IC_{50} (μM)
41.8

Figure 4.

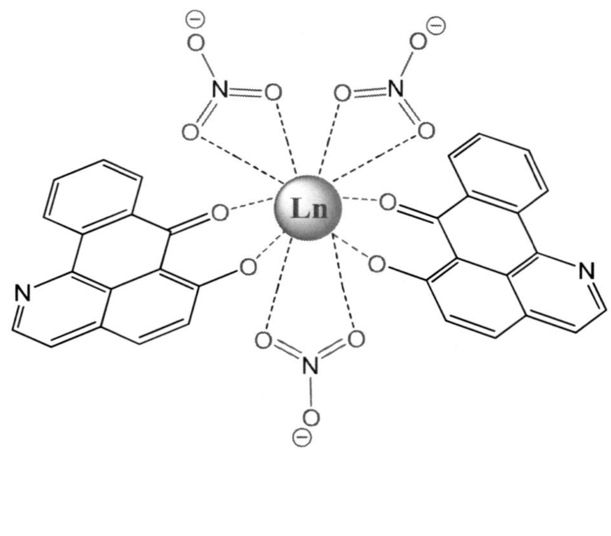

Figure 5.

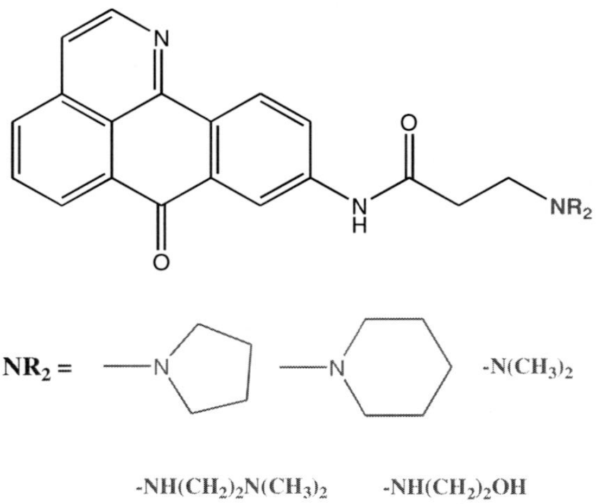

Figure 6.

Recently has been discovered that an oxoisoaporphine derivative with an amino group at the position 6 called **Lakshminine** (Figure 7) (6-amino-1-aza-5-methoxy-7*H*-dibenzo[*de,h*]quinolin-7-one) has presented an interesting antiproliferative activity. This rare natural product has now been synthesized in order to have sufficient amounts for biological testing, its 4-amino isomer, their 6- and 4-nitro precursors, the intermediate 5-methoxy-7*H*-dibenzo[*de,h*]quinolin-7-one, and the unsubstituted skeleton were tested against normal human fibroblasts and three human solid tumor cell line s. Only the 4-nitro precursor showed marginal antiproliferative activity [34].

In this sense, 5-Bromo-oxoisoaporphine (**A**) (Figure 8) [35] was synthesized *via* reduction of 4-bromophenylacetonitrile; the resulting 4-bromophenylethylamine underwent condensation with phthalic anhydride to give the corresponding phthalimide, which underwent Friedel-Crafts ring-closing reaction to give 2-(7-bromoisoquinolin-1-yl)benzoic acid, which underwent intramolecular cyclization to give the bromo derivative. Later **A** and a new derivative (**B**) were tested with inhibitory activity on proliferation of HepG2, MGC80-3, SK-OV-3 and various human tumor cell strains such as T-24, and results show that **B** has obviously anti-tumor activity *in vitro* with better potential medicinal value [36].

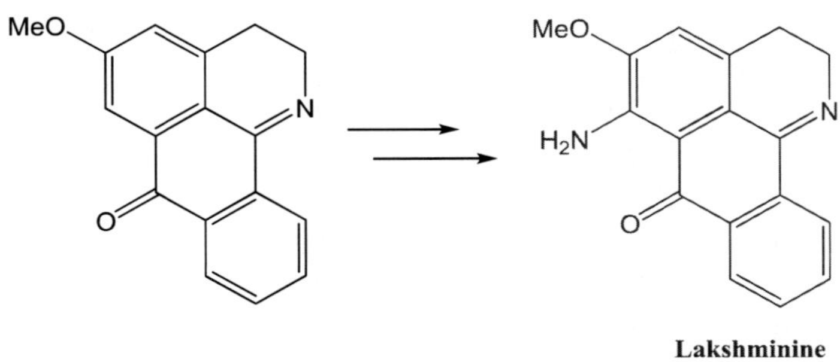

Lakshminine

Figure 7.

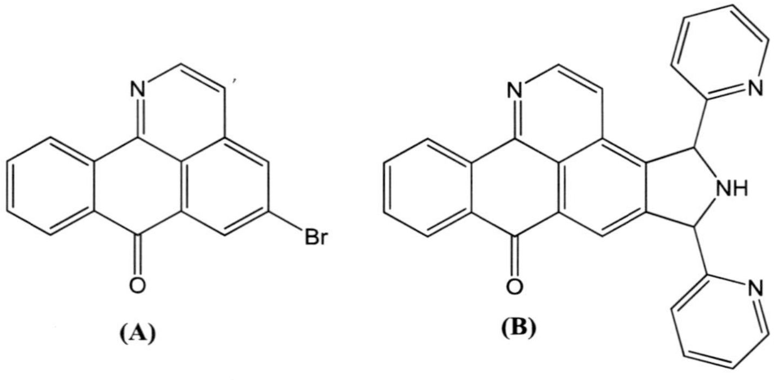

Figure 8.

Therefore, we have reported some examples of antineoplastic agents by using oxoisoaporphine derivatives, wherein the use of both metal and alkaloid as organ ometallic complex has been a novel alternative against several cancer cell line s, affording a new pharmacological tool to be considered. Our suggestion in this aspect is to continue the *in vivo* assays in order to find an oxoisoaporphine derivative with low cytotoxicity that can be used by enteral administration, and to be used as novel organometallic tool for treatment against cancer and similar neoplastic disease.

B. Use of Oxoisoaporphines in the Treatment of Mental Disorders and Neuro-Degenerative Behavior

Some therapeutic alternatives for treatment of behavioral diseases as depression or neuro-degenerative disorders like Parkinson's and Alzheimer's disease have been the use of natural products obtained from roots and bark of trees and shrubs. Nevertheless, the small quantities of these substances in the above mentioned sources and the high cost of isolation in pure state of some analogues with different stereochemistry; they make viable the possibility of obtaining these compounds for alternative synthetic routes. Many chemical structures are involved on the market of the psychotropic drugs, hallucinogenic and in treatment of psychiatric diseases. However, many of the side effects that present the above mentioned substances come from the ignorance of the potentialities of the administrated drugs that often produce addiction in the patient without necessarily relieving the treated ailment.

Thus, the important source of alkaloids of varied chemical structure provides alternatives that they can contribute to the drugs availability for treatment of psychiatric diseases. In this sense, one of the most important groups of alkaloids are the classified as isoquinolines that involve oxoisoaporphines. Although its chemical structure is very different from the alkaloids known with psychotropic activities as Harmaline and Harmine [37], oxoisoaporphines, isomers of oxoaporphines, have been described for a few years ago with interesting pharmacological activities as acetylcholinesterase (AChE), butylcholinesterase (BuChE) and monoamino oxidase A (MAO-A) inhibitors, responsible for the motor activity, cognitive and the anxiety -depression. In this sense, we will see some examples of the interesting proposal of the use of these alkaloids that represent an excellent treatment route for its low cost of production of certain derivatives, its purity and low cytotoxicity either in its parenteral administration or as nanoencapsulated systems.

In this sense, a series of 9-aminoalkanamido-1-azabenzanthrone derivatives (Figure 9) [R = 1-pyrrolidinyl, NMe_2, n = 1, 2, 3; R = NEt_2, $NH(CH_2)_2NMe_2$, $NH(CH_2)_2OH$, n = 2], and corresponding quaternary methiodide salts, such as I [R = 1-methyl-1-pyrrolidinium, $N+Me_3.I-$, n = 1, 2, 3; R = $N+(Me)Et_2.I-$, n = 2], were designed and synthesized as acetylcholinesterase (AChE) or butyrylcholinesterase (BuChE) inhibitors. Such compounds exhibited high AChE inhibitory activity with IC_{50} values in the nanomolar range and high selectivity for AChE over BuChE (45- to 1980-fold). According to this research, the cationic compounds with quaternary nitrogen showed higher inhibitory effects on AChE. The compounds with pyrrolidine at the end of side chain possessed higher inhibitory activity. But the compounds with variation of chain length showed less influence on inhibitory activity [38].

As showed in Figure 9, the oxoisoaporphine derivatives with large chain attached at the ring D can have some variation respect to the functional groups such as amide, carboxylic acid, etc. and its final pharmacological reactivity. However, when there is a change in the position of the ring D to afford an oxoaporphine, the pharmacological data on AChE and BuChE experiments would change drastically. In this sense, novel derivatives of oxoaporphine alkaloids (4-carboxylic amide-7-oxo-7*H*-dibenzo[*de,g*]quinoline, Ar-$CONH(CH_2)_nNR$) and their quaternary methiodide salts (Ar-$CONH(CH_2)_nN^+(CH_3)RI^-$) were designed and synthesized as acetylcholinesterase (AChE) and/or butyrylcholinesterase (BuChE) inhibitors. The AChE inhibition potency of synthetic oxoaporphine derivatives was decreased about 2-3 orders of magnitude as compared with that of oxoisoaporphine derivatives [39].

Sometimes, minimal changes in the position of a chain on the oxoisoaporohine framework are not necessary to increase the biological potential of a compound. In order to test this concept, 3-substituted and 4-substituted oxoisoaporphine derivatives were synthesized. It was found that all these synthetic compounds had IC50 values at micro or nano molar range for cholinesterase inhibition, and most of them could inhibit amyloid β (Aβ) self-induced aggregation with inhibition ratio from 31.8% to 57.6%. Thus, the results from cell toxicity study indicated that most quaternary methiodide salts had higher IC50 values than the corresponding non-quaternary compounds. These results are novel and it could be an excellent information in the design of new selective drugs in the cholinesterase inhibition and, therefore, in the treatment of Alzheimer's disease [40]. Furthermore, dual binding site acetylcholinesterase (AChE) inhibitors have been designed, synthesized, and tested for their ability to inhibit AChE, butyrylcholinesterase (BChE), AChE-induced and self-induced β-amyloid (Aβ) aggregation. For this propose, oxoisoaporphine and a tacrine or its congener, connected through an oligomethylene linker containing an amine group at variable position. These hybrids exhibit high AChE inhibitory activity with IC50 values in the nanomolar range in most cases. Moreover, some hybrids of this series, particularly those bearing a tetrahydroacridine moiety, exhibit a significant in vitro inhibitory activity toward the AChE-induced and self-induced Aβ aggregation, which makes them promising anti-Alzheimer drug candidates and, therefore, to promote a new alternative in this pharmaceutical field [41,42].

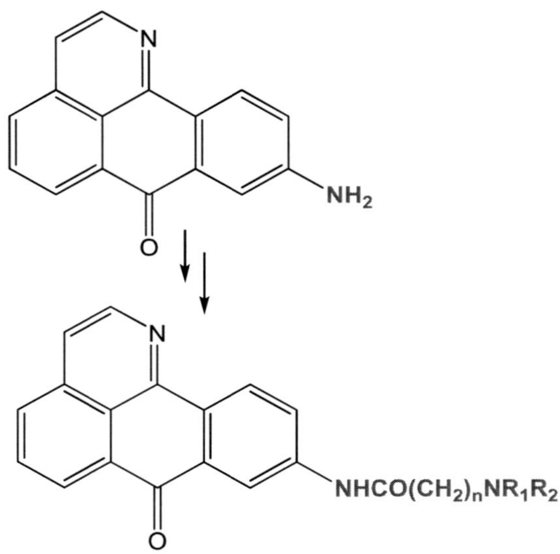

Figure 9.

On the other hand, some oxoisoaporphine derivatives (**OXO 1**, **OXO 2**) showed an unusual monoamino oxidase inhibition, being only selective for the isoform A (MAO-A), and whose low concentration of the compounds (up to pM concentration) to allowed to regulate and to reduce the depression in animal testing under *in vitro* experiments [43]. Due to that there are not references about this activity in this type of alkaloids, as showed in 2,3-dyhidro-oxoisoaporphine derivatives (Figure 10), such results were protected by means of a Spanish patent and its international preliminary [44]. This allowed to this type of alkaloids to be considered as possible antidepressant drug through the MAO-A inhibition. Its low cytotoxicity and easy formation by synthetic routes of the different derivatives have afforded an unusual interest for this small group of alkaloids [45].

Although the administration of this alkaloid in animal testing is via parental and its low solubility in aqueous medium makes it difficult to release the drug in the living system, at the time there exists several supramolecular systems to release drug of controlled manner. Thus, the use of synthetic polymer as poly-ε-caprolactone (PCL) helps in studies of drug delivery in many liposoluble compounds and to increase its bioavailability in the organism. Refered to this, oxoisoaporphine nanocapsules for treating depression were synthesized by using PCL in order to afford nanocapsule systems that incorporate this alkaloid in high yield (> 95%) that allow controlled and timed release the compounds by using 2,3-dihydro-7H-dibenzo[de,h]quinolin-7-one derivatives and their oxidized analogues. The nanocapsules were characterized by the zeta potential method and encapsulation efficiency (%). The results of this novel proposal have afforded an excellent tool in order to release this drug insoluble in water and to avoid its elimination by the organism.

Respect to a oxoisoaporphine derivative whose selectivity was refered as MAO-A inhibitor, 2,3-dyhidro-5-methoxy-7*H*-dibenzo[*de,h*]quinolin-7-one, **OXO 3**, this derivative was used in the Porsolt swimming forced test as a nanoencapsulated solution and administered *via* parenteral [46].

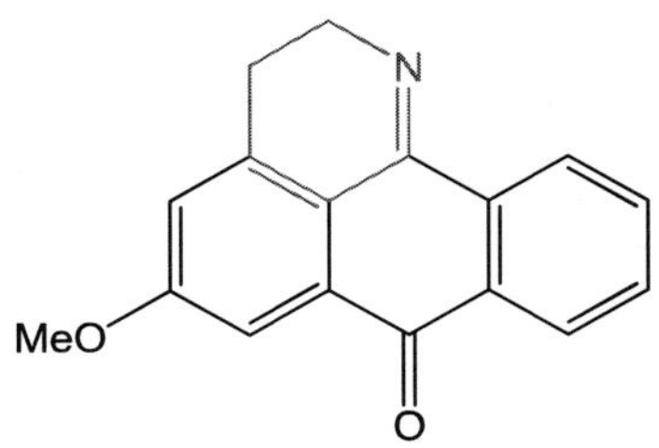

OXO 1

IC_{50} = 13.65 nM

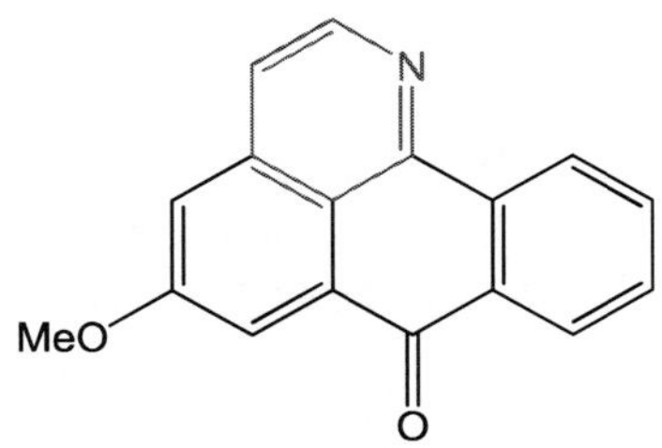

OXO 2

IC_{50} = 0.83 nM

Figure 10.

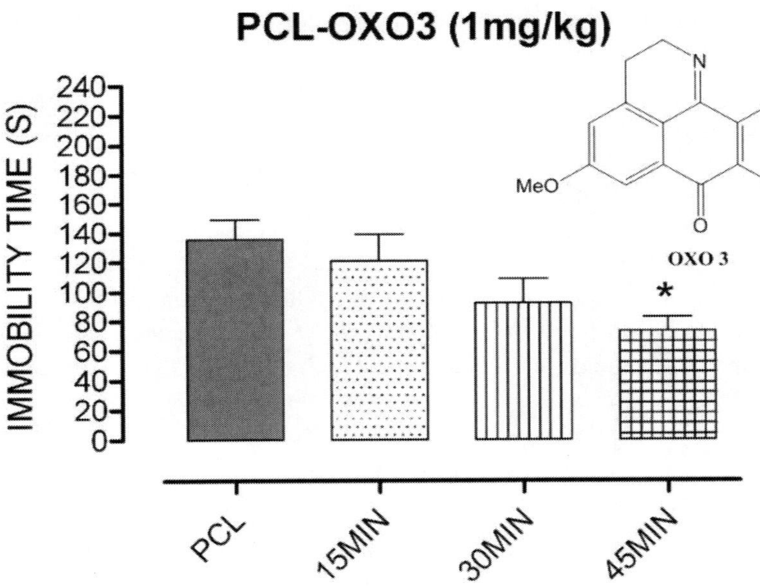

Figure 11.

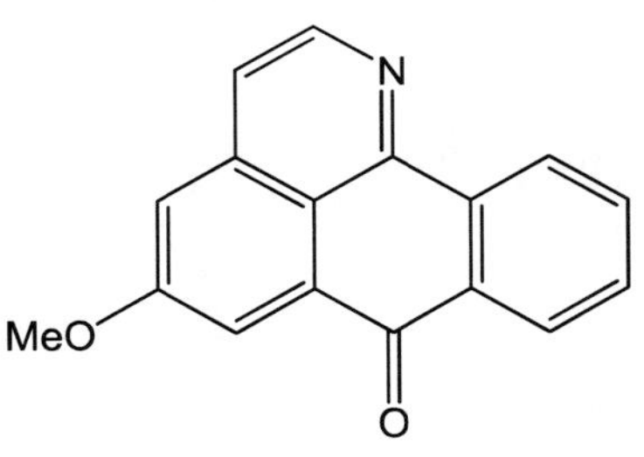

OXO 4

Figure 12.

The results confirmed the tendency in order to release the drug the controlled manner in the time and affording an antidepressive effect. In this sense, the immobility time, commonly used to measure the effectiveness of antidepressants, is important in 45 min (Figure 11).

Therefore, the pharmacological data respect to the control would mean to use a minimal amount of this drug and to achieve the same effect with minor side effects as cytotoxicity, and to increase the possibility of across the blood -brain barrier (BBB).

On the other hand, antidepressant -like effects of intra-cerebroventricularly administration of oxoisoaporphine derivatives in post stroke -depressive like behavior were examinated through despair swimming and tail suspension models. For this aim, acute ischemic stroke was induced by bilateral common carotid arteries occlusion which significantly changed the normal behaviors of male balb/c mice. Results show that some 2,3-dihydro- and oxoisoaporphine derivatives modified the abnormality in the behaviors through decreasing in the immobility time in tail suspension and despair swimming models and increasing in the swimming and climbing times in despair swimming model. Therefore, these alkaloids can be used in the treatment of post strokedepressive like behavior in acute ischemic stroke patients, in agreement with our previous results of oxoisoaporphine derivatives as novel antidepressant [47].

Finally, we have assayed the *in vivo* antidepressant effect of a natural phenolic compound, *trans*-resveratrol, and an oxoisoaporphine derivative, 5-methoxy-7*H*-dibenzo[*de,h*]quinolin-7-one, **OXO 4** (Figure 12). The antidepressant-like effect of *trans*-resveratrol and **OXO 4** was evaluated through a Forced Swimming Test (FST), and they were compared with reference antidepressants: imipramine, desipramine, nomifensine, bupropion, nisoxetine, citalopram and moclobemide. In this sense, we evaluate **OXO 4** at 15, 30 and 45 min. affording the mayor reduction at 30 min after the administration.

Thus, our results suggest that **OXO 4** has a great antidepressant effect non-reported for this type of isoquinoline alkaloids, but in agreement with the reports given above [48].

CONCLUSION

According to the information brought by the varied studies of antineoplastic activities and behavioral disorders, including neuro-degenerative diseases, using like drugs study oxoisoaporphine derivatives, we have verified that different functional groups, moieties semi- and completely oxidized of the alkaloid chemical structure, they compromise important pharmacological properties that can be justified by interactions like hydrogen bonding, hydrophobic or π-stacking. In spite of being a small group of isoquinoline alkaloids that had been isolated previously of oriental natural sources, already were known the properties of energy transfer and of the use as industrial dyes. Probably these new therapeutic properties of oxoisoaporphines in the applied health scope will be an interesting alternative to be considered for treatment of the different carcinogenic cellular lines, knowing that precedents exist which bring the use of oxoisoaporphines as a real possibility in the photodynamic therapy.

On the other hand, the new projections of these compounds as antidepressant and anxiolytic of specific activity and long action to be encapsulated in inert polymers that prolong the half life of the drug, as well as its specific reactivity as cholinesterase inhibitors (AChE/BuChE) related to the therapeutic action against both the Parkinson's and Alzheimer's diseases, it offers us a scale of possibilities of low cost and of high efficiency in future treatments.

The investigations carried out still keep on developing the search of new compounds that extend the period of action of an antineoplastic drug or replacing the compounds that, at present, generate undesirable side effect as addiction of psychotropic substances and control of mood disorder s as depression. It is expected that these references included in this chapter help to guide clearly and precisely in the study of the chemical and pharmacological properties of this small group of alkaloids of big projection that have been developed during the last 5 years.

REFERENCES

[1] Okamoto, Y., Tanaka, S., Kitayama, K., Isomoto, M., Masaishi, M., Yanagawa, H. & Kunitomo, J. (1971). Studies on the alkaloids of menispermaceous plants. CCLXI. Alkaloids of Menispermum dauricum DC. 8. *Yakugaku Zasshi*, *91*, 684-687.

[2] Hou, C. Y. & Xue-Hong. (1985). Studies on the Chemical constituents of *Menispermum dauricum* DC. *Yaoxue Xuebao*, *20* (2), 112-117.

[3] Hou, C. Y. & Xue, H. (1984). Studies on the chemical constituents of *Menispermum dauricum* DC. *Yaoxue Xuebao*, *19*(6), 471-472.

[4] Kunitomo, J. & Satoh, M. (1982). Structure of Menisporphine: A new type of isoquinoline alkaloid. *Chem. Pharm. Bull.*, *30*, 2659-2660.

[5] Takani, M., Takasu, Y. & Takahashi, K. (1983). Studies on Constituents of Medicinal Plants. XXIII. Constituents of the Vines of *Menispermum dauricum* DC. (2) *Chem. Pharm. Bull.*, *31*, 3091-3093.

[6] Kunitomo, J., Satoh, M. & Shingu, T. (1983). Structure and synthesis of menisporphine, a new type of isoquinoline alkaloid: Alkaloids of *Menispermum dauricum* DC. (9) (1983). *Tetrahedron*, *39*, 3261-3265.

[7] Hu, S.-M., Xu, S.-X., Yao, X.-S., Cui, C.-B., Tezuka, T. & Kikuchi, K. (1993). Dauricoside, a New Glycosidal Alkaloid Having an Inhibitory Activity against Blood-Platelet Aggregation. *Chem. Pharm. Bull.*, *41*, 1866-1868.

[8] Kunitomo, J., Kaede, S. & Satoh, M. (1985). The Structure of 2, 3-Dihydromenisporphine and the Synthesis of Dauriporphine, Oxoisoaporphine Alkaloids from *Menispermum dauricum* DC. *Chem. Pharm. Bull.*, *33*, 2778-2782.

[9] Takani, M., Takasu, Y. & Takahashi, K. (1983). Studies on Constituents of Medicinal Plants. XXIII. Constituents of the Vines of *Menispermum dauricum* DC. (2) *Chem. Pharm. Bull.*, *31*, 3091-3093.

[10] Kunitomo, J., Kaede, S. & Satoh, M. (1985). The Structure of 2,3-Dihydromenisporphine and the Synthesis of Dauriporphine, Oxoisoaporphine Alkaloids from *Menispermum dauricum* DC. *Chem. Pharm. Bull.*, *33*, 2778-2782.

[11] Sugimoto, Y., Babiker, H. A. A., Inanaga, S., Kato, M. & Isogai, A. (1999). Oxoisoaporphines from *Menispermum dauricum*. *Phytochemistry, 52*, 1431-1435.

[12] Kunitomo, J-I. & Miyata, Y. (1986). Further Evidence for the Structure of Bianfugecine. *Heterocycles*, *24*, 437-440.

[13] Min, Y. D., Choi, S. U. & Lee, K. R. (2006). Aporphine alkaloids and their reversal activity of multidrug resistance (MDR) from the stems and rhizomes of *Sinomenium acutum*. *Arch. Pharm. Res.*, *29*(8), 627-632.

[14] Yu, B-W., Meng, L-H., Chen, J-Y., Zhou, T-X., Cheng, K-F., Ding, J. & Qin, G-W. (2001). Cytotoxic oxoisoaporphine alkaloids from *Menispermum dauricum*. *J. Nat. Prod.*, *64* (7), 968-970.

[15] Killmer, L., Vogt, F. G., Freyer, A. J., Menachery, M. D. & Adelman, C. M. (2003). Lakshminine, a New Rare Oxoisoaporphine Alkaloid from Sciadotenia toxifera, and Structural Revisions of Telazoline and Teladiazoline, Two Related Oxoaporphines from *Telitoxicum peruvian*um and *T. glaziovii*. *J. Nat. Prod.*, *66*, 115-118.

[16] Kunz, M. A., Koberle, K. & Kochendoerfer, G. (1938). Azabenzanthrone dyes. DE 655592 19380119.

[17] Allen, C. F. H., Crawford, J. V., Sprague, R. H., Webster, E. R. & Wilson, C. V. (1950). 3-Azabenzanthrone dyes. *J. Am. Chem. Soc. 72*, 585-588.

[18] Sekar, N. (2001). 1-Azabenzanthrone colorants. *Colourage*, *48*(1), 54-57.

[19] Srinivasan, K. V. & Seshadri, S. (1987). Synthesis of disperse dyes derived from 1-cyano-2-substituted-3-azabenzanthrones. *Dyes and Pigments*, *8*(3), 225-30.

[20] Pieri, G., Carlini, F. M., Paffoni, C. & Boffa, G. (1976). Azabenzanthrone fluorescent dyes. Ger. Offen., DE 2617321 A1 19761104.

[21] De la Fuente, J. R., Kciuk, G., Aliaga, C. & Bobrowski, K. (2014). Spectral and Kinetic Properties of Radical Cations Derived from Oxoisoaporphines: Relevance to Electron-Transfer Processes Involving Phytoalexins. *J. Phys. Chem. A.*, *118*(21), 3775-3786.

[22] Aliaga, C., Ceron-Neculpan, M., Saitz, C., Jullian, C., Sobarzo-Sanchez, E., De la Fuente, J. R. (2011). Oxoisoaporphines: regioselective deuterium labeling involving the metastable hydrogenated photoproduct anions. *J. Photochem. Photobiol.* A: *Chemistry*, *222*(2-3), 360-365.

[23] De la Fuente, J. R., Aliaga, C., Poblete, C., Zapata, G., Jullian, C., Saitz, C., Canete, A., Kciuk, G., Sobarzo-Sanchez, E. & Bobrowski, K. (2009). Photoreduction of Oxoisoaporphines by Amines: Laser Flash and Steady-State Photolysis, Pulse Radiolysis, and TD-DFT Studies. *J. Phys. Chem. A.*, *113*(27), 7737-7747.

[24] Sobarzo-Sanchez, E., Gonzalez, P., Valdes, C., Sanchez, G. & Hidalgo, M. E. (2012). Applied biological and physicochemical activity of isoquinoline alkaloids: oxoisoaporphine and boldine. *Molecules*, *17*, 10958-10970.

[25] González-Díaz, H., Prado-Prado, F., Sobarzo-Sánchez, E., Haddad, M., Maurel Chevalley, S., Valentin, A., Quetin-Leclercq, J., Dea-Ayuela, M. A., Gomez-Muños, M. T., Munteanu, C. R., Torres-Labandeira, J. J., García-Mera, X., Tapia, R. A. & Ubeira. F. M. (2011). NL MIND-BEST: a web server for ligands & proteins discovery; theoretic-experimental study of proteins of Giardia lamblia and new compounds active against *Plasmodium falciparum*. *J. Theor. Biol.*, *276*, 229-249.

[26] Sobarzo-Sánchez, E., Bilbao-Ramos, P., Dea-Ayuela, M., González-Díaz, H., Yañez, M., Uriarte, E., Santana, L., Martínez-Sernández, V., Bolás-Fernández, F. & Ubeira, F. M. (2013). Synthetic Oxoisoaporphine Alkaloids: In Vitro, In Vivo and In Silico Assessment of Antileishmanial Activities. *Plos One*, *8*(10), e77560.

[27] Castro-Castillo, V., Suarez-Rozas, C., Castro-Loiza, N., Theoduloz, C. & Cassels, Bruce K. (2013). Annulation of substituted anthracene-9,10-diones yields promising selectively antiproliferative compounds. *Eur. J. Med. Chem.*, *62*, 688-692.

[28] Gomez-Jeria, J. S. (2013). Flores-Catalan, M. Quantum-chemical modeling of the relationships between molecular structure and in vitro multi-step, multimechanistic

drug effects. HIV -1 replication inhibition and inhibition of cell proliferation as examples. *Can. Chem. Trans.*, *1*(3), 215-217.

[29] Chen, Z., Liang, H., Liu, Y. & Qin, Q. (2013). 1-Azabenzanthrone-platinum (II) complex, and synthetic method and application thereof as cell proliferation inhibitor. *Faming Zhuanli Shenqing*, CN 103450281 A 20131218.

[30] Chen, Z., Liang, H., Liu, Y., Qin, Q. & Zhang, C. (2013). 6-Hydroxyoxoisoaporphine rare earth complexes, synthetic method and application thereof. *Faming Zhuanli Shenqing*, CN 103450236 A 20131218.

[31] Chen, Z., Liang, H., Liu, Y., Yang, G. & Qin, Q. (2014). A 6-aminooxoisoaporphine platinum (II) complex synthesis method and application. *Faming Zhuanli Shenqing*, CN 103524564 A 20140122.

[32] Yu, B.-Wu., Meng, L.-H., Chen, J.-Y., Zhou, T.-X., Cheng, K.-F., Ding, J. & Qin, G.-W. (2001). Cytotoxic oxoisoaporphine alkaloids from *Menispermum dauricum*. *J. Nat. Prod.*, *64*(7), 968-970.

[33] Tang, H., Wang, X.-D., Wei, Y.-B., Huang, S.-L., Huang, Z.-S., Tan, J.-H., An, L.-K., Wu, J.-Y., Sun -Chi Chan, A. & Gu, L.-Q. (2008). Oxoisoaporphine alkaloid derivatives: Synthesis, DNA binding affinity and cytotoxicity. *Eur. J. Med. Chem.*, *43*(5), 973-980.

[34] Castro-Castillo, V., Rebolledo-Fuentes, M., Theoduloz, C. & Cassels, B. K. (2010). Synthesis of Lakshminine and Antiproliferative Testing of Related Oxoisoaporphines. *J. Nat. Prod.*, *73*(11), 1951-1953.

[35] Chen, Z., Liang, H., Liu, Y., Tang, H., Zhang, C. & Qin, Q. (2013). 5-Bromo-oxoisoaporphine as antitumor drug and its preparation, pharmaceutical compositions and use in the treatment of cancer. *Faming Zhuanli Shenqing*, CN 103044326 A 20130417.

[36] Chen, Z., Liang, H., Liu, Y. & Qin, Q. (2013). Oxo-isoaporphine alkaloid derivative, its synthetic method and application. *Faming Zhuanli Shenqing*, CN 103130803 A 20130605.

[37] Zhao, T., Zheng, S. S., Zhang, B. F., Li, Y. Y., Bligh, S. W., Wang, C. H. & Wang, Z. T. (2012). Metabolic pathways of the psychotropic-carboline alkaloids, harmaline and harmine, by liquid chromatography /mass spectrometry and NMR spectroscopy. *Food Chem.*, *15*, 134(2), 1096-105.

[38] Tang, H., Ning, F.-X., Wei, Y.-B., Huang, S.-L., Huang, Z.-S., Chan, A. S.-C. & Gu, L.-Q. (2007). Derivatives of oxoisoaporphine alkaloids: A novel class of selective acetylcholinesterase inhibitor s. *Bioorg. Med. Chem. Lett.*, *17*(13), 3765-3768.

[39] Tang, H., Wei, Y.-B., Zhang, C., Ning, F.-X., Qiao, W., Huang, S.-L., Ma, L., Huang, Z.-S. & Gu, L.-Q. (2009). Synthesis, biological evaluation and molecular modeling of oxoisoaporphine and oxoaporphine derivatives as new dual inhibitors of acetylcholinesterase /butyrylcholinesterase. *Eur. J. Med. Chem.*, *44*(6), 2523-2532.

[40] Li, Y.-P., Ning, F.-X., Yang, M.-B., Li, Y.-C., Nie, M.-H., Ou, T.-M., Tan, J.-H., Huang, S.-L., Li, D., Gu, L.-Q. & Huang, Z. S. (2011). Syntheses and characterization of novel oxoisoaporphine derivatives as dual inhibitors for cholinesterases and amyloid beta aggregation. *Eur. J. Med. Chem.*, *46*(5), 1572-1581.

[41] Tang, H., Zhao, L.-Z., Zhao, H.-T., Huang, S.-L., Zhong, S.-M., Qin, J.-K. & Chen, Z.-F., Huang, Z.-S. (2011). Liang, H. Hybrids of oxoisoaporphine-tacrine congeners:

Novel acetylcholinesterase and acetylcholinesterase-induced β-amyloid aggregation inhibitors. *Eur. J. Med. Chem.*, *46*(10), 4970-4979.

[42] Tang, H., Zhao, H.-T., Zhong, S.-M., Wang, Z.-Y., Chen, Z.-F. & Liang, H. (2012). Novel oxoisoaporphine-based inhibitors of acetyl- and butyrylcholinesterase and acetylcholinesterase -induced beta-amyloid aggregation. *Bioorg. Med. Chem. Lett.*, *22*(6), 2257-2261.

[43] Prado-Prado, F., Garcia-Mera, X., Escobar, M., Sobarzo-Sanchez, E., Yanez, M., Riera-Fernandez, P. & Gonzalez-Diaz, H. (2011). 2D MI-DRAGON: A new predictor for protein-ligands interactions and theoretic-experimental studies of US FDA drug-target network, oxoisoaporphine inhibitors for MAO-A and human parasite proteins. *Eur. J. Med. Chem.*, *46*(12), 5838-5851.

[44] Sobarzo-Sanchez, E., Yanez Jato, M., Orallo Cambeiro, F., Uriarte Villares, E. & Cano Rubio, E. (2009). Use of oxoisoaporphines and the derivatives thereof as selective inhibitors of monoamino oxidase A. *PCT Int. Appl.*, WO 2009034216 A1 20090319.

[45] Molina, E., Sobarzo-Sanchez, E., Speck-Planche, A., Matos, M. J., Uriarte, E., Santana, L., Yanez, M. & Orallo, F. (2012). Monoamino oxidase a: an interesting pharmacological target for the development of multi-target QSAR. *Mini Rev. Med. Chem.*, *12*(10), 947-958.

[46] Sobarzo Sanchez, E., Cebey Lopez, M., Fontenla, J. A., Otero Espinar, F., Torres Labandeira, J. J., Gonzalez-Díaz, H., Grillo, R. & Fraceto, L. F. (2013). Oxoisoaporphine nanocapsules comprising polymers and oils for treating depression. *PCT Int. Appl.*, WO 2013050637 A1 20130411.

[47] Nabavi, S. F., Sobarzo-Sanchez, E., Nabavi, S. M., Daglia, M., Moghaddam, A. H. & Silva, A. G. (2013). Behavioral Effects of 2,3-Dihydro- and Oxoisoaporphine Derivatives in Post Stroke-Depressive Like Behavior in Male Balb/c Mice. *Curr. Top. Med. Chem.*, *13*(17), 2127-2133.

[48] Lopez, M. C., Fontenla, J. A., Uriarte, E., Santana, L. & Sobarzo-Sanchez, E. (2014). Comparison of the Antidepressive Effects of Trans-Resveratrol and 5-Methoxy-7*H*-Dibenzo[*de,h*]Quinolin-7-One. *Curr. Top. Med. Chem.*, *14*(2), 234-238.

In: Alkaloids
Editor: Eduardo Sobarzo-Sánchez

ISBN: 978-1-63482-074-5

Chapter 10

DAPHNIPHYLLUM ALKALOIDS

Haifeng Wu[1,*], Xiaopo Zhang[1], Lisheng Ding[2], Shilin Chen[1,3] and Xudong Xu[1,†]

[1]Institute of Medicinal Plant Development, Chinese Academy of Medical Sciences and Peking Union Medical College, Beijing, China

[2]Chengdu Institute of Biology, Chinese Academy of Sciences, Chengdu, China

[3]Institute of Chinses Materia Medica, China Academy of Chinese Medical Sciences, Beijing, China

ABSTRACT

The unique polycyclic fused ring systems of *Daphniphyllum* alkaloids, along with their extensive bioactivities, make this family of alkaloids especially attractive targets for total synthesis and biogenetic studies. In particular, successive discoveries of new alkaloids with unprecedented skeletons in recent years have made great contribution to structural diversities of alkaloids elaborated by plants of the genus *Daphniphyllum*. By the end of 2008, more than 200 alkaloids belonging to 14 different skeletal types have been isolated from different parts of plants of thirteen *Daphniphyllum* species. These alkaloids show cytotoxic, antioxidant, vasorelaxant and anti-platelet activating factor (PAF) effects. The plausible biosynthetic pathways for *Daphniphyllum* alkaloids have been proposed and biomimetic total syntheses of some alkaloids completed. To provide an update of the previous reviews published in 2009, chemical structures, biosynthesis, syntheses and pharmacological activities of *Daphniphyllum* alkaloids reported in the latest years are presented in this chapter. Between 2009 and 2012, additional fifty-one novel alkaloids have been isolated and identified. Among them, some possess unprecedented frameworks. Several state -of-the-art total syntheses through cascade reactions were completed.

[*] Haifeng Wu: Institute of Medicinal Plant Development, Chinese Academy of Medical Sciences and Peking Union Medical College, Beijing 100193, China. E-mail: wwwtony505@yahoo.com.cn. Mailing Address: No.151, Malianwa North Road, Haidian District, Beijing, PR China. Postal code: 100193, tel.: +86 10 5783 3296, fax: +86 10 5783 3296.

[†] Xudong Xu: Institute of Medicinal Plant Development, Chinese Academy of Medical Sciences and Peking Union Medical College, Beijing 100193, China. E-mail: xdxu@implad.ac.cn.

Keywords: alkaloids, *Daphniphyllum* biosynthesis, total synthesis, pharmacological activities

1. Introduction

Daphniphyllum alkaloids, triterpenoid-alkaloids with highly variable polycyclic skeletons, are characteristic constituents of the genus *Daphniphyllum*. *Daphniphyllum*, referring to Daphne and leaf in Greek, is the sole genus in the Daphniphyllaceae family, including about 30 species of dioecious evergreen trees and shrubs mainly distributed from India to Japan and from central China to New Guinea [1]. About ten species were found in China, some of which are traditionally used in the treatment of asthma, rheumatism and snakebite, etc. The isolation of the first C_{30}-type *Daphniphyllum* alkaloid, daphnimacrine from *D. macropodum*, was reported by Yagi in 1909 [2]; however, the structure remained unresolved. The systematic investigation of *Daphniphyllum* alkaloids began in the 1960s. Three new alkaloids, daphnicaline, daphnicadine and daphnicamine were isolated from the seeds of *D. calycinum*, Niu-Er-Fon (in Chinese) by Fang et al. [3]. Niu-Er-Fon is a poisonous plant growing in southern China. The authors measured the melting points and proposed the molecular formulae of these alkaloids. Subsequently, investigation on the Hong Kong species of *D. calycinum* and *D. glaucescens* resulted in the isolation of another three new alkaloids, calycine, glaucescine and glaucescinine [4]. ^{1}H-NMR spectrum of calycine was reported. In 1966, the structures of several alkaloids from *D. macropodum* were determined with X-ray analysis of their hydrobromide by Hirata et al. [5-7]. Since then, more than 200 *Daphniphyllum* alkaloids have been isolated from the genus *Daphniphyllum* by the end of 2008. The initial five skeletal types have been expanded to 14 main types. These alkaloids show cytotoxic, antioxidant, vasorelaxant and anti-platelet activating factor (PAF) effects, etc. The unusual ring systems of *Daphniphyllum* alkaloids, together with their biological aspects, make this family of alkaloids especially attractive targets for total synthesis and biogenetic studies. In 1986, the first total synthesis of a *Daphniphyllum* alkaloid, (+)-methyl homodaphniphyllate, was completed by Heathcock et al. [8]. Subsequently, a series of total syntheses were reported [9-12]. A biosynthetic pathway for *Daphniphyllum* alkaloids was also proposed and several biomimetic total syntheses of some *Daphniphyllum* alkaloids conducted by Heathcock and coworkers [13-19]. Considering structural peculiarities of *Daphniphyllum* alkaloids, Yamamura et al. successively reviewed the isolation, structural elucidation, chemical and biological properties of *Daphniphyllum* alkaloids in 1975 and 1986 [20, 21]. Thereafter, the application of modern chromatographic and spectroscopic techniques greatly facilitated the discoveries of *Daphniphyllum* alkaloids. In 2003, Kobayashi and Morita summarized reports published during the 15-year period between 1987 and 2002 [22]. From then on, 66 new alkaloids from eleven species of *Daphniphyllum* had been reported until 2007, and about two thirds of these alkaloids were isolated from the Chinese species of *Daphniphyllum* by Chinese researchers. Li and Guo published a mini review highlighting progress over the period [23]. In 2009, the two comprehensive reviews were successively published [24, 25]. Both thoroughly summarized all reported *Daphniphyllum* alkaloids. Since 2009, additional 51 novel alkaloids have been isolated from eight species of *Daphniphyllum* genus.

Some of these alkaloids belong to the known skeleton, while others possess unprecedented frameworks. In addition, several state -of-the-art total syntheses through cascade reactions were also completed. To provide an update of the previous reviews by Kobayashi [25], this chapter focuses on isolation, biosynthesis, synthesis and biological activities of *Daphniphyllum* alkaloids reported in the past few years (2009–2012).

2. Structure of Daphniphyllum Alkaloids

Daphniphyllum alkaloids possess complex polycyclic ring systems represented by fused pentacyclic, hexacyclic, heptacyclic, octacyclic skeleton or aza-adamantane nucleus. Most of these alkaloids were derived from six molecule of the mevalonic acid pathway *via* a squalene-like intermediate except for a few originated from amino acid s. According to structural classification proposed by Kobayashi, *Daphniphyllum* alkaloids are divided into 14 main structural types including daphniphyllines, secodaphniphyllines, yuzurimines, daphnilactone A, daphnilactone B, yuzurines, bukittinggines, daphnezomines, daphnicyclidins, daphmanidins, daphniglaucins, calyciphyllines, paxdaphnines, daphlongeranines A and B, as shown in Figure 1. Classification of recently isolated alkaloids basically follows that of previous reviews, but the newly obtained skeletons have been added (Table 1).

Table 1. Daphniphyllum alkaloids isolated between 2009 and 2012

No.	Compounds	Sources	Parts	References
1	11-hydroxycodaphniphylline	*D. subverticillatum*	leaves	[26]
2	daphnezomine V	*D. humile*	leaves and branches	[27]
3	homodaphniphyllate	*D. calycinum*	stems and leaves	[28]
4	calycinumine A	*D. calycinum*	twigs	[29]
5	9,10-epoxycalycinine A	*D. calycinum*	stems and leaves	[28]
6	macropodumine J	*D. macropodum*	bark	[30]
7	macropodumine K	*D. macropodum*	bark	[30]
8	4,21-deacetyl-deoxyyuzurimine	*D. macropodum*	bark	[31]
9	daphnezomine T	*D. humile*	leaves and branches	[27]
10	daphhimalenine A	*D. himalense*	leaves	[32]
11	daphhimalenine B	*D. himalense*	leaves	[32]
12	daphangustifoline B	*D. angustifolium*	whole plants	(He, Zhou et al. 2011)
13	daphhimalenine C	*D. himalense*	leaves	[33]
14	17-hydroxydaphnigraciline	*D. subverticillatum*	leaves	[26]
15	daphlongamine I	*D. longeracemosum*	fruits	[34]
16	daphlongamine J	*D. longeracemosum*	fruits	[34]
17	daphlongeranine F	*D. longeracemosum*	fruits	[35]
18	daphlongeranine D	*D. longeracemosum*	fruits	[36]
19	daphlongeranine E	*D. longeracemosum*	fruits	[36]
20	daphmalenine A	*D. himalense*	leaves	[37]
21	daphmalenine B	*D. himalense*	leaves	[37]

Table 1. (Continued)

No.	Compounds	Sources	Parts	References
22	daphnioldhamine A	*D. oldhami*	fruits	[38]
23	daphmacromine A	*D. macropodum*	leaves and stems	[39]
24	daphmacromine B	*D. macropodum*	leaves and stems	[39]
25	daphmacromine C	*D. macropodum*	leaves and stems	[39]
26	daphmacromine D	*D. macropodum*	leaves and stems	[39]
27	daphmacromine E	*D. macropodum*	leaves and stems	[39]
28	daphmacromine F	*D. macropodum*	leaves and stems	[39]
29	daphmacromine G	*D. macropodum*	leaves and stems	[39]
30	daphmacromine H	*D. macropodum*	leaves and stems	[39]
31	daphmacromine I	*D. macropodum*	leaves and stems	[39]
32	daphmacromine J	*D. macropodum*	leaves and stems	[39]
33	dapholdhamine A	*D. oldhami*	leaves	[40]
34	angustimine	*D. angustifolium*	twigs	[41]
35	dapholdhamine B	*D. oldhami*	leaves	[40]
36	daphnezomine U	*D. humile*	leaves and branches	[27]
37	daphlongeranine C	*D. longeracemosum*	fruits	[36]
38	daphnezomine L methyl ester	*D. subverticillatum*	leaves	[26]
39	calycinumine B	*D. calycinum*	twigs	[29]
40	dapholdhamine C	*D. oldhami*	leaves	[40]
41	dapholdhamine D	*D. oldhami*	leaves	[40]
42	caldaphnidine H	*D. calycinum*	leaves and stems	[42]
43	angustifolimine	D. angustifolium	twigs	[41]
44	daphangustifoline A	*D. angustifolium.*	whole plants	[43]
45	daphenylline	*D. longeracemosum*	fruits	[44]
46	subdaphnidine A	*D. subverticillatum*	leaves	[26]
47	daphlongamine E	*D. longeracemosum*	leaves	[45]
48	daphlongamine F	*D. longeracemosum*	leaves	[45]
49	daphlongamine G	*D. longeracemosum*	leaves	[45]
50	daphhimalenine D	*D. himalense*	leaves	[33]
51	daphlongamine H	*D. longeracemosum*	leaves	[45]

2.1. Daphniphylline-Type Alkaloids

Daphniphylline-type alkaloids possess 22 carbon core with or without the C8 side chain. The C8 unit consists of 6-oxabicyclo[3.2.1]octane or 2,8- dioxabicyclo[3.2.1]octane. Three novel daphniphylline-type alkaloids, 11-hydroxycodaphniphylline (1) [26], daphnezomines V (2) [27] and homodaphniphyllate (3) [28] were isolated recently (Figure 2).

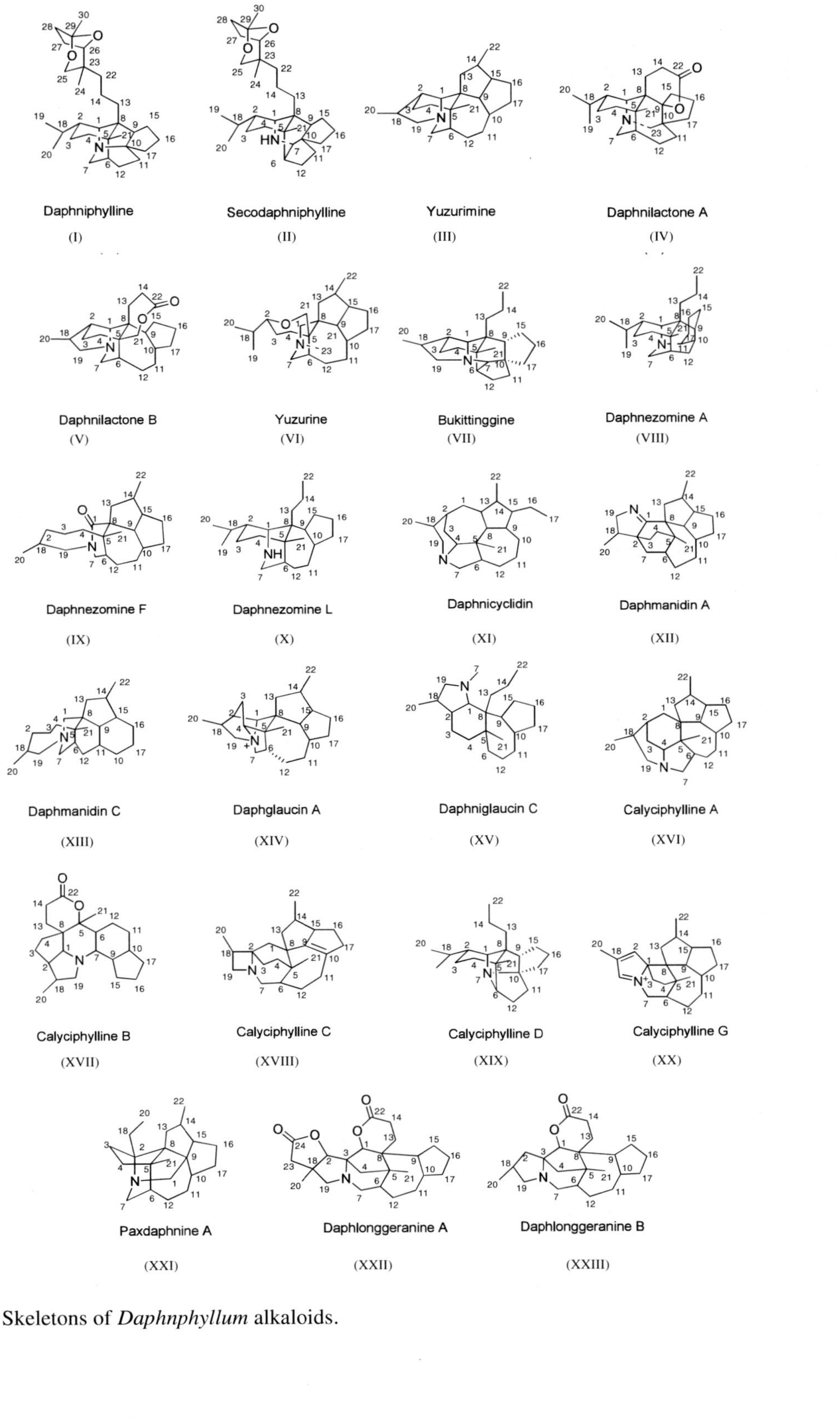

Figure 1. Skeletons of *Daphnphyllum* alkaloids.

2.2. Secodaphniphylline-Type Alkaloids

This group of alkaloids is closely related to daphniphylline-type alkaloids, possessing a C_{22} skeleton with or without the C_8 side chain.

The distinct difference is that the C-10–N linkage in daphniphylline-type alkaloids is broken and C-10–C-7 bond is formed in secodaphnan-type alkaloids. No secodaphniphylline-type alkaloid has been obtained in recent years.

2.3. Yuzurimine-Type Alkaloids

Yuzurimine-type alkaloids contain 22 carbon atoms and C-22 often is carboxyl group. Recently, ten new yuzurimine-type alkaloids, calycinumine A (**4**) [29], 9,10-epoxycalycinine A (**5**) [28], macropodumines J (**6**) and K (**7**) [30], 4,21-deacetyl-deoxyyuzurimine (**8**) [31], daphnezomine T (**9**) [27], daphhimalenine A (**10**) and B (**11**) [32], daphangustifoline B (**12**) [43] and daphhimalenine C (**13**) [33] have been reported (Figure 3). Calycinumine A (**4**) is the first example of C-22-nor yuzurimine-type alkaloid and its structure was confirmed by a single crystal X-ray diffraction. Macropodumine J (**6**) contains a CN group which is relatively rare in naturally occurring alkaloids. Daphhimalenine A (**10**) with rearrangement C-21 skeleton, contains a unique 1-azabicyclo[5.2.1]decane ring system with the cleavage of C-1 and C-8 bond. The absolute configuration of Daphhimalenine A (**10**) was assigned by computational methods. Daphhimalenine B (**11**) was biogenetically related to daphhimalenine A (**10**). Daphangustifoline B (**12**) is the first alkaloid with a benzyl group in the genus of *Daphniphyllum*.

2.4. Daphnilactone A-Type Alkaloids

The skeleton of daphnilactone A-type alkaloids seems to be constructed by the insertion of a C_1 unit into a nitrogen –carbon bond in the daphniphylline-type skeleton followed by lactonization.

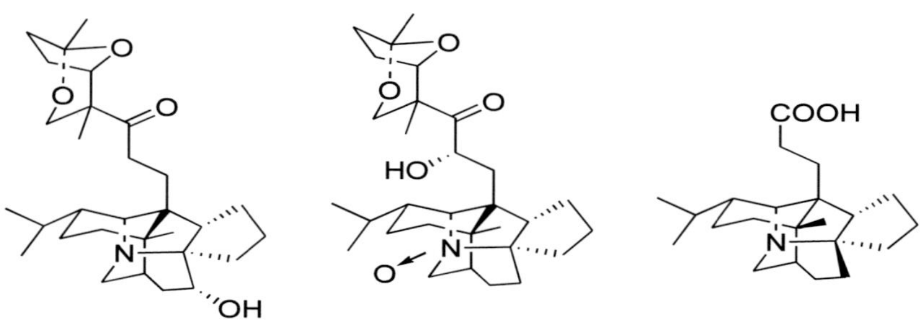

1 **2** **3**

Figure 2. Daphniphylline-type alkaloids.

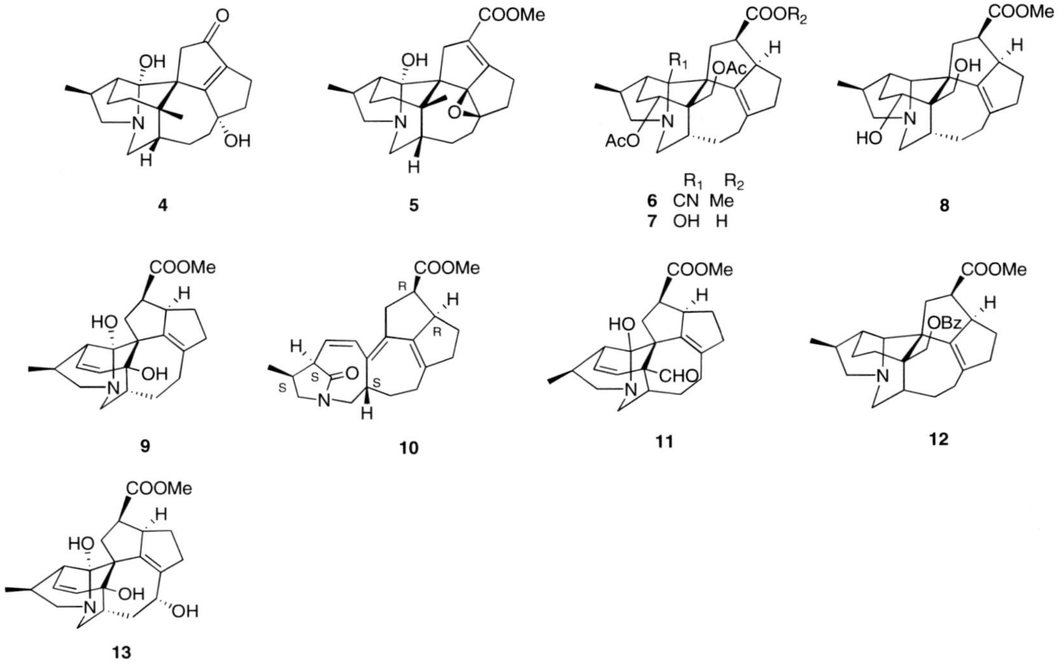

Figure 3. Yuzurimine-type alkaloids.

2.5. Daphnilactone B-Type Alkaloids

Daphnilactone B type alkaloids are a small group of C-14,15-seco-yuzurimines. Most of them form a lactone between the carbonyl at C-22 and the hydroxy attached to C-21.

2.6. Yuzurine-Type Alkaloids

Different from other *Daphniphyllum* alkaloids with 22 C-atoms, yuzurine-type alkaloids have 23 carbons in their frameworks with a methyl group attached to the nitrogen atom. In the latest few years, nineteen yuzurine-type alkaloids, 17-hydroxydaphnigraciline (**14**) [26], daphlongamines I (**15**) and J (**16**) [34], daphlongeranine F (**17**) [35], daphlongeranines D (**18**) and E (**19**) [36], daphmalenines A (**20**) and B (**21**) [37], daphnioldhamine A (**22**) [38], and daphmacromines A–J (**23–32**) [39] have been obtained (Figure 4). Daphmalenines A (**20**) and B (**21**) are biogenetically related yuzurine-type *Daphniphyllum* alkaloids of the rare (14*R*, 15*S*) series, possessing an unusual penta- or tetracyclic ring system, respectively. Daphmalenine B (**21**) is the first *seco*-10, 17-yuzurine type *Daphniphyllum* alkaloid.

The absolute configurations of daphmalenines A (**20**) and B (**21**) were determined on X-ray diffraction by using the Flack parameter and computational methods, respectively. Daphnioldhamine A (**22**) is the first *Daphniphyllum* alkaloid with transannular effect and is easily tautomerized under acidic or alkaline conditions (Figure 5).

The structure of Daphmacromine A (**23**) was confirmed by single-crystal X-ray diffraction.

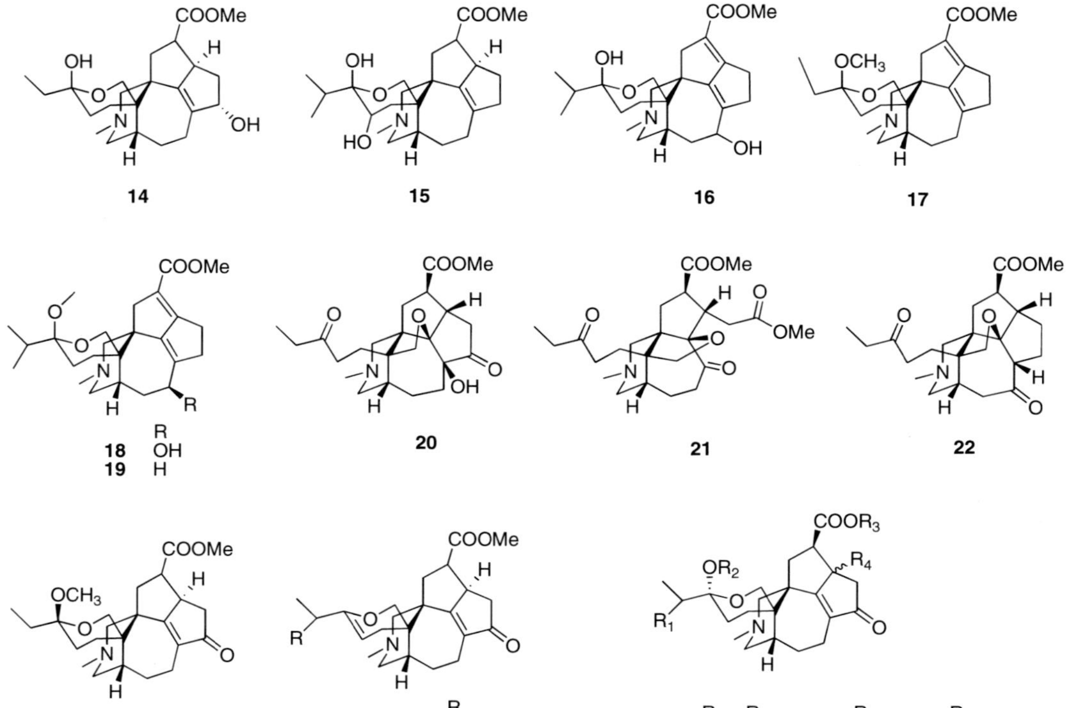

	R_1	R_2	R_3	R_4
23	H	CH_3	CH_2CH_3	α-H
24	H	CH_3	CH_2CH_3	β-H
25	H	CH_2CH_3	CH_3	α-H
26	H	CH_2CH_3	CH_3	β-H
27	H	CH_3	CH_3	α-H
29	H	CH_3	CH_3	β-H
30	CH_3	H	CH_3	β-H

Figure 4. Yuzurine-type alkaloids.

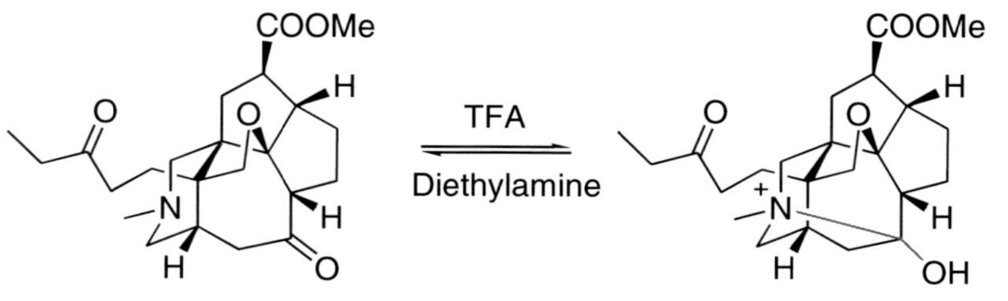

22

Figure 5. Chemical transformation of **22**.

2.7. Bukittinggine-Type Alkaloids

Two new bukittinggin-type alkaloids, dapholdhamine A (**33**) [40] and angustimine (**34**) [41], were isolated (Figure 6). Dapholdhamine A (**33**) is closely related to bukittinggine. Angustimine (**34**) is an intramolecular salt featuring an unprecedented hexacyclic fused skeleton through the cleavage of a C-6–C-7 bond and the formation of a C-6–N bond.

2.8. Daphnezomine

2.8.1. Daphnezomine A–Type Alkaloids

Daphnezomine A-type alkaloids possess unique aza-adamantne core with an amino ketal bridge. Dapholdhamine B (**35**) was isolated from *D. oldhami* [40] (Figure 7).

2.8.2. Daphnezomine F–Type Alkaloids

Daphnezomine F-type alkaloids possess characteristic 1-azabicyclo[5.2.2]undecane ring system. Two new daphnezomine F-type alkaloids, daphnezomine U (**36**) [27] and daphlongeranine C (**37**) [36], were obtained (Figure 8). Daphlongeranine C (**37**) is the first daphnezomine F-type alkaloid obtained from *D. longeracemosum.*

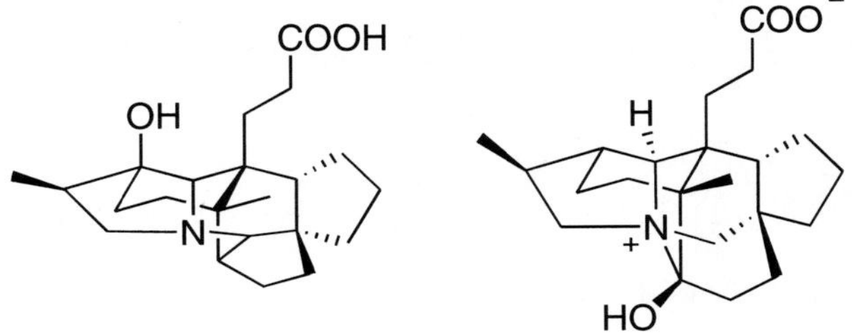

33 **34**

Figure 6. Bukittinggine-type alkaloids.

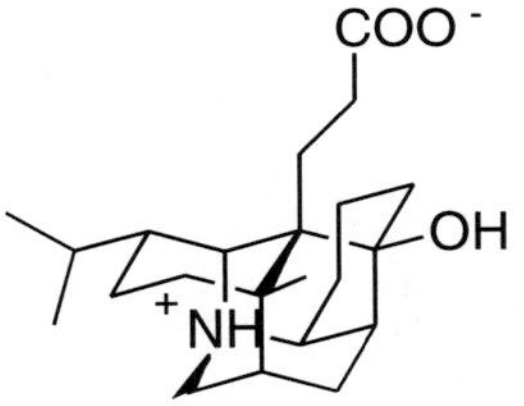

35

Figure 7. Daphnezomine A-type alkaloid.

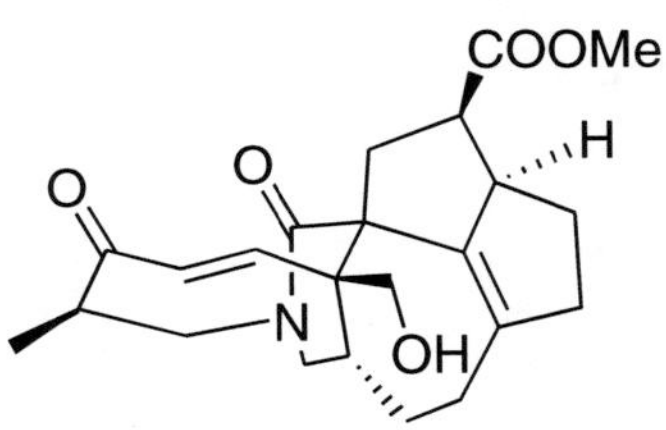

36

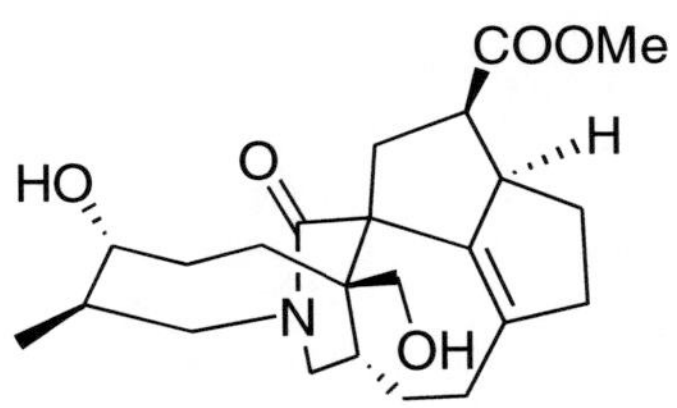

37

Figure 8. Daphnezomine F-type alkaloids.

2.8.3. Daphnezomine L-Type Alkaloids

Two new daphnezomine L-type alkaloids, daphnezomine L methyl ester (**38**), [26] and calycinumine B (**39**) [29], were reported (Figure 9). Daphnezomine L methyl ester (**38**) is structurally close to a hypothetical biogenetic intermediate between the secodaphnane and daphnane skeletons. Calycinumine B (**39**) features an unprecedented heteroatom-containing adamantane -like western hemisphere of daphnezomine L-type alkaloid.

2.9. Daphnicyclidins

Daphnicyclidin-type alkaloids are a group of rare C-22 fused hexa- or pentacyclic nor-Daphniphyllum alkaloids. Four new daphnicyclidin-type alkaloids, dapholdhamines C (40) and D (41) [40], caldaphnidine H (42) [42] and angustifolimine (43) [41] were reported (Figure 10). Angustifolimine (43) represents the second diamino Daphniphyllum alkaloid.

2.10. Daphmanidines

2.10.1. Daphmanidine-A Type Alkaloids

Daphmanidine-A type alkaloids contain C_{22} core with the formation of imine between C-1 and N atom and breakup of C-7–N bond.

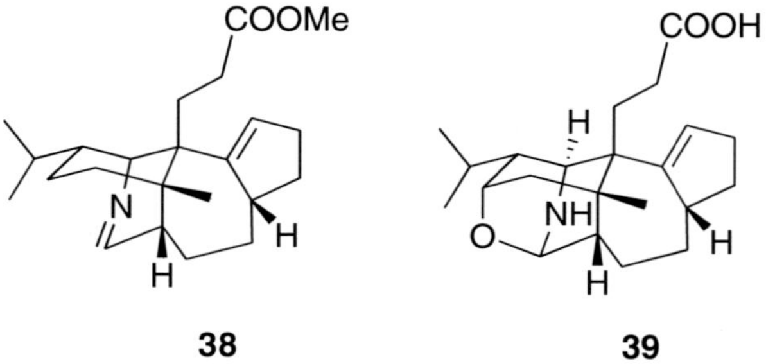

Figure 9. Daphnezomines L-type alkaloids.

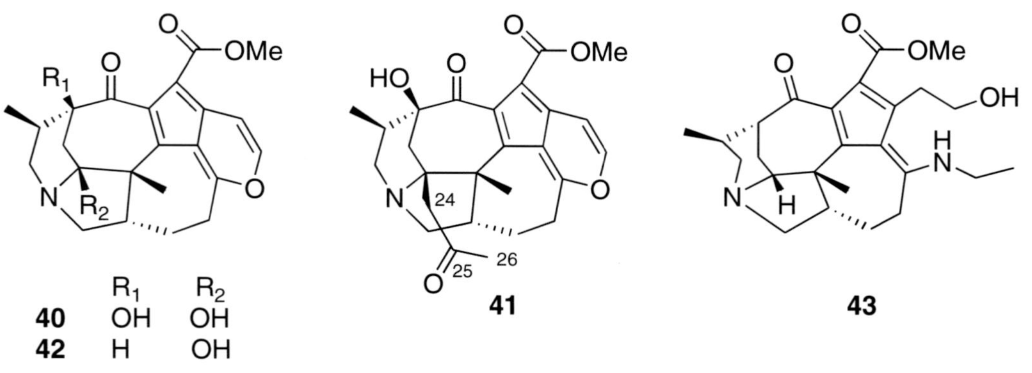

Figure 10. Daphnicyclidin-type alkaloids.

2.10.2. Daphmanidines C and D

This group of alkaloids consist of an 1-azabicyclo[5.2.2]undecane, a hexahydronaphthalen-1-one, and a cyclopentane.

2.11. Daphniglaucins

2.11.1. Daphniglaucin A-Type Alkaloids

A new daphniglaucin A-type alkaloid, daphangustifoline A (**44**) was isolated from the whole plant of *D. angustifolium* [43] (Figure 11).

2.11.2. Daphniglaucin C-Type Alkaloids

This group of alkaloids is biogenetically originated from yuzurimine and daphnilactone type alkaloids by cleavage of the C-6–C-7 bond. All of them have a tetracyclic ring system comprising an octahydroindole and a hexahydroazulene rings.

2.12. Calyciphyllines

2.12.1. Calyciphylline A-Type Alkloids

Calyciphylline A-type alkaloids are a group of rare C-22 fused hexacyclic nor-*Daphniphyllum* alkaloids, biosynthesized by the fission of C-1–N bond of yuzurimine-type alkaloid and the formation of a new bond between C-4 and nitrogen. Calyciphylline A-type alkaloids possess a 6/5/5/7/5/6-membered ring system and an oxo group on C-1.

Six new calyciphylline A-type alkaloids, daphenylline (**45**) [44], subdaphnidine A (**46**) [26], daphlongamines E–G (**47**-**49**) [45] and daphhimalenine D (**50**) [33] were reported (Figure 12). Daphenylline (**45**), possessing an unprecedented rearranged 22-nor-calyciphylline skeleton, has an expanded neohexatomic ring with C-13 connected to C-1 instead of C-8 as usual. The absolute configuration of daphenylline (**45**) was elucidated on the basis of computational approaches.

2.12.2. Calyciphylline B-Type Alkloids

Calyciphylline B type alkaloids are a group of rare C_{22} fused-pentacyclic yuzurimine type nor-*Daphniphyllum* alkaloids. Most of them contain a six-member lactone ring. A new calyciphylline B type alkaloid, daphlongamines H (**51**) was reported (Figure 13) [45].

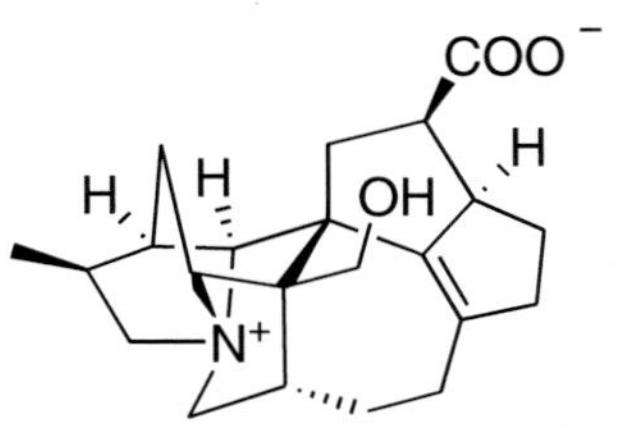

44

Figure 11. Daphniglaucin A-type alkaloids.

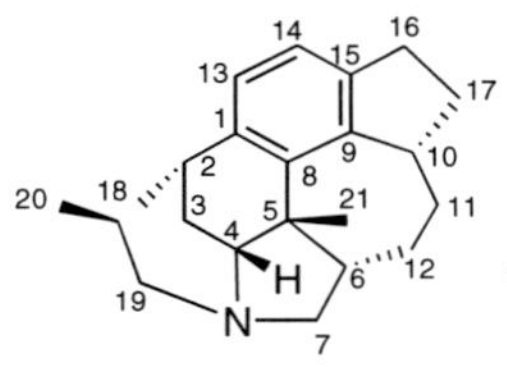

45

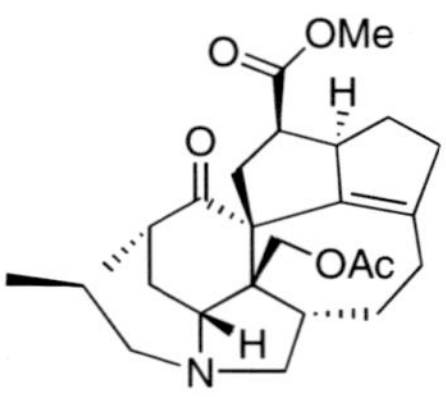

46

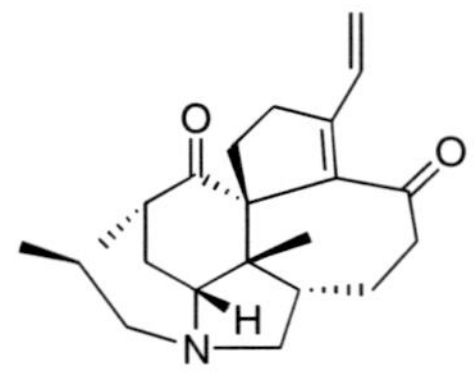

47

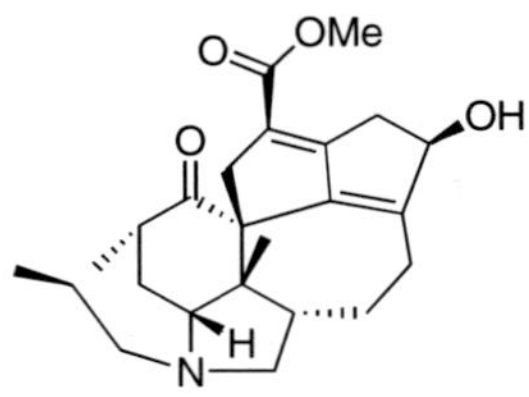

48

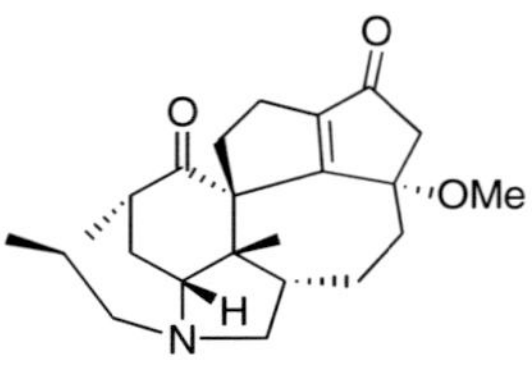

49

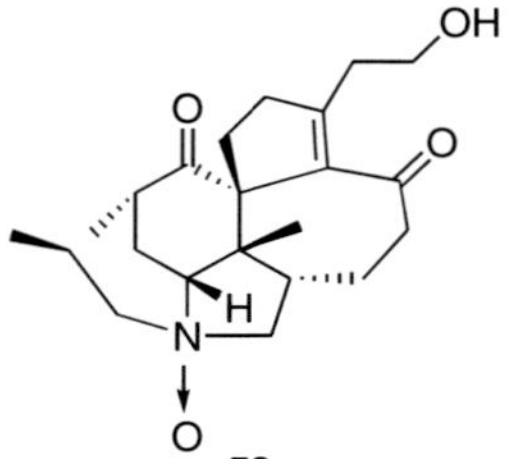

50

Figure 12. Calyciphylline A-type alkaloid.

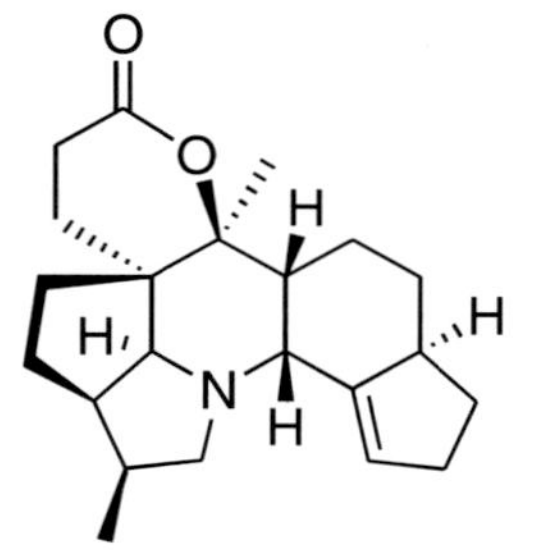

51

Figure 13. Calyciphylline B-type alkaloids.

2.12.3. Calyciphylline C-T Type Alkloids

Calyciphylline C type alkaloids are a group of $_{C22}$ hexacyclic alkaloids, containing a 4-azatricyclo[5.2.2.$0^{1,4}$]undecane ring system with an oxo group on C-1.

2.12.4. Calyciphylline D-Type Alkloids

The carbon skeletons of calyciphyllines D were elucidated to be a fused-pentacyclic system containing a 8-azatricyclo[4.2.1.$0^{4,8}$]-nonane ring.

2.12.5. Calyciphylline G

Calyciphylline G consists of fused hexacyclic system containing a 5-azatricyclo [6.2.1.$0^{1,5}$]undecane ring with a 2*H*-pyrrolium ring.

2.13. Paxdaphnine A-Type Alkaloids

Paxdaphnine A alkaloids are a group of 19-nor-*Daphniphyllum* alkaloids.

2.14. Daphlongeranines A and B

Daphlongeranines B and A possess a unique heptacyclic ring system with one or two lactones, respectively.

3. Biosynthesis

From a biogenetic point of view, *Daphniphyllum* alkaloids are classified into triterpenoid-alkaloids originated from mevalonic acid (MVA, **53**) that construct triterpene skeletons via a squalene-like intermediate [46]. Suzuki and Yamamura conducted isotope tracer experiments to demonstrate that daphniphylline (**54**) and codaphniphylline (**55**) were biosynthesized from six molecules of MVA through a squalene-like intermediate [47] and daphnilactone B (**56**), one of the major C_{22}-type *Daphniphyllum* alkaloids, was generated from four equivalents of MVA [48] (Figure 14). Heathcock et al. proposed a biosynthetic pathway of the *Daphniphyllum* alkaloids [13] (Figure 15). Squalene (**52**) is chemo-selectively oxidized to dialdehyde (**57**). The resulting dialdehyde (**57**) condenses with a primary amine, followed by cyclization and addition/elimination, affording the 6, 5-fused ring system (**60**) that contains a dihydropyridine ring. Diels–Alder cycloaddition followed by an ene-like cyclization completes the "aza"-cyclase phase that gives the carbon skeleton (**62**) of the *Daphniphyllum* alkaloids. Alkaloid **62** was named proto-daphniphylline.

A series variation of proto-daphniphylline, including fission of C–C and/or C–N bonds followed by oxidation, methylation, rearrangements, cyclizations, affords more than fourteen other skeletons above mentioned. Several plausible biosyntheses proposed recently are presented.

Daphnezomine T (**9**) is the first alkaloid without a branched C_1 unit at C-5.

Biogenetically, daphnezomine T (**9**) might be generated from an intermediate like pordamacrine B, which could be derived from yuzurimine by an elimination of acetic acid from C-3–C-4 and hydrolysis of acetyl ester at C-21, via oxidative loss of C_1 branch at C-5 (Figure 16) [27].

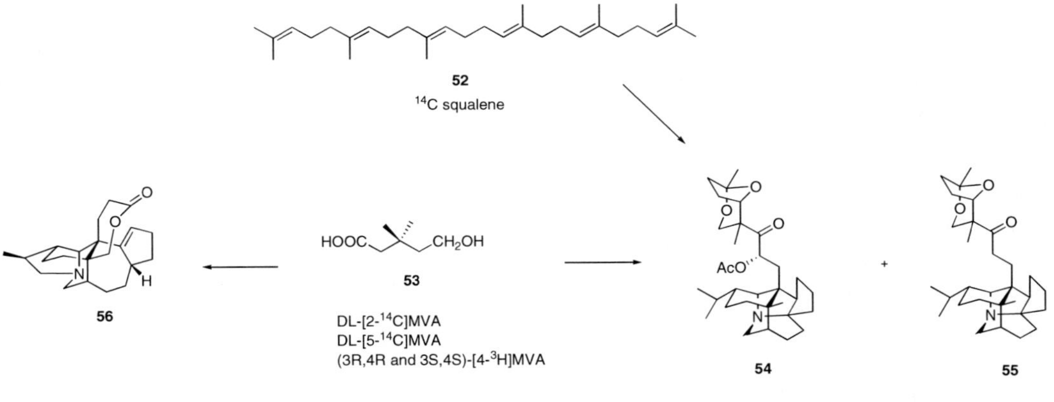

Figure 13. Feeding experiments with labeled mevalonic acid (**53**) and squalene (**52**) into daphniphylline (**54**), codaphniphylline (**55**) and daphnilactoneB (**56**).

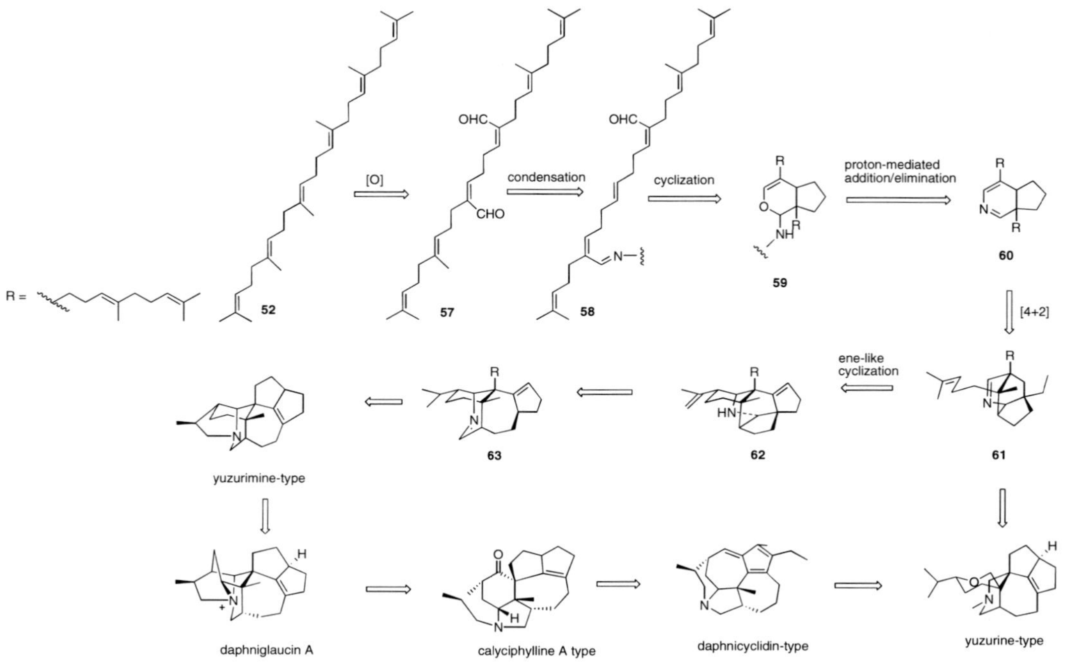

Figure 15. Plausible biosynthetic pathway for the production of *Daphniphyllum* alkaloids.

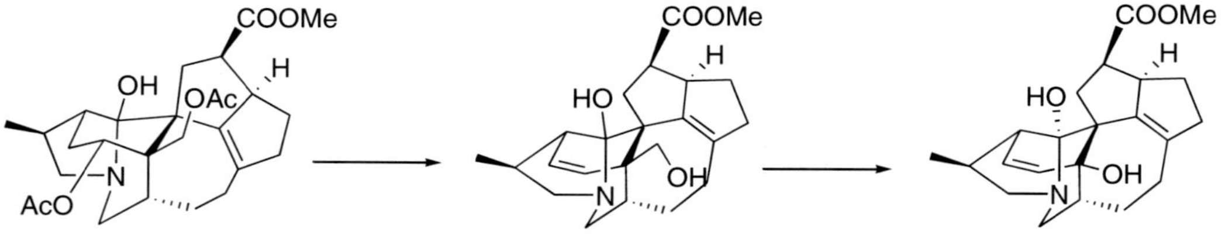

Figure 16. Plausible biogenetic pathway of daphnezomine T (**9**).

Daphenylline (**45**) might be generated from daphnilongeranin C which could be reduced and dehydrated to form a ring expanded intermediate **65** via the Wagner-Meerwein rearrangement. Then the intermediate **65** could be involved in the elimination of the carboxyl group at C-21 by decarboxylase to generate **66** or **67** with a double bond between C-13 and C-14 or between C-14 and C-15, followed by one or two steps of syn-[1,3] sigmatropic rearrangement to yield daphenylline (**45**) (Figure 17) [44].

Daphhimalenine B (**11**) might undergo oxidation to give intermediate **i**, and then oxidative decarboxylation of **i** and subsequent oxidation might yield daphnezomine T. After dehydration, daphnezomine T could further convert to the key carbonium intermediate **ii**, which finally underwent a carbocation rearrangement to afford 1-azabicyclo[5.2.1]decane ring system in daphhimalenine A (**10**) (Figure 18) [32].

Daphmalenines A (**20**) and B (**21**) could be originated from common imine intermediate A which involves an alternative route during the process of the formation of the C-14 and C-15 bond (Figure 19) [37].

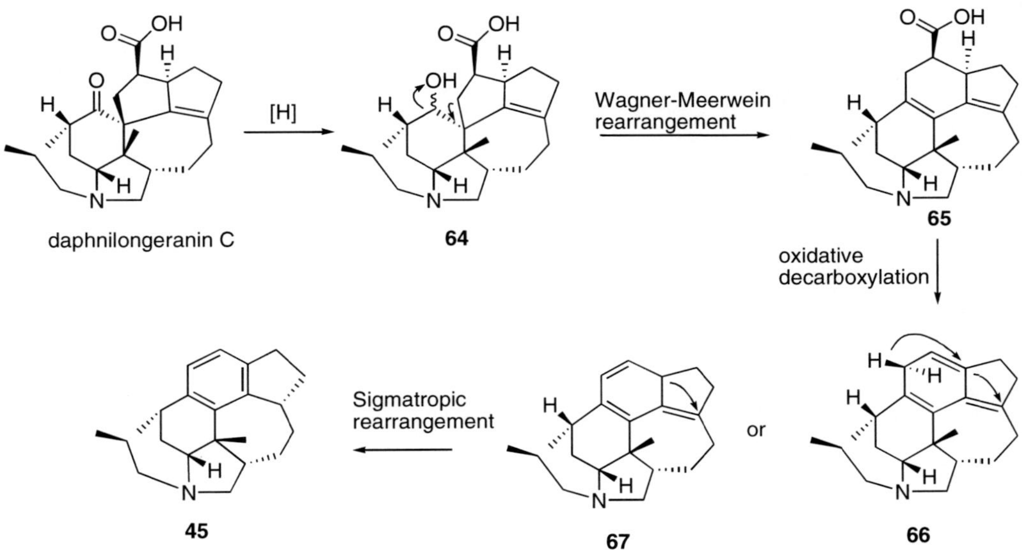

Figure 17. Plausible Biosynthetic Pathway of daphenylline (**45**).

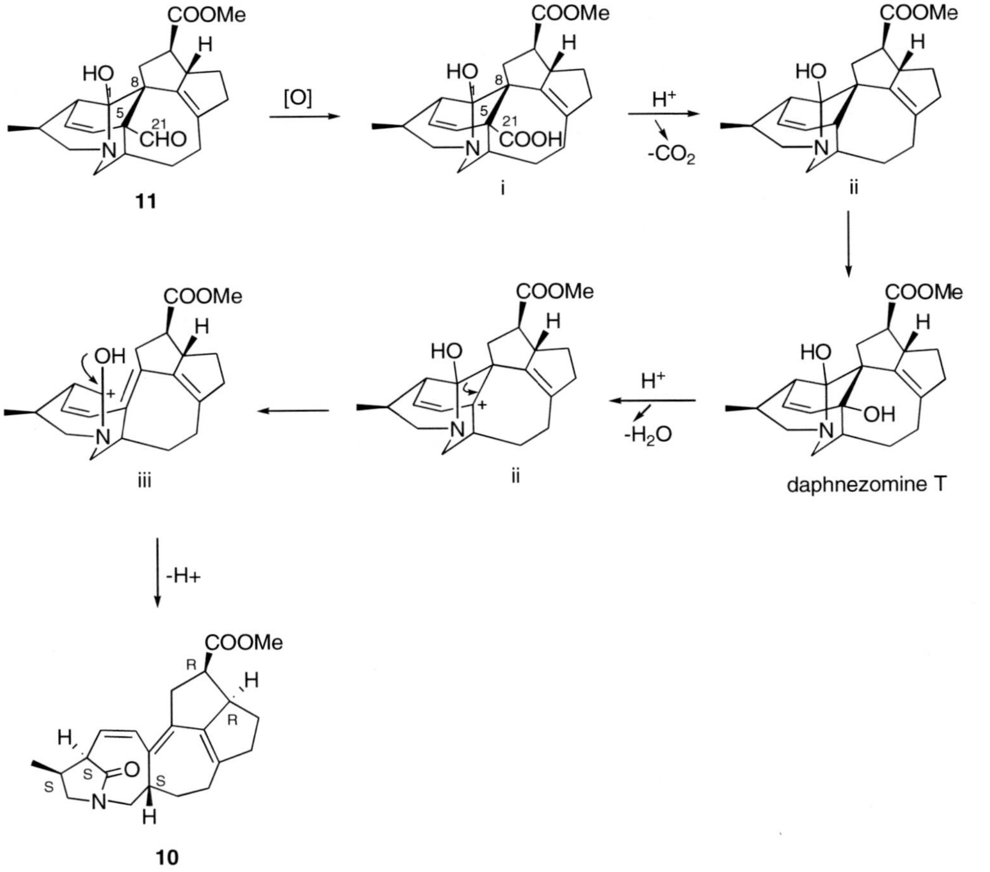

Figure 18. Biogenetic Pathway Proposed for daphhimalenine A (**10**), daphhimalenine B (**11**) and daphnezomine T.

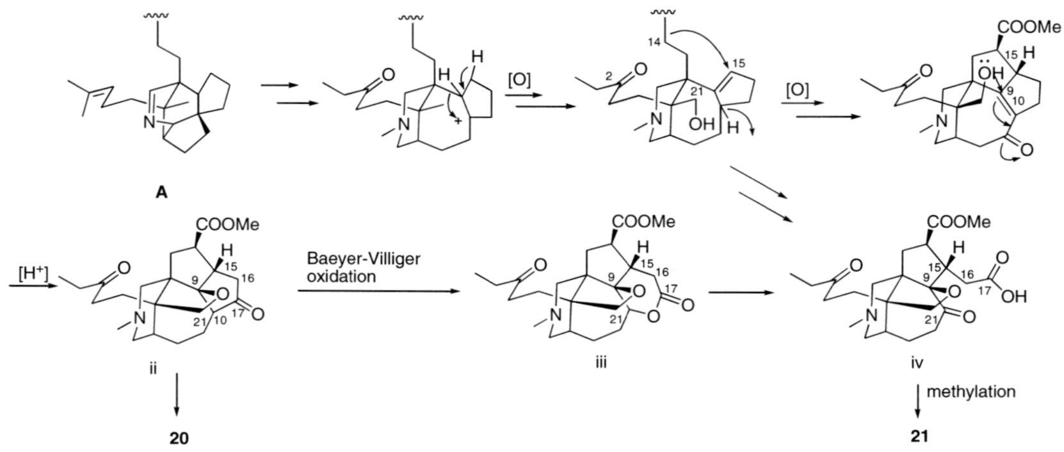

Figure 19. Plausible biosynthetic pathway of daphmalenines A (**20**) and B (**21**).

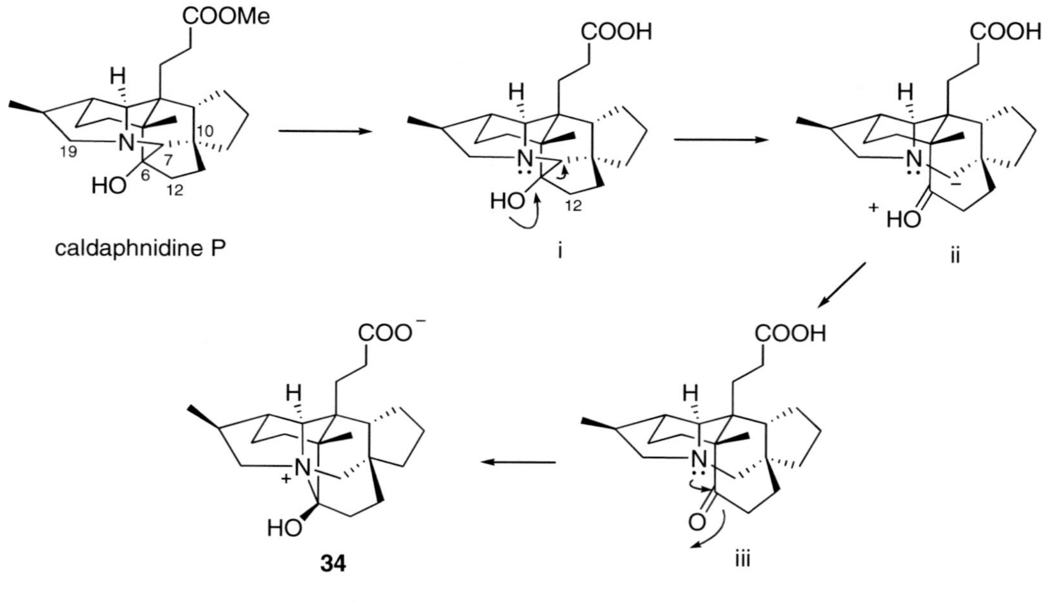

Figure 20. Plausible biosynthetic pathway of angustimine (**34**).

The biosynthetic origin of angustimine (**34**) could be traced back to a bukittinggine-type alkaloid, caldaphnidine P, which after hydrolysiswould give intermediate i. The intermediate i would yield **34** via a cascade of chemical reactions by involving the key steps of a C-6–C-7 bond cleavage (ii) and a C-6–N bond formation (iii) (Figure 20) [41].

A plausible biogenetic pathway of daphnioldhamine A (**22**) was proposed [38]. Daphnioldhamine A might be generated from a common intermediate A (**61**), which involved oxidative reaction to form the ether linkage between C-21 and C-9, and the C=O group at C-11 (Figure 21).

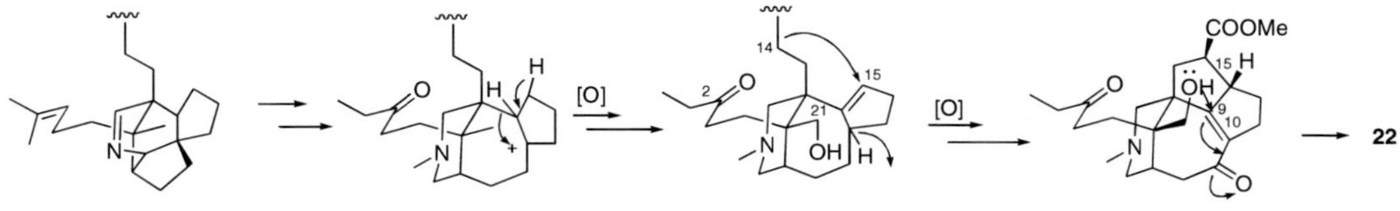

61

Figure 21. Plausible biogenetic pathway of daphnioldhamine A (**22**).

4. Chemical Synthesis

The intriguing structures of *Daphniphyllum* alkaloids have been challenging organic chemists since the first *Daphniphyllum* alkaloid was unambiguously elucidated in the 1960s. When and how the nitrogen atom(s) in these molecules is (are) incorporated in Nature can steer the synthetic chemist towards a biomimetic approach. Heathcoc's biomimetic synthesis of the daphniphylline alkaloids is state -of-the-art work with a characteristic elegance and simplicity reminiscent of nature's approach to formidable structures [49].

Cascade reactions are useful method for the construction of polycyclic skeletons, which are important cores for biological activities [50]. Recently, several publications involving intramolecular cycloaddition reactions are reported.

Weiss et al. reported a first total synthesis of daphmanidin A type alkaloid, (+)-daphmanidin E [51]. The key features of the strategy involve rapid access to an enantiomerically pure bicyclo-[2.2.2]octadione and elaboration around its periphery through the implementation of two Claisen rearrangements, a diastereoselective hydroboration, and a cobalt -catalyzed alkyl-Heck cyclization (Scheme 1).

Darses et al. developed a synthetic strategy for the construction of the [7–5–5] all-carbon tricyclic core common in the *Daphniphyllum* alkaloid family using a key intramolecular Pauson–Khand reaction [52]. A subsequent base -mediated double-bond migration and a regio- and stereoselective radical late stage allylic oxygenation completed the synthesis of the model DEF tricyclic ring system of the biologically active daphniyunnine D (Scheme 2).

Ovchinnikova et al. reported a new type of template-assisted cyclization for the formation of crownophane bearing the azaheterocyclic fragment with the same core structure intermediate (**61**) of 2-azabicyclo[2.2.2]oct-2-ene (Scheme 1) which take place when chalcone podand is allowed to react with aminoenone in the presence of alkali ions (Scheme 3) [53].

In addition, an asymmetric synthesis of the ABCD ring system of daphnilactone B is described [54]. The synthesis features a tandem, double intramolecular, [4+2]/[3+2] cycloaddition of a highly functionalized, enantiomerically enriched nitroalkene to generate a pentacyclic nitroso acetal. The cycloaddition establishes six contiguous stereogenic centers including the critical CD ring junction that bears two quaternary stereogenic centers. Hydrogenolysis of the nitroso acetal followed by amide reduction and cyclization provided the AB rings. The methyl substituent on the A ring was installed in the correct configuration via hydrogenation of an exocyclic olefin in the final step.

A tandem one-pot reaction of an aldehyde with a primary amine involving condensation and then cyclization (N-alkylation), followed by intramolecular dipolar cycloaddition of the resulting nitrone or azomethine ylide, provides a synthesis of bridged tricyclic amines.

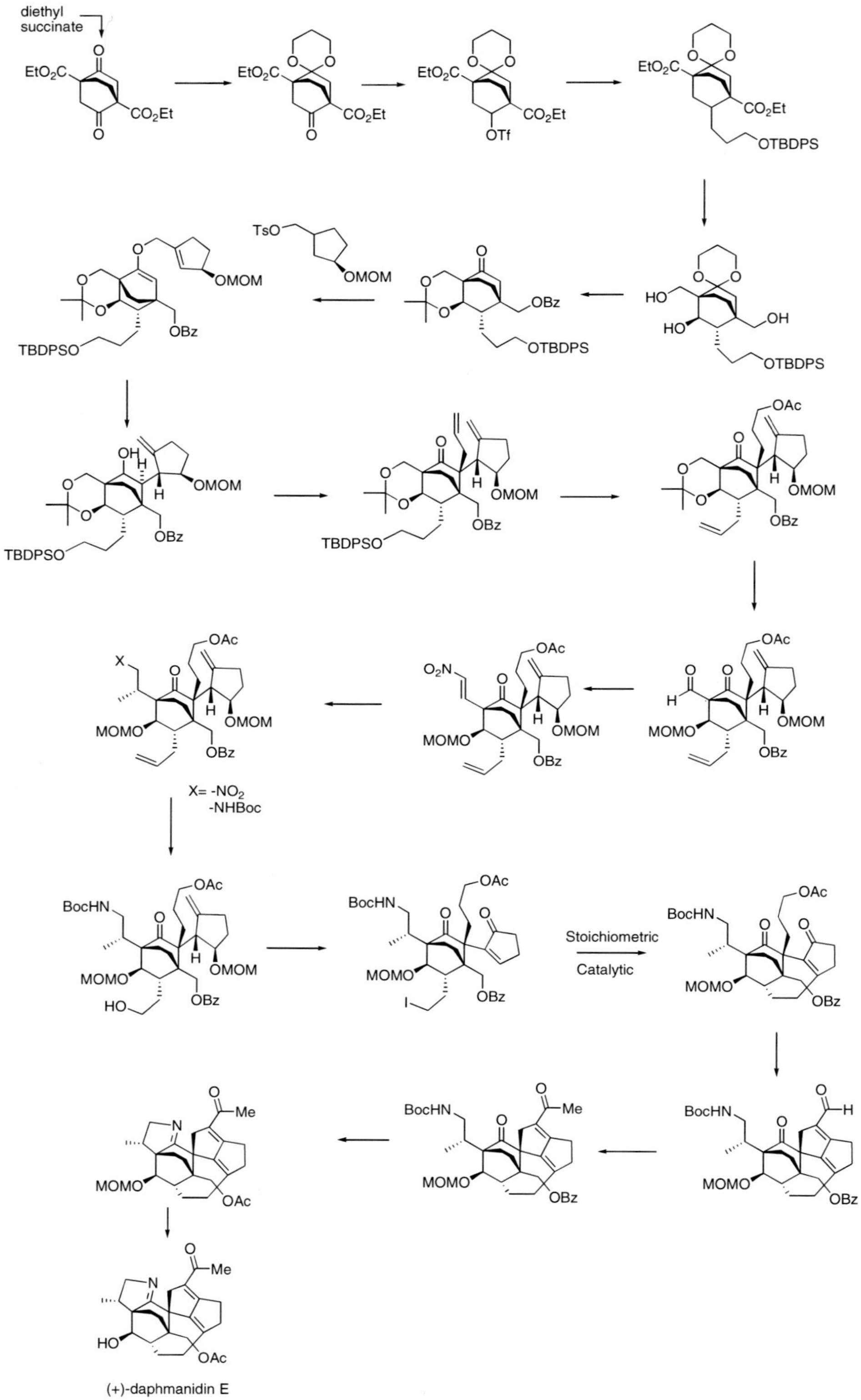

Scheme 1. Synthesis of (+)-daphmanidin E.

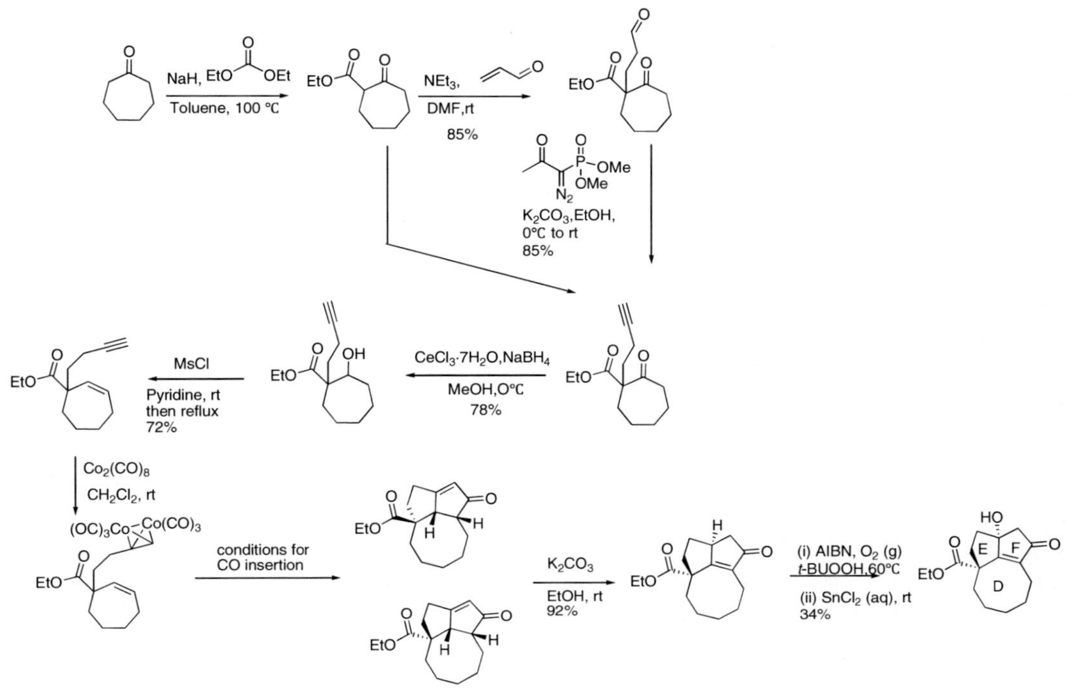

Scheme 2. Synthesis of the tricyclic core.

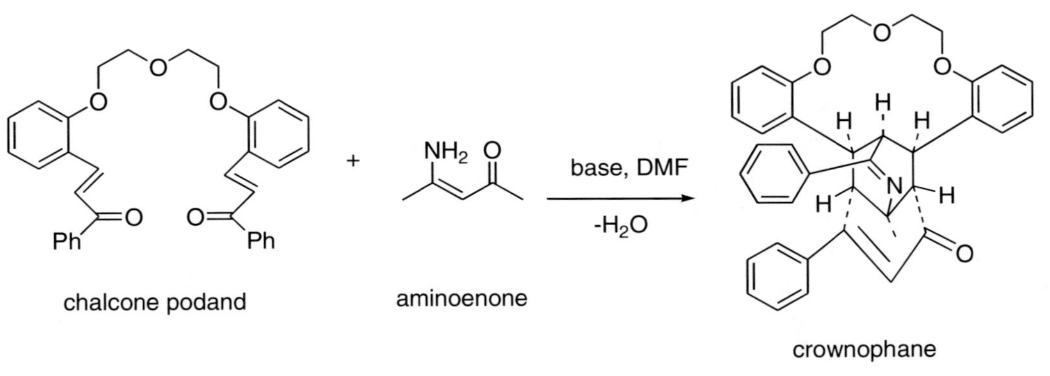

Scheme 3. Synthesis of crownophane.

The reaction was most successful using hydroxylamine, and when the dipolarophile was an unsaturated ester, subsequent reduction of the N-O bond and cyclization to the lactam provided the core ring system of the yuzurimine, daphnilactone B, and bukittinggine type Daphniphyllum alkaloids [55]. Asymmetric construction of rings A-D of *Daphnicyclidin*-Type alkaloids was reported [56].

Coldham et al. reported on a cascade reaction sequences involving condensation, cyclization, intramolecular dipolar cycloaddition for preparation of fused tricyclic heterocycles which can be converted in two steps to a 1,3-oxazine present in some yuzurimine-type *Daphniphyllum* alkaloids under some conditions [57]. In another study, addition of hydroxylamine to substituted 4-chlorobutanals gives intermediate nitrones that undergo tandem cyclization and then intramolecular dipolar cycloaddition to give the core ring system of the yuzurimine-type natural products. Ring-opening of the isoxazolidines gives amino alcohols that can be converted to 1,3-oxazines, representing the tetracyclic core of alkaloids such as daphcalycic acid and daphcalycine [58].

5. Pharmalogical Activities

Daphangustifoline B (**12**) showed weak inhibition of tumor growth against the HL-60, MCF-7, and A549 cell line s [43]. Daphmacromine A-J (**23**-**32**) in vitro displayed pesticide activities against brine shrimp (*Artemia salina*) at 100 mg/L, and the corrected lethality ranged from 40.98% to 70.90% [39]. Daphnilongeridine showed cytotoxicity against several tumor cell lines at IC_{50} values in the range 2.4-9.7μM and against the HMEC human microvascular endothelial cell line with an IC_{50} of 2.7 μM [26]. It is likely that the *Daphniphyllum* alkaloids play an important role in the protection of the plants which produce them against attacking insects [30]. Daphtedinine C exhibited the most potent insecticidal effect against *Plutella xylostella* (score 99) and a moderate effect against *Heliothis virescens* (score 50), while deoxyyuzurimine (4) and daphnicyclidin D (7) showed weak insecticidal effects against *Aphis gossypii* (score 50).

Conclusion

So far, more than 250 alkaloids have been reported from 15 species of *Daphniphyllum* genus. It is highly likely that further phytochemical investigations on the other species will result in much more isolations of *Daphniphyllum* alkaloids with structural variation s.

Total syntheses of *Daphniphyllum* alkaloids are still ongoing, and the biosynthesis of *Daphniphyllum* alkaloids requires thorough studies to elucidate the intermediates and the relevant enzymes. The biological activities of *Daphniphyllum* alkaloids should be extensively investigated. In addition, plants of the genus *Daphniphyllum* produce the structurally diverse and complex alkaloids, which should play an important role for the plant itself from the evolutionary perspective. Biological/ecological role of these alkaloids in the life cycle of the plant should be paid much more attentions.

References

[1] K. Kubitzki, *Daphniphyllaceae Flowering Plants Eudicots*, In: K. Kubitzki (Ed.), Springer Berlin Heidelberg, 2007, pp. 127-128.

[2] S. Yagi, *Kyoto Igaku Zasshi* 6 (1909) 208-222.

[3] S. D. Fang, W. Zhou, Y. Chen, J. H. Chu, Alkaloids of the poisonous plant Niu-Er-Fon *Daphniphyllum calycinum* Benth, *Acta Chim. Sinica* 30 (1964) 270-274.

[4] H. R. Arthur, R. P. K. Chan, S. N. Loo, Alkaloids of *Daphniphyllum calycinum* and *D. glaucescens* of Hong Kong, *Phytochemistry* 4 (1965) 627-629.

[5] N. Sakabe, H. Irikawa, H. Sakurai, Y. Hirata, Isolation of 3 new alkaloids from *Daphniphyllum macropodum* Miquel, *Tetrahedron Lett.* 7 (1966) 963-964.

[6] N. Sakabe, Y. Hirata, X-ray structure determination of a new type alkaloid, daphniphylline hydrobromide, *Tetrahedron Lett.* 7 (1966) 965-968.

[7] H. Sakurai, N. Sakabe, Y. Hirata, X-ray structure determination of yuzurimine hydrobromide, *Tetrahedron Lett.* 7 (1966) 6309-6314.

[8] C. H. Heathcock, S. K. Davidsen, S. Mills, M. A. Sanner, Total synthesis of (.+-.)-methyl homodaphniphyllate, *J. Am. Chem. Soc.* 108 (1986) 5650-5651.

[9] R. B. Ruggeri, K. F. McClure, C. H. Heathcock, Daphniphyllum alkaloids. Part 5. Total synthesis of (.+-.)-daphnilactone A: a novel fragmentation reaction, *J. Am. Chem. Soc.* 111 (1989) 1530-1531.

[10] J. A. Stafford, C. H. Heathcock, Daphniphyllum alkaloids. Part 8. Asymmetric total synthesis of (-)-secodaphniphylline, *The Journal of Organic Chemistry* 55 (1990) 5433-5434.

[11] C. H. Heathcock, S. K. Davidsen, S. G. Mills, M. A. Sanner, Daphniphyllum alkaloids. 10. Classical total synthesis of methyl homodaphniphyllate, *J. Org. Chem.* 57 (1992) 2531-2544.

[12] C. H. Heathcock, S. Piettre, R. B. Ruggeri, J. A. Ragan, J. C. Kath, Daphniphyllum alkaloids. 12. A proposed biosynthesis of the pentacylic skeleton. proto-Daphniphylline, *The Journal of Organic Chemistry* 57 (1992) 2554-2566.

[13] C. H. Heathcock, Nature knows best: An amazing reaction cascade is uncovered by design and discovery, *Proc. Natl. Acad. Sci. US* 93 (1996) 14323-14327.

[14] C. H. Heathcock, M. M. Hansen, R. B. Ruggeri, J. C. Kath, *Daphniphyllum* alkaloids.11. Biomimetic total synthesis of methyl homosecodaphniphyllate. Development of the tetracylization reaction, *J. Org. Chem.* 57 (1992) 2544-2553.

[15] C. H. Heathcock, D. Joe, *Daphniphyllum* Alkaloids. 17. A possibly biomimetic transformation of the secodaphnane to the Daphnane Skeleton[1], *Journal of Organic Chemistry* 60 (1995) 1131-1142.

[16] C. H. Heathcock, S. Piettre, R. B. Ruggeri, J. A. Ragan, J. C. Kath, *Daphniphyllum* alkaloids. 12. A proposed biosynthesis of the pentacyclic skeleton. *proto*-Daphniphylline[1], *J. Org. Chem.* 57 (1992) 2554-2566.

[17] C. H. Heathcock, S. Piettre, J. Kath, Mechanic investigations of a biomimetic polycyclization process that leads to the *Daphniphyllum* alkaloids, *Pure and Applied Chemistry* 62 (1990) 1911-1920.

[18] R. B. Ruggeri, C. H. Heathcock, A proposal for biosynthesis of the *Daphniphyllum* alkaloids, *Pure and Applied Chemistry* 61 (1989) 289-292.

[19] S. Piettre, C. H. Heathcock, Total synthesis of *proto*-daphniphylline, *Science* 248 (1990) 1532-1534.

[20] S. Yamamura, Y. Hirata, *The Alkaloids,* Academic Press, New York, 1975.

[21] S. Yamamura, In: *The Alkaloids,* Academic Press, New York, 1986.

[22] J. Kobayashi, H. Morita, In: *The Alkaloids,* Academic Press, New York, 2003.

[23] Z. Y. Li, Y. W. Guo, Progress in the study of Daphniphyllum alkaloids, *Chin. J. Org. Chem.* 27 (2007) 565-575.

[24] M. Dong, M. L. Zhang, Q. W. Shi, Y. C. Gu, H. Kiyota, The Daphniphyllum Alkaloids, *Current Organic Chemistry* 13 (2009) 646-682.

[25] J. I. Kobayashi, T. Kubota, The *Daphniphyllum* alkaloids, *Natural Product Reports* 26 (2009) 936-962.

[26] C. R. Zhang, H. B. Liu, T. Feng, J. Y. Zhu, M. Y. Geng, J. M. Yue, Alkaloids from the leaves of *Daphniphyllum subverticillatum*, *J. Nat. Prod.* 72 (2009) 1669-1672.

[27] T. Kubota, T. Suzuki, K. I. Ishiuchi, T. Kuhara, J. I. Kobayashi, Daphnezomines T-V, alkaloids from *Daphniphyllum humile*, *Chem. Pharm. Bull.* 57 (2009) 504-507.

[28] Z. Y. Li, S. Y. Peng, L. Fang, Y. M. Yang, Y. W. Guo, Two new *Daphniphyllum* alkaloids from the stems and leaves of *Daphniphyllum calycinum*, *Chem. Biodivers.* 6 (2009) 105-110.

[29] C. R. Zhang, H. B. Liu, S. H. Dong, J. Y. Zhu, Y. Wu, J. M. Yue, Calycinumines A and B, two novel alkaloids from *Daphniphyllum calycinum*, *Org. Lett.* 11 (2009) 4692-4695.

[30] Z. Y. Li, Y. C. Gu, D. Irwin, J. Sheridan, J. Clough, P. Chen, S. Y. Peng, Y. M. Yang, Y. W. Guo, Further *Daphniphyllum* alkaloids with insecticidal activity from the bark of *Daphniphyllum macropodum* MIQ, *Chemistry and Biodiversity* 6 (2009) 1744-1750.

[31] Z. Y. Li, H. G. Xu, Z. Z. Zhao, Y. W. Guo, Two new *Daphniphyllum* alkaloids from *Daphniphyllum macropodum* Miq, *J. Asian Nat. Prod. Res.* 11 (2009) 153-158.

[32] Y. Zhang, Y. T. Di, Q. Zhang, S. Z. Mu, C. J. Tan, X. Fang, H. P. He, S. L. Li, X. J. Hao, Daphhimalenine A, a new alkaloid with an unprecedented skeleton, from *Daphniphyllum himalense*, *Org. Lett.* 11 (2009) 5414-5417.

[33] Y. Zhang, H. P. He, L. L. Guo, S. F. Li, Y. T. Di, X. J. Hao, Daphhimalenines C and D. New Alkaloids from *Daphniphyllum himalense*, *Z. Naturforsch.* 67 (2012) 407-410.

[34] C. S. Li, Y. T. Di, J. Guo, Q. Zhang, X. Fang, X. J. Hao, Daphlongamines I and J. New yuzurine-type alkaloids from *Daphniphyllum longeracemosum*, Zeitschrift Fur Naturforschung Section B-A, *Journal of Chemical Sciences* 65 (2010) 1406-1408.

[35] T. Q. Yang, Y. T. Di, H. P. He, X. J. Hao, A new alkaloid from the fruits of *Daphniphyllum longeracemosum* (Daphniphyllaceae), *Yunnan Zhiwu Yanjiu* 32 (2010) 561-563.

[36] T. Q. Yang, Y. T. Di, H. P. He, Q. Zhang, Y. Zhang, X. J. Hao, Further alkaloids from the fruits of *Daphniphyllum longeracemosum*, *Helv. Chim. Acta* 94 (2011) 397-403.

[37] Y. Zhang, Y. T. Di, H. P. He, S. F. Li, Y. Lu, N. B. Gong, X. J. Hao, Daphmalenines A and B: Two new alkaloids with unusual skeletons from *Daphniphyllum himalense*, *Eur. J. Org. Chem.* 2011 (2011) 4103-4107.

[38] C. J. Tan, Y. H. Wang, Y. T. Di, H. P. He, S. Z. Mu, S. F. Li, Y. Zhang, X. J. Hao, Daphnioldhamine A, a novel alkaloid from *Daphniphyllum oldhami*, *Tetrahedron Lett.* 53 (2012) 2588-2591.

[39] M. M. Cao, Y. Zhang, H. P. He, S. F. Li, S. D. Huang, D. Z. Chen, G. H. Tang, S. L. Li, Y. T. Di, X. J. Hao, Daphmacromines A-J, alkaloids from *Daphniphyllum macropodum*, *J. Nat. Prod.* 75 (2012) 1076-1082.

[40] Y. Zhang, Y. T. Di, S. Z. Mu, C. S. Li, Q. Zhang, C. J. Tan, Z. Zhang, X. Fang, X. J. Hao, Dapholdhamines A-D, alkaloids from *Daphniphyllum oldhami*, *J. Nat. Prod.* 72 (2009) 1325-1327.

[41] C. R. Zhang, C. Q. Fan, S. H. Dong, H. B. Liu, W. B. Zhou, Y. Wu, J. M. Yue, Angustimine and Angustifolimine: Two new alkaloids from *Daphniphyllum angustifolium*, *Org. Lett.* 13 (2011) 2440-2443.

[42] Z. J. Zhan, G. W. Rao, X. R. Hou, C. P. Li, W. G. Shan, Alkaloids from the leaves and stems of *Daphniphyllum calycinum*, *Helv. Chim. Acta* 92 (2009) 1562-1567.

[43] T. He, Y. Zhou, Y. H. Wang, S. Z. Mu, X. J. Hao, Two new akaloids from *Daphniphyllum angustifolium* Hutch, *Helv. Chim. Acta* 94 (2011) 1019-1023.

[44] Q. Zhang, Y. T. Di, C. S. Li, X. Fang, C. J. Tan, Z. Zhang, Y. Zhang, H. P. He, S. L. Li, X. J. Hao, Daphenylline, a new alkaloid with an unusual skeleton, from *Daphniphyllum longeracemosum*, *Org. Lett.* 11 (2009) 2357-2359.

[45] C. S. Li, Y. T. Di, Q. Zhang, Y. Zhang, C. J. Tan, X. J. Hao, Alkaloids from the leaves of *Daphniphyllum longeracemosum* Rosenth, *Helv. Chim. Acta* 92 (2009) 653-659.

[46] E. C. Cherney, P. S. Baran, Terpenoid-alkaloids: Their biosynthetic twist of fate and total synthesis, *Isr. J. Chem.* 51 (2011) 391-405.

[47] K. T. Suzuki, S. Okuda, H. Niwa, M. Toda, Y. Hirata, S. Yamamura, Biosynthesis of *Daphniphyllum* alkaloids, *Tetrahedron Lett.* 14 (1973) 799-802.

[48] H. Niwa, Y. Hirata, K. T. Suzuki, S. Yamamura, Biosynthesis of daphnilactone-B, *Tetrahedron Letters* 14 (1973) 2129-2132.

[49] J. Kim, M. Movassaghi, Biogenetically inspired syntheses of alkaloid natural products, *Chemical Society Reviews* 38 (2009) 3035-3050.

[50] M. Ihara, Cascade reactions for syntheses of heterocycles, *Arkivoc vii* (2006) 416-438.

[51] M. E. Weiss, E. M. Carreira, Total synthesis of (+)-daphmanidin E, *Angew. Chem. Int. Ed.* 50 (2011) 11501-11505.

[52] B. Darses, I. N. Michaelides, F. Sladojevich, J. W. Ward, P. R. Rzepa, D. J. Dixon, Expedient construction of the [7-5-5] all-carbon tricyclic core of the *Daphniphyllum* alkaloids daphnilongeranin B and daphniyunnine D, *Org. Lett.* 14 (2012) 1684-1687.

[53] I. G. Ovchinnikova, O. V. Fedorova, Eugenia G. Matochkina, M. I. Kodess, A. A. Tumashov, P. A. Slepukhin, Gennady L. Rusinov, V. N. Charushin, The first example of cascade synthesis of alkaloid-like subunit incorporated into crown ethers, *Macroheterocycles* 3 (2012) 108-113.

[54] S. E. Denmark, R. Y. Baiazitov, S. T. Nguyen, Tandem double intramolecular [4+2]/[3+2] cycloadditions of nitroalkenes: construction of the pentacyclic core structure of daphnilactone B, *Tetrahedron* 65 (2009) 6535-6548.

[55] I. Coldham, A. J. M. Burrell, H. D. S. Guerrand, N. Oram, Cascade cyclization, dipolar cycloaddition to bridged tricyclic amines related to the *Daphniphyllum* alkaloids, *Org. Lett.* 13 (2011) 1267-1269.

[56] T. B. Dunn, J. M. Ellis, C. C. Kofink, J. R. Manning, L. E. Overman, Asymmetric construction of rings A-D of daphnicyclidin-type alkaloids, *Org. Letters* 11 (2009) 5658-5661.

[57] I. Coldham, A. J. M. Burrell, L. Watson, N. Oram, N. G. Martin, Synthesis of fused tricyclic heterocycles by condensation, cyclization, dipolar cycloaddition cascade of α-benzenesulfonyl and α-phenylthio substituted aldehydes, *Heterocycles* 84 (2012) 597-613.

[58] I. Coldham, L. Watson, H. Adams, N. G. Martin, Synthesis of the core ring system of the yuzurimine-type *Daphniphyllum* alkaloids by cascade condensation, cyclization, cycloaddition chemistry, *J. Org. Chem.* 76 (2011) 2360-2366.

About the Editor

Dr. Eduardo Sobarzo-Sánchez
Professor
Department of Pharmaceutical Technology
Faculty of Pharmacy
University of Santiago de Compostela
15782 Santiago de Compostela, Spain
e.sobarzo@usc.es

INDEX

#

A

B

C

D